Composition *IN* Black *AND* White

Composition *in* Black *and* White

THE LIFE OF PHILIPPA SCHUYLER

Kathryn Talalay

New York *Oxford*
OXFORD UNIVERSITY PRESS
1995

Oxford University Press

Oxford New York
Athens Auckland Bangkok Bombay
Calcutta Cape Town Dar es Salaam Delhi
Florence Hong Kong Istanbul Karachi
Kuala Lumpur Madras Madrid Melbourne
Mexico City Nairobi Paris Singapore
Taipei Tokyo Toronto

and associated companies in
Berlin Ibadan

Published by Oxford University Press, Inc.,
198 Madison Avenue, New York, New York 10016

Library of Congress Cataloging-in-Publication Data
Talalay, Kathryn M.
Composition in black and white : the life of Philippa Schuyler /
Kathryn Talalay.
p. cm. Includes index.
ISBN 0-19-509608-8
1. Schuyler, Philippa. 2. Pianists—United States—Biography.
I. Title.
ML417.S42T35 1995
786.2'092—dc20 [B] 94-49016

1 3 5 7 9 8 6 4 2

Printed in the United States of America
on acid-free paper

In Memoriam

With unbounded affection and deep admiration
to an extraordinary man,
whose kindness, creativity, and spirit
were unmatched.
I am blessed to have shared
with him the writing of this book,
and I cherish our times together.

To my father

ANSELM TALALAY

1912–1994

with love forever

PREFACE

Over a dozen years have passed since I first discovered, by chance, a slim red file, hidden in a rare book room. The words "Philippa Duke Schuyler" were embossed on the cover. As I opened the file, a haunting image stared back at me.

I sat down on the floor and began turning the pages—more photos, newspaper articles, some music compositions, letters. An increasingly complex tapestry began to emerge: One of her parents was white, the other black. A child prodigy, Philippa had been compared to Mozart. By age eleven she had penned over two hundred musical works.

Each clipping helped to compose a larger portrait—that of a gifted and beautiful young woman who spent her life in travel to more than seventy-five countries, performing for kings and queens, in small African towns and quaint Dutch villages, at Town Hall in New York and the Palais des Beaux-Arts in Brussels. Then I came across a final item—a tattered and yellowed article written in 1966. Under Philippa's veneer, a journalist had discerned "an atmosphere of childlike emotional fragility. . . . 'I'm not sure what I am,'" she admitted to the reporter.

How had I not heard of Philippa Schuyler? News of her death had made the front page of the *New York Times*. The 1939 World's Fair dedicated a day to her. She was profiled in the *New Yorker*. Why had such a bright star burned out? And, most intriguing, where were all those music manuscripts, those thousands of lost notes?

The red file had been donated to the Indiana University School of Music Library—where I was a faculty member—by Joseph and Mary Myers, a couple from North Carolina. I tracked them down and although they knew nothing regarding the whereabouts of Philippa's music, they invited me to their home, a rambling, century-old white house in Lexington, to reminisce about an "unusual woman."

The visit was imbued with an air of mystery. Joseph and Mary wanted to perform a seance in order to get in touch with Philippa's spirit. A colleague who had traveled with me nervously accepted an offer to sleep under a newly built "pyramid" in their backyard.

Although the Myers knew Philippa only briefly, their insights were keen and their generosity to a stranger remarkable. Mary told me that the entire Schuyler family was deceased, but that "a man named Mitchell" might know something about the Schuyler estate. Excited by the lead, I grabbed my pen to write down his address: "Somewhere in New York," Joseph said.

I returned to Bloomington determined to find Mr. Mitchell. But the Manhattan directory alone lists over 750. Night after night on the university WATS line, I called number after number. Sensing the enormity of the project, I turned superstitious and decided to go to the end of the alphabet. "Well, thank you anyway. Sorry to have bothered you," became my mantra.

Nine months later, at about 11 p.m., having promised myself this would be my final attempt, I made one last round of calls. A sleepy-voiced woman answered the phone. Oh yes, she knew the Schuylers, but who was I? Yes, she had some of Philippa's music in her home, but most of the Schuyler papers had been donated to the Schomburg Center. The woman at the other end of the line was Mrs. Carolyn H. Mitchell, executrix of the Schuyler estate. I am eternally grateful to Carolyn for not hanging up on me at that very late hour.

Over the next few years, with the help of a sabbatical and a Rockefeller grant, I began to trace the life of an unusual woman. With permission from the gracious Carolyn Mitchell, I sorted through sixty-odd cartons at the Schomburg Center for Research in Black Culture. While wading through letters, diaries, photographs, and manuscripts sandwiched between ballet slippers and swimming medals, I began to realize that hidden in the Schuyler memorabilia lay a life far removed from what I had expected. Music was only a part. Philippa's place in history, her complex relationship to her biracial provenance, the "grand experiment" of her parents, George and Josephine, who believed that the solution to America's race problems lay in miscegenation—all this, and more, left Philippa searching the world for her place in it.

My own travels would take me across America and Europe, interviewing her friends, family, and acquaintances. Some erstwhile lovers shunned me, saying they "never wanted to hear that name ever again." Some on the white side of the family would not speak to me, but others welcomed a view of their "family skeletons." A man I met in California told me things he had never repeated to another living soul. And a woman who braved the threats of her husband to meet me in secret provided valuable missing pieces to Philippa's enigmatic past.

Soon Philippa became the adopted daughter in our family. My father, in particular, took an extraordinary interest, and in time he became actively involved in deciphering Philippa's mysteries. We spent days, nights, and months together, thinking, talking, reconstructing her yearly travels through faded stamps on her various passports. I wrote, he read, I rewrote, we edited. For years my parents' house was a mess—papers in every corner, scraps covered with ideas my father had written down at midnight scattered on the floor of their bedroom, my own books, drafts, and thoughts piled high on their capacious dining-room table. When I was living in New York or Bloomington, my father and I spoke every day. Sometimes he would call me very early in the morning with an (often crazy) idea: would I mind researching it? In turn, I would think nothing of calling him from abroad to wake him at two in the morning after having interviewed a key figure in the Schuyler saga. He was as interested in her as I was.

During the final stages of this project, my father became quite ill. I am eternally grateful, however, that he lived long enough to witness the signing of my contract with Oxford University Press. He died a week later.

❖ ❖ ❖

Philippa exerted an extraordinary influence on the culture of her times. For many African Americans growing up in the 1940s and 1950s, she was a role model. Not only to aspiring musicians, encouraged that a black woman could be so successful in the white world of classical music, but also to others who later became doctors and lawyers. Philippa was the first "colored girl" to achieve national prominence at so young an age. One woman told me she decided to become a writer after hearing Philippa play: It was easier than learning all those notes. But it was Philippa who had inspired her on her journey. This was repeated to me time and again.

What struck me also, time and again, was the ubiquitous refrain from so many African Americans of her generation: "Of course we knew Philippa Schuyler. . . . But whatever happened to her?"

Here, then, is her alluring, complex, tragic, sometimes disturbing American saga—for all those who knew her, and for all those who should know her.

New York
February 1995

K. T.

ACKNOWLEDGMENTS

Many wonderful people contributed to the creation of this book. If there are omissions, the fault lies with me: I am forever indebted to each and every person who made *Composition in Black and White* possible.

I am very grateful to the Rockefeller Foundation for awarding me a Scholar-in-Residency at the Schomburg Center for Research in Black Culture, which enabled me to finish the research and begin writing. I am also grateful to Indiana University for granting me the time and money to pursue research on the Schuylers. In particular I owe many thanks to Dr. David E. Fenske, head of the Music Library at Indiana University, not only for providing moral support but also for approving my numerous and extended leaves—and for trusting that one day I would *finish* the manuscript.

Many people corresponded with me over the years, breathing life into Philippa's memory. Along with their reminiscences, some graciously enclosed photographs from personal collections; others exchanged letters with me for months, answering my endless questions. I extend immeasurable thanks to Taha Baasher; Marjorie and Ralphaella Banks; Elizabeth Watson Blanchard; Dr. John Blacking, who knew Philippa in South Africa when she was "passing"; Sister C. A. Carroll; the late Dr. Gilbert Chase and his wife Kathleen; Lena Corbett; Dr. Dominique-René de Lerma; Lori Heise; Sabina L. Ehlers; Virginia Elwood-Akers, for alerting me to her book on women war correspondents in Vietnam; Walter Goldman; Joseph Mitchell (who also spoke with me on several occasions); John MacKenzie for his poetic insights; Laura Jane Musser, who tirelessly wrote to me about the family; Robert E. Ohlman, Jr.; Ernie Pereira for answering yet another of my interminable letters; Dr. Michael W. Peplow; Edith Kermit Roosevelt; Lillian Schiff; Charlotte Churchill Starr; Dr. Eileen Southern, who published my first article on Philippa; Dr. Harry McKinley Williams, whose dissertation on George Schuyler saved me hours of research; and Denise Wontner.

Much of the primary research material on the Schuylers as well as hundreds of photographs are housed at the Schomburg Center for Research in Black Culture in New York. I would especially like to thank Howard Dodson, chief of the Schomburg Center, for his encouragement and interest. The staff at the Schomburg was exceptional. I extend my sincerest thanks in particular to Berlena Robinson for her enthusiasm and endless assistance; to Diana Lachatanere, James Briggs Murray, and four former staff members, Deborah Carter, Robert Morris, Susan Davis, and Deborah Willis—for their magnificent help, day after day, and for photocopying an obscene amount of material. To Mary Yearwood and Jim Huffman from the Schomburg's Photographs and Prints, for their extreme patience as I foraged through boxes of photographs—and left them with dozens to duplicate. I am very grateful to Carolyn Davis and her staff at the Syracuse University Library, Department of Special Collections; to the staff of the Beinecke Rare Book and Manuscript Library at Yale University, where George Schuyler's papers are part of the Yale Collection of American Literature; and to Columbia University, New York, for allowing me access to *Reminiscences of George Samuel Schuyler*, part of their Oral History Collection. I would also like to acknowledge Barbara Layne Hicks, Monica Smith, Joseph Solomon, executor of the Carl Van Vechten Estate, Robert Speller, and Ed Weber at the University of Michigan, Ann Arbor, for granting me permission to use photographs from their collections.

Having the good fortune to talk with people who knew or remembered the Schuylers was quintessential in preparing this biography. My interviews ran the gamut from brief telephone conversations to weekend visits with people who kindly invited me into their homes. For their singular contributions to this book I am profoundly appreciative to Olive Abbott; Vincent Baker of Baker Enterprises; Dr. John Henrick Clarke; the Cincinnati Cogdells—Gaston, his mother, the late Susie May, and Gaston's daughter, the great-grandniece of Josephine, whose physical resemblance to Jody is uncanny; Eschilia Cosi in Rome, who lovingly showed me the hole Philippa's high heels had worn in the marble floor under the damper pedal; Leonard de Paur, whom I interviewed in the hospital; the late Elton Fax; André Gascht in Brussels; Gerd Gamborg, with whom I spent four wonderful days in Oslo, Norway; Robert Hill; the late Kathleen Houston; Jean Blackwell Hutson, a veritable encyclopedia of information; Delilah Jackson; Dr. Hylan Lewis; Joseph and Mary Myers; Klaus George Roy; Dr. Yohanan Ramati, who took time out from his exceptionally busy international schedule to meet with me at a crowded Penn Station diner; Oriel Schadel from Herder and Company, Rome; Robert Speller and his wife; Philippa's extraordinary piano teacher Pauline Apanowitz Styler and her daughter Lori; and Lineke and Theo Snijders van Eyk, in Woerden, Holland (and Gus Müller, who found their elusive address). I interviewed two people under the promise that they remain anonymous. To both of them I owe much: I understand how difficult it was to come forward.

Several people helped with the research and writing of this biography. In particular the late Jeff Remington, who died at much too young an age, assisted me with Philippa's later life: Jeff's prose is impressed on the pages of this book; and his companion Margaret Swift, whose unerring sense of style and encouragement are dear to me. Katie Daley was there practically from the start helping to pull together vast amounts of material covering the early years of Philippa's life. I will be eternally grateful to her for setting me on the correct path. Dr. Norman Harris was also there at the beginning, giving me a perspective on George I otherwise would not have had. I thank Dr. Robert Harrist for his research; Geneva Riley, who was my assistant while I was a Scholar-in-Residence at the Schomburg Center; Barbara Gilbert, who transcribed the "unexpurgated" version of Josephine's diaries; and Sandy Siebenschuh, who carefully edited an initial draft. Dr. Arnold Rampersad read an early version and provided invaluable advice. Selma Epstein, a pianist from Maryland, asked if I would edit some of Philippa's piano works, several of which she has subsequently performed, from New York to Hong Kong. I thank her for taking Philippa around the world, again.

At Oxford University Press, my eternal gratitude to Elizabeth Maguire for her pivotal role in getting this book published; to Elda Rotor for her very conscientious work; to Carole Schwager and Philip Reynolds for their eagle-eyed copyediting.

To Faith Childs, for her encouragement. To Dr. Austin Caswell and Dr. David Baker, both at Indiana University, for their warm friendship and support. To Richard Newman, at Harvard's W. E. B. Du Bois Institute, without whom I would have been lost. And to my wonderful friends who have listened to me babble over the years about the Schuylers—and never once *looked* bored.

Carolyn Mitchell, executrix of the Schuyler estate, has more than anyone stood by Philippa and me during the long years of research and writing, always supportive, encouraging, and generous. She stands alone in my gratitude, and I give her my most cherished and sincere acknowledgment.

Finally, to my family—to my aunt and uncle, Dorothy and Lewis Cullman, for their extraordinary support and generosity; to my sister Nina Callahan, for her patience and sharp insights, and for her considerable contributions to understanding Philippa's early childhood development; to her husband, Eddie Callahan, for his support; to my twin sister, Dr. Lauren Talalay, for her hours and hours of work, her critical mind, and for daily abusing MCI; to her husband, Dr. Steven Bank, a forensic psychologist, for spending invaluable "couch time" with Philippa and me, and for making me laugh; to Minnie Gilleylen, with love; to Lucille, for her seal of approval; and to my husband Frank Ponzio, for going to bed with two women every night for all these many years—and never once complaining.

And foremost to my exceptional parents: my incredible mother, for always having the time to read, edit, talk, cajole, criticize, comfort, and dream. And to the memory of my beloved father, without whom this book would never have been.

CONTENTS

Notes on the Text

The words *mulatto* and *octoroon* have been eschewed except when their use was absolutely necessary. *Negro* and *colored*, however, are used throughout as standard terms of the day. *African American* and *black* appear less frequently.

The names of countries are those in use at the time of Philippa's visits, for example, the Belgian Congo rather than modern-day Zaire.

Composition *IN* Black *AND* White

Prelude

On Tuesday, May 9, 1967, at 18:10 hours local time, a helicopter assigned to the 282nd Aviation Company, Vietnam, crashed into the ocean approximately ten miles north of Da Nang. The aircraft, a Lycoming UH–1D, was returning from Hue Citadel on a routine support mission to Da Nang's Marble Mountain Airfield, its home base. Approximately seventy-five yards from shore, the helicopter capsized, disappearing into seventy feet of water.

While the UH–1D is authorized to carry up to seven passengers when transporting VIPs, it is often loaded according to weight capacity. Of the sixteen persons on board this flight, three perished: twenty-year-old Pfc. Michael L. Elmy, an only child, from Pontiac, Michigan; Doan Van Lien, a small Vietnamese boy being evacuated with seven other orphans; and thirty-five-year-old Philippa Duke Schuyler, a gifted musician and journalist, listed on the manifest as "Magazine Writer—Union Leader, Manchester, N.H." The others escaped unscathed or with relatively minor injuries.

The aircraft was recovered from Da Nang Bay, and an accident investigation board was immediately appointed to interview all American survivors plus the unit's instructor. No material failure of any kind was found. Nor was weather deemed a factor. Visibility ranged from seven to fifteen miles. The temperature was eighty-three degrees Fahrenheit, surface wind was ten knots gusting to fifteen, and sunset was not to occur for another hour.

While the straight-line distance between Hue and Da Nang is little more than fifty miles, the terrain is treacherous. A narrow east–west mountain range, with peaks rising to 4,500 feet, bisects the route just north of Da Nang Bay. At that time, the V-shaped pass through the mountains cradled a Hawk missile site. Air traffic in and out of Marble Mountain was heavy as a rule, and the recommended procedure for aircraft returning to Da Nang from the north was to descend below the traffic pattern to 500 feet above the jungle canopy.

Captain Hosey, an experienced flyer who had qualified as a helicopter pilot only a year earlier, lifted the aircraft off at 17:40. He picked up the craft before starting the journey; it hovered at 90 percent of capacity, indicating no particular overload. Copilot Toews called Hue Radio for a weather check on Da Nang. Receiving a favorable report, Hosey followed the contour of the beach while climbing gradually to 3,000 feet.

Before entering the mountain pass, the captain admonished the crew to watch the children: "See that they are strapped in, because when we are going over the missile site we are going to 'autorotate'."[1]

There were not enough seatbelts to go around, however, and one of the children sat on Philippa's lap while the others braced themselves between strapped-in adults.

Autorotation occurs when the power of the engine is so cut and the aspect of the rotor blades so changed that the aircraft can no longer sustain its altitude. The helicopter falls and the upward-rushing air begins to spin the rotor like a windmill. Under proper conditions the helicopter regains the capability to glide, much like a fixed-wing aircraft that has lost power. The sensation experienced by the passenger, however, is that of an extended roller coaster descent. A minute can feel like eternity.

The aircraft descended 1,500 feet in probably less than a minute. At about 2,000 feet, the air-driven rotor began to overspeed. The commander tried a number of corrective maneuvers, but the engine continued to fluctuate wildly. The craft never recovered lift; it struck the water at a forward speed of perhaps thirty knots. Shortly before the impact, Captain Hosey, realizing that he had lost control, told Toews to send out a Mayday.

This rather unusual flight method puzzled the copilot, bothered the instructor, and obviously disturbed members of the accident investigation board. Toews offered the information that "Captain Hosey was more abrupt on the controls than normal." The unit's instructor testified that Hosey's method of descent was uncalled for—there had been plenty of time to descend under normal power.

Perhaps the most disturbing question by one of the board members was asked of the commander: "Concerning the woman reporter, do you think you could have in any way tried to fly the aircraft to show her a little bit more of what the situation was like over there?" Not convinced by Hosey's denial, the board member asked Henry Garcia, the crew chief, "Was Captain Hosey's flying technique different on the day of the accident from the other times you flew with him?" Garcia answered, "This is the first time I have flown with him that he low-leveled."

Haunted by the deaths of the three people, particularly Philippa's, Toews revealed that before loading the aircraft in Hue, he heard someone say to Captain Hosey: "If you can convince her [Philippa] not to come back, we will get you anything you want." Toews added that the comment seemed in jest because people were laughing and smiling.

Unless equipped with pontoons, helicopters do not float well. Yet no life vests were carried on board, and the survivors had only sketchy recollections of escaping from the aircraft. Hosey remembered trying to free himself from the seatbelt and swimming to the top.

> Upon reaching the top I shouted for everyone to swim to shore. I had then two of the kids and I started for shore. Passing a floating object I put one of them on it. At this time the lady passenger was shouting that she couldn't swim.[2] I shouted back and called her to at least try. Going a little further I found another floating object and put the other boy on it. I started back for the lady and just didn't feel I could make it. I turned back towards the shore. A few yards out I called that I *couldn't* make it [to shore either]. Pfc. Quigley [the gunner] shouted back to me that I *had* to. The next thing I remember was getting on a Marine helicopter and being taken to the hospital.

Captain Hosey had sustained a head injury, generalized body contusions, and abrasions.

"Right after Toews got on the radio to call Mayday we hit," Quigley testified.

> I released the seat belt of the kid sitting with me and put my arms around his head. We hit the water very quickly. After I came up from under the water I saw the ship lying with both skids up in the air. The little boy came up before me and he was crying. About this time Captain Hosey came up. He was one of the last to get his head above water. He also had a kid with him; I think he had two of them. Anyway, by this time the kid [I was with] was holding onto my neck and I couldn't get anywhere. So I pushed him away, went back towards the ship, and got a piece of the helicopter that was floating. I gave it to him. I pulled him for a while, and just could not go any further. So I left him afloat and started for the shore myself. The girl reporter cried out, "I can't swim." Captain Hosey yelled back, "Try. Kick your feet and move your hands." Then Captain Hosey yelled that he couldn't make it. I yelled back, "You *can* make it, sir, we've got to." By this time Toews and Garcia were on shore with a couple of kids and they were yelling for me to come in. But I felt I couldn't. Finally, I reached the rocks and lay down. After that I don't remember a thing until the H-34 picked us up and took me off to the hospital.

Quigley was treated for swallowing aviation fuel.

"I don't remember anything after impact until I came to, underwater, upside-down, still locked in my seat harness," Toews testified.

> I think the aircraft came apart on impact. I don't remember getting out. I dumped my harness and metal chest protector. My gloves and helmet must have been torn off. They were not on me when I surfaced.
>
> The aircraft was not visible. Garcia was pulling one Vietnamese to shore when I surfaced. Several Vietnamese were further out. I took one child to shore. Three more made it by themselves. At this time I saw Captain Hosey and Quigley making their way to shore. Three Vietnamese [children] were in the water with them; two were holding onto the debris. I went back to get the third one. I did not see the American woman or [Elmy], the maintenance man.
>
> I sent Garcia to try to reach the highway to get help. I tried to collect the rest of the people. I asked Captain Hosey to take command if he was all right. He said "I'll never be all right." Then I asked if he was all right physically. He said "Yes."

> When I caught up with Garcia he had written a partial SOS in the sand. Two flights of Marine H-34s went over us, but it appeared that they did not see us. As we began to go to the highway, an Army 0-1 came over and circled. . . . Shortly thereafter a Marine helicopter picked up the people on the point and took off. Then a 282nd [Company] helicopter picked up Garcia and myself.

Michael Elmy was first declared missing. His body was subsequently found trapped in the wreckage.

Lt. Frederick D. Gregory, commander of the rescue mission, arrived forty-five minutes after the fatal crash. By this time all the survivors had been evacuated to the hospital in Da Nang for observation. "There were many personal articles floating in the water; however, since the security of the area was questionable, we were unable to recover any of it. . . . I was advised that three passengers were still missing. My crew quickly located Philippa's floating body. . . . Every effort was made to revive her."[3]

❖ ❖ ❖

Three weeks earlier, on April 15, 1967, Philippa Duke Schuyler had given a piano concert over South Vietnamese television. She performed her composition *Normandie* (a set of variations based on a fifteenth-century tune), her own piano transcriptions of Gershwin's *Rhapsody in Blue* and *An American in Paris*, and Copland's *Scherzo humoristique (The Cat and the Mouse)*. The television station was perched high atop Monkey Mountain on a breathtaking site overlooking Da Nang, a city that both "lured and mesmerized" Philippa: "Hue was romantic, Saigon raucous, [but] Da Nang infinitely sad," she had written.[4]

Philippa had also been involved in missions of mercy, evacuating young children from a Catholic orphanage in Hue to Da Nang, where they could continue their education in greater safety. She was ferrying the last group of eight boys when she died. A book she had been writing on Vietnam was almost complete. It would be published posthumously under the title *Good Men Die*.

Three times Philippa had planned to return to America, and three times she had postponed it. Josephine Schuyler, concerned about her daughter's well-being in a war-torn country and having booked a concert for her in America on the twelfth of May, tried to convince Philippa to stick to her original plans. But Philippa was inflexible. Bolstered by her latest visit to an astrologer (a bent she had followed throughout her life), who had predicted that she would emerge from "the mouth of the dragon" on Tuesday night (May 9), Philippa had once again rearranged her travel plans.

❖ ❖ ❖

On May 17, 1967—dressed in her favorite gold concert gown—Philippa lay in state at St. Charles Borromeo Roman Catholic Church in Harlem. John Lindsay, then mayor of New York, paid his respects, as did hundreds of other friends and acquaintances. President and Mrs. Johnson sent a basket of white and red flowers.

The following morning, her funeral cortege traveled south along Seventh Avenue to 110th Street, east to Fifth Avenue, and down Fifth to Sixtieth Street, where it joined schoolchildren given the morning off. The procession moved slowly, paced by the marching of 150 Roman Catholic children, nuns, and friends, to St. Patrick's Cathedral. The only sound that morning on the usually busy thoroughfare was the doleful tapping of the drums by the six Harlem youths who led the procession to the cathedral.

Traffic stopped; hushed shoppers and business people lined the sidewalks, listening to the muffled roll of the drums. To add to the tribute, bells solemnly tolled from St. Thomas Episcopal Church at Fifty-third Street. Along the route, other churches paid their respects—St. Andrew's, St. Joseph's, and All Souls.

At about eleven-thirty, the procession reached St. Patrick's, where several hundred people gathered on the cathedral steps. The flowers were removed ceremoniously and black papal cloth draped over the silver casket. Six U.S. servicemen carried the coffin inside, where two thousand mourners stood to celebrate the Pontifical Requiem Mass, the highest Catholic tribute in death. His Eminence Cardinal Francis Spellman presided.

Philippa's death was reported on the front page of the *New York Times* and noted in papers all over the world. Posthumously, she received some of the highest awards—the Harriet Cohen International Piano Medal (presented by Benjamin Britten), the Catholic War Veterans Award, the Mary Church Terrell Award, an Overseas Press Club Plaque, and the Distinguished American Award. On the day of her death, Senator Karl Mundt of North Dakota read an eloquent eulogy into the *Congressional Record.*[5]

❖ ❖ ❖

Born in New York City during the Great Depression, Philippa was the child of a racially mixed marriage between George Schuyler, a noted black journalist and writer, and Josephine Cogdell, the artistic and spirited youngest child of a wealthy white Texas family.

Philippa first attracted public attention when her parents discovered that she could read and write at the age of two and a half. She was playing Mozart at four, composing by five. She had spectacular success as a child prodigy, concertizing widely and winning numerous awards.

But as Philippa approached adulthood, the acceptance of her youthful genius was replaced by America's preoccupation with the color of her skin. She had to go abroad to find recognition as a musician, and to earn a livelihood. While in her twenties, Philippa added a second, although subordinate, career to her music: She became a roving journalist, reporting on world events from the many countries in which she concertized. From then on, her twin keyboards—the piano and the typewriter—became constant companions.

Doubleness pervaded Philippa's existence: from her two careers as a writer and musician to her broad acceptance overseas as an "American" artist but her rejection at home. Her sophisticated public persona hid a childlike private one, and her racial makeup, a kind of chiaroscuro, signified her own duality.

Behind this scrim of awards and medals, of travel and applause, lay the tragedy of promise denied. Philippa's life was a mixture of accolade and heart-break; an unending struggle for perfection, for recognition, and for acceptance. Neither white nor black, she searched all her life for an identity, for a home.

I

From Texas to Harlem with Love

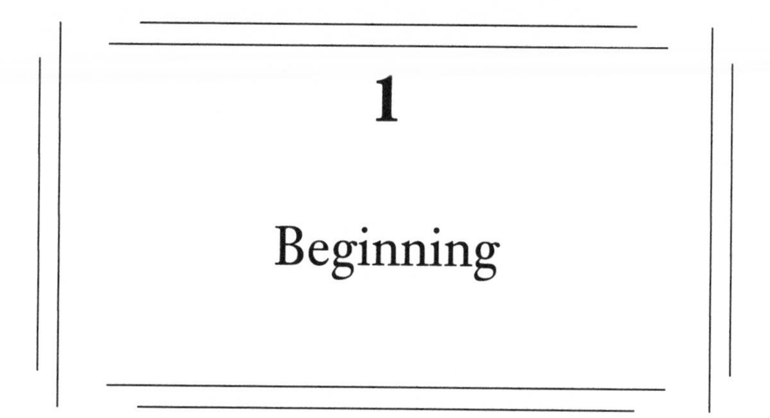

1

Beginning

At 6:35 a.m., on August 2, 1931, a hot and humid Sunday, Philippa Duke Schuyler was born at home, in Harlem. She weighed a healthy 7¼ pounds, yelled loudly when slapped, opened and shut her fists, and waved her arms. George Schuyler, who was still recovering from the lingering effects of malaria contracted while on assignment in Liberia earlier that year, stayed at home until eight, when he had to leave to fulfill a series of lecture engagements, beginning at Brookwood College in Katonah, New York. George's absence during the early days of Philippa's life was a sad necessity, the inevitable consequence of a black writer's effort to earn a living in America.

Their apartment was on the fourth floor of the elite Park Lincoln at 321 Edgecombe Avenue, a quiet, tree-lined thoroughfare stretching from 145th to 155th with no intervening side streets. Opposite was Colonial Park, where weathered benches and tall maples lined the broad sidewalk. As a small child, Philippa would spend many hours playing in the park or sitting with her mother on these benches, watching the passing parade of Harlem.

The Edgecombe apartments—all brown and yellow and white brick, sturdy, immaculately kept, many with recessed entrances and colorful marquees, and tended by doormen in gold braid—announced the opulence of Sugar Hill. Harlemites had coined the sobriquet "Sugar Hill" not only because the street was perched on a bluff far above the shivering hunger of Harlem but also because sugar was slang for money. One had to have money—or means of securing it—to live on these heights.

Contrary to appearances, not *all* residents were affluent. "Working-class strivers," who were forced to throw rent parties or take in boarders to meet the monthly deadline, had slipped into the neighborhood. "In some of the more expensive apartment buildings, pimps and racketeers mingled with the upper crust—though only in the inescapable democracy of the lobbies and elevators."[1]

Although George was of the working class, he was not one of the "strivers," nor did he throw rent parties; it was his status as a journalist that qualified him to be a denizen of Sugar Hill. Schuyler was a steady contributor of editorials and features to the *Pittsburgh Courier*, one of America's oldest black newspapers; he wrote for H. L. Mencken's *American Mercury* and for the *Saturday Evening Post;* he freelanced successfully for other magazines and was often invited to lecture.

❖ ❖ ❖

Philippa was born at the depth of the Great Depression. Black America, and Harlem in particular, was hard hit. The Harlem Renaissance, when the "Negro was in vogue," clearly was a thing of the past. Twelve million Americans were unemployed, and in Harlem itself, a six-square-mile city within the city, almost half the people were out of work while a mere 9 percent had government relief jobs. The tuberculosis rate was five times higher than in white Manhattan, and overall two black mothers and two black babies died for every white mother and infant. The community's single medical facility, Harlem General Hospital, with 273 beds and 50 bassinets, served 200,000.

Yet through all of this, Sugar Hill proudly maintained an aura of opulence. A few buildings up from the Schuylers' was 409 Edgecombe, home to the most "distinguished Hillites," including W. E. B. Du Bois and Roy Wilkins. In fact, Walter White's apartment, on the thirteenth floor, was facetiously known as the "White House of Harlem" because many notables from the world of politics, literature, and theater crossed its threshold.[2]

The Schuylers' apartment had three small rooms, with only a modicum of natural light. But the cheerful decoration compensated for this lack. Reflecting Josephine's taste, the living-room walls were apple green, the chairs green and orange, the bookcase vermillion with touches of blue, red, and purple. A turquoise scarf fell from one corner of the bookshelves across the multicolored volumes that marched up the walls. Patches of sunlight struck the dagger, the sword, the harp, and the pouch that George Schuyler had brought from Liberia.[3] Baby Philippa's bedroom had pictures of nymphs and satyrs painted by West Coast artist John Garth, and at the head of her bed was a painting of Christ—without the loincloth, like the life-sized image of the naked savior that hung in Michelangelo's historic home in Rome.

Philippa's birth was mentioned in all the black newspapers. It was widely held that she was the product of the first interracial celebrity marriage of the twentieth century (except for Jack Johnson's, the heavyweight boxing champion). From the moment she was born, many eyes were on her.

Josephine in particular, but George also, kept detailed scrapbooks in the form of "letters" to Philippa during the early years of her life.[4] "This Diary is for you. In case something should happen to your mother, it will be necessary that you know her love," read the verso of the title page of their first scrapbook. Also pasted on that page were two newspaper clippings. One shows a handsome mustachioed gentleman wearing a Dutch hat. The caption reads: "Colonel Schuyler, New York aristocrat of Revolutionary War fame, had a mulatto son, named Chalk, by one of his female slaves. . . . The boy was given a good education, a fine well-stocked farm, and married to a white woman by the colonel." The other is an etching of New York (then New Amsterdam) from the 1650s. It is captioned: "The Negroes helped build the city, the forts, and the defenses against the Indians. Some were slaves; others were free workers and mechanics; others landowners and planters; and at least one was a physician. There was very little color prejudice under the Dutch."

❖ ❖ ❖

From the day of her birth, Philippa wasted little time: In her fourth week, she began to crawl. It took her fifteen minutes to move eighteen inches, Josephine noted dutifully in their scrapbook on September 4. Four months later, Philippa was sitting and standing on her own, and in twice that time, she had taken her first unsteady steps.

George's homecomings were infrequent during that first year, but he could at least share with his wife, in a bittersweet way, the daily progress of his little girl through the scrapbooks—when she first uttered "mama," when she first stood up on her own, when she first laughed. He probably laughed when he read Josephine's marginalia about their daughter: "I now perceive that you can pick up your teething ring deliberately and put it in your mouth. You recognize it as your property. Thus, sad to relate, you are already on the way to becoming a capitalist. This must be remedied, for it would never do for two socialists to raise a capitalist."[5] Despite frequent absences, George was able to pinpoint, almost to the day, the progress in Philippa's vocabulary. Two months past her first birthday, Philippa had a grand total of four phrases: "Joe" (for "Josephine"); "Daddy"; "God damn"; and "How do!" Perhaps more for George's benefit than Philippa's, Josephine recorded, "It is after all not a bad vocabulary. Most people use little more in their daily communication than a greeting and an oath."[6]

When George was home, he shared writing in the scrapbook. Unlike Josephine's handwriting, which was large and scrawly, George's was small, neat, and tight, almost as if he were trying to squeeze in as much as possible during these brief visits with Philippa. Always the reporter, his overwhelming joy is betrayed only by punctuation:

> Philippa! It is the evening of October 4, 1932, and you are two days over 14 months of age! You are an excellent mimic. You clap your hands, snap your fingers, drum, dance, pat your stomach and rub your head all in direct imitation of your father. You play hide and seek with him and have a most jolly time. Yesterday . . . for the first time you walked on the sidewalk with your father and ran after dead leaves which you collected and brought to him. Today your mother took you out and you walked three blocks! Last night your mother went to the theater (*Ol' Man Satan*) and you stayed home with your father who put you to bed!

Josephine steadfastly attributed Philippa's amazing progress to two causes: "hybrid vigor" and a diet of raw food. The Cogdells were farmers and cattle ranchers. It is not surprising, then, that Josephine would have heard talk around the dinner table about hybridization, the crossing of independent strains of plants and animals. Professor Edward Murray East had reported in 1919 that when inbred strains of corn were crossed, they produced a generation with notable increase in size and productivity. His thesis led to the successful use of hybridization by animal and plant breeders.

Josephine was convinced that hybridization might well apply to human beings. She was therefore greatly intrigued when, as a young girl, she had come across a reference to the descendants of the mutineers of the *Bounty* evincing hybrid vigor, or "heterosis," as a result of miscegenation. (Later, Harry Shapiro would find that the males of the first generation born on Pitcairn Island after the mutiny showed an average increase of about two and a half inches in stature compared to their British sailor fathers or to the Tahitian men whose women they had married. What was more, their fecundity was one of the highest on record anywhere; these couples averaged 11.4 children.)[7]

Perhaps the most avant-garde of Josephine's theories were her dietary ones. As early as fourteen years old, she had experimented with nutrition and now, as a young adult, firmly believed that all foods must be eaten raw, for cooking destroyed their valuable vitamin content. Even meats, especially liver and brains, were never cooked but simply run under hot water for several minutes to kill any lurking germs. No alcohol, tobacco, sugar, or anything artificial was allowed in the house. Philippa was being raised, and happily so, on the rather bizarre combination of mother's milk, cod liver oil, wheat germ, *un*pasteurized milk, and many fruits.

Weaned at twelve months, Phil grew strong on her mother's "scientifically" prepared diet, and as a child she was rarely sick. Josephine began to experiment more and more with nutrition as Philippa grew. Most of her experiments were beneficial, and her discovery in December 1934 that vitamin C could prevent colds appears to have predated the Nobel Prize winner Linus Pauling's work. Other of her dietary adventures, however, such as overdosing her daughter with vitamins A and D when she feared Phil had been exposed to measles, adversely affected her daughter, but cleared up as soon as her mother stopped the experiment. The daily doses of vitamin C and cod liver oil, however, remained a constant.[8]

George was equally intent on his daughter, but he was also awed by the fact that at almost thirty-seven years old he was suddenly the father of a beautiful little girl who was the color of "lightly done toast with dark liquid eyes of a fawn, and eyelashes like the black glistening stems of maiden hair ferns, turned back almost to meet your eyebrows."[9]

Philippa was a happy baby with a sunny disposition. "You are always laughing and possessed of an exceptional sense of humor," wrote George at the beginning of the 1933 scrapbook, when she was one year and seven months old. He surrounded the text with snapshots of a plump, round, smiling face, sporting sparse hair which stood up ungracefully in clumps. At the time it did not appear that beauty was in her stars. But by the end of that year, Josephine began commenting on how much better looking her daughter was becoming: "You are slim and brown and trim with crinkly brown hair and gorgeous jewel-like black eyes. You have not the conventional prettiness of little girls—your beauty is handsome and smart rather than pretty. You do not photograph well at this age. You become stiff and self-conscious while your beauty lies a great deal in your sparkle and grace and change of expression which is not caught by the Kodak."[10]

Philippa's very early world consisted primarily of the multicolored apartment on Sugar Hill; at five weeks a trip to see the A. Philip Randolphs, thus initiating her into the world of the intelligentsia; daily sunbaths on their Harlem roof (an essential health ritual, according to Josephine); multiple trips to the motion pictures that she "watched with growing curiosity";[11] and the omnipresence of an awestruck mother. As she gazed upon her sleeping fourteen-month-old daughter, Josephine wrote: "There lies the meaning and fulfillment of existence. To continually clothe the ancient seed of man in fresh, sweet flesh which when it loses its freshness will bring forth other new and fresh fruit. This is the beginning and end of existence."[12]

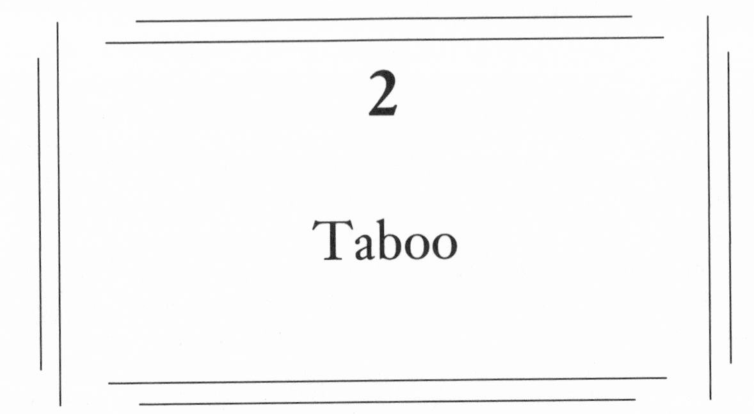

2

Taboo

Recording events was second nature to Josephine. She had faithfully kept a diary since her early teens—the diary of an indulged child from a wealthy and powerful Texas family. She was the youngest of seven and her closest sibling was almost ten years older. Whatever has been preserved of her diaries is less noteworthy for its insights than for her remarkable ability to recall and record conversations.

She continued to write when she married in Texas, at the age of sixteen or seventeen, Jack Lewis, a dapper and much older traveling salesman. The marriage, motivated by a desire to liberate herself from the Cogdell family, was unhappy and short-lived. Josephine moved to California to continue her education—to study painting, to take dancing lessons with Ruth Saint Denis and Elmira Morisini, to study Chinese philosophy, and to attend lectures by the English poet John Cowper Powys. She had a brief fling at being one of Mack Sennett's bathing beauties, and she modeled for pinups—quite demure by today's standards but undoubtedly considered risqué in the pre-bikini era.

In 1921, while living in San Francisco, she met a young painter, John Garth, and fell in love with him. They had a stormy life together for the next five years, mostly supported by Josephine's allowance, spread thin to pay for both their apartment and John's studio. What appeared to hold their alliance together was her appreciation of his art and his belief in her talent as a writer.

In the spring of 1927, Josephine's life with John Garth had, however, reached an emotional nadir. Haunted by jealousy and doubt, she was becoming increasingly despondent, and she rarely left their San Francisco apartment. They

---

Chapters 2 through 4 are culled (and paraphrased) from Josephine's extensive 1927–28 diaries, most of which are notated in dialogue. The diaries are part of the Philippa Duke Schuyler Collection, Schomburg Center for Research in Black Culture, New York Public Library.

decided on a temporary separation. Josephine would go to New York for a few months to pursue her writing career.

Her father, who despite her sometimes delinquent behavior still adored his last-born,[1] was receptive to her moving to New York and willing to continue financial support, on the condition that her sister Lena accompany her at least on the initial trip.

❖ ❖ ❖

Josephine and John had a tearful parting at the San Francisco railway depot. Before going east, she visited Lena in San Diego and stopped off in Texas. In her frequent letters to John, she continued to pour her heart out.

Shortly after her arrival in New York, "Jody"—as she was known to her friends—found a studio apartment in Greenwich Village. She began to establish a social life of sorts, meeting eligible men whom she would, in retrospect, consider pitifully unsatisfactory. Yet out of sheer ennui some of these relationships led to intimacies that were perfunctory enough to allow her to rationalize that she was not really being unfaithful to John Garth.

In the relative comfort of her Village apartment she continued to write every day, working on stories and keeping her daily journal. Having assumed a pen name, Heba Jannath, she had submitted a novel to a publisher and was awaiting a response.

On Wednesday, July 27, 1927, with great trepidation, Josephine went to the office of the *Messenger*, a left-wing black publication to which she had contributed poetry and prose since 1923, to meet its editor, George Schuyler. She had been reading and clipping his iconoclastic and satirical articles for some time, and he had been screening her pieces without suspecting that she was a white southerner.

The day was unseasonably hot, the temperature soaring into the nineties. Josephine fussed all morning about what to wear. Green was her favorite color, but she finally decided to wear her blue crepe suit because it matched her eyes.

The crepe stuck to her skin as she hailed a taxi to take her uptown. At 2305 Seventh Avenue she got out, glanced at the converted brownstone, and with a well-rehearsed bravado, climbed the three long flights of stairs. She hesitated a moment, conscious of small beads of perspiration collecting on her upper lip. Quickly she wiped them away and walked confidently and unannounced into George Schuyler's office.

"He was stunning," she would later write. "His black skin gleamed like satinwood and his hands were as long and graceful as the wings of a raven." Though she had thoroughly prepared herself for the encounter and in later years would admit that one of the subliminal urges that brought her to New York was the desire to meet George Schuyler, a firestorm began raging in her mind as she stood in the doorway.

Regaining her composure, Josephine demurely introduced herself as one of the *Messenger*'s San Francisco contributors. She had always admired his work and wanted to meet him.

George was not in an accommodating mood. He had a hectic afternoon ahead of him with several deadlines to meet. As the door opened, Schuyler looked up. For a moment all he could focus on was a pair of lavender blue eyes, surrounded by delicately arched eyebrows and long lashes. He stared at the vision, so out of place in his dusty office. She was beautiful and seductively shapely, he thought, noticing her lovely bosom and elegant ankles, the golden hair piled on top of her head, the porcelain white skin, the red lips. One can well believe that for a brief spell Schuyler thought he was dreaming.

They started to talk about the magazine and drifted naturally into other areas on reading and writing, literature and politics, art and travel. It was late afternoon and George's deadlines had long since been forgotten. By six o'clock they were still engaged in a spirited exchange of ideas when Schuyler suggested they dine at Tubbs, one of the finer restaurants in Harlem, at 140th and Lenox. The evening ended at the Savoy, where they danced until closing time.

The attraction between George and Josephine was immediate, and in a rare confession, George would later write, "There is a certain affinity between individuals of opposite colors. The fascination of the unknown is so alluring that mutual stimulation is inevitable."[2]

On a Sunday night in mid-August, Josephine joined Schuyler again at the office and they once more went to the Savoy. Coming home in the cab he said very quietly and diffidently, "I wonder . . . would you mind . . . if I kissed you?" It was said so seriously and courteously that her heart was touched. She had felt mellow as he held her in his arms on the dance floor, more aware of his body yet less fearful of him and curious because his attitude was so restrained. He seemed so much more of a gentleman than the white men in her life. So when he asked to kiss her, she said coyly that he might try. He placed his arm around her and pressed her lips softly to his. He did not part them as whites do, she observed. It was a modest and refined salute. "His lips are softer and more sensuous than white lips," she thought.

"You certainly know how to kiss," she heard him say. He kept his arms around her but did not attempt to kiss her again.

They began to see each other frequently, she going up to Harlem more often than he coming down to the Village. Everything about Harlem thrilled her, but she tempered her enthusiasm, for George appeared politely bored by what seemed to her astonishing and novel:

> The octoroon choruses at the Lafayette, the black sheiks at Small's, the expert amateur dancing . . . at the Savoy, the Curb Market along the 8th Avenue "L," with its strange West Indian roots and flare of tropical fruits which bloomed even when snow had whitened the face of Harlem, the displays of chitterlings and pig-snouts in the restaurant windows; Strivers Row, where the colored aristocracy lived in stately houses . . . and three blocks away the dirty tenements of 142nd Street with their shrieking swarms of black, brown, yellow and pale ivory children; and the foreign Negroes speaking French, Spanish, and Dutch and, strangely enough, looking their nationalities; the British West Indians with the Africanized Cockney

and thoroughly British temperament; the rural Southern Negroes so entirely different from the urban Southern Negro and yet more like him than like their Northern rural cousins.[3]

Josephine continued to see some of her friends in the Village, but their conversations seemed "less and less clever and more like childish exhibitionism." By contrast, she was impressed by George Schuyler's caustic wit and infallible logic. She hung on his every word. "He is so bawdy in some of the tales he tells, but at the same time so aloof and impersonal."

Returning from Harlem in bright morning light three weeks after their first encounter, she was too excited to sleep. Opening her diary she wrote:

It is Nine o'clock Wednesday morning, Aug. 17. I just got in from Harlem. I rode back on the "L," the air was brisk and pure, the city spread beneath the tracks cut crisply out of sunlight & shadow. I am full of happiness, of peace. . . . Something marvelous has happened . . . and by all moral and social logic I should now be feeling disgust, regret and contrition. I feel none of these. I cannot for the life of me conjure up a single shred of shame. This morning George possessed me. My blood is still ringing from it. Somehow, strangely enough, I feel ennobled. I cannot explain why except that a complete satisfaction fills me with wholesomeness and gratitude towards life.

She stopped writing, lost in deep thought. She was twenty-seven and had made love before, though not to so many men that she was unable to clearly remember every one of them or count them on the fingers of one hand. But they all seemed anemic next to George. She was still lost in reverie when a special-delivery letter arrived. It was from George, who must have typed and mailed it as soon as he reached the office.

Using a somewhat stilted (if not Victorian) prose, Schuyler wrote to express the pleasure her company had given him the night before and that morning; that he wished her to know he considered it an honor to have "embraced" her; that Josephine was as proficient at love as she was in other things; that her amorous accomplishments had greatly surprised and gratified him; that her cooperation had been more spirited than any he had yet met with and that he hoped she had no regrets. Playing his cards very carefully, George assured her that their relationship, if she so desired, could return to pure friendship and that he would never again refer to their intimacy, for he valued their intellectual communion too highly ever to wish to jeopardize it.

In a warm glow after reading from the letter, Josephine stretched out on the couch and tried to relive each tiny incident of the previous eighteen hours: "How bawdy the music had been at the Savoy. How Fess Williams had waved his magic baton over the dancers, converting them into gaily gliding pans and shepherdesses. So like Pan did Williams look with his dark African face, his large . . . proud, ugly mouth . . . and his arms spread out like the black wings of a soaring eagle. Funny too, how he and George might almost be brothers, they are the same build and both carry themselves with the same nonchalant air, the same dapper ease." Jody recalled with a quiver that as she danced with Schuyler last night her "flesh burned under his touch though he made no move to be covertly amorous."

"After dancing until two," Josephine's diary continues,

> we went to our favorite eating place and as usual had it all to ourselves. I was so sleepy I could hardly hold my eyes open. He noticed it and when I said "I dread that long trip to the Village," asked very softly, "Well, why take it?" I guess I looked surprised for he added: "Won't you do me the honor to be my guest for the rest of the night?" A sort of shock that was not unpleasant went thru me. . . . "If you mean that I am really to be your guest and nothing else," I said and felt immediately embarrassed for questioning his motives. "I mean what I say," he answered quietly.
>
> We took a taxi to his place . . . went thru the big empty marble lobby and walked up . . . to his . . . apartment. . . . His room was . . . almost painfully spotless and orderly. . . . He hung up my light wrap and hat and offered me his bath robe. I said I would just sleep in my slip. He turned the bed down and took his robe & pajamas to the bathroom. . . . I removed my dress and got into bed. My slip covered me modestly from armpits to knees. I felt excited but almost dead with weariness. He returned and put out the bed light. I turned my face to the wall and he got into bed. He did not try to touch me and almost at once I fell . . . into deep sleep.
>
> It was about an hour later I think when I awoke, aware that he had left the bed. He sat at the window where the day was just breaking, smoking his pipe. The smoke had awakened me.
>
> "Have I run you out of your bed," I asked. . . .
>
> "I couldn't sleep," he said.
>
> "If you don't come back I shall get up and go home because you have to work today and need some sleep," I threatened. He laid his pipe down and came back to bed. Stretching his arm out, I laid my head on it, for it seemed prudish to avoid touching him, and we both went back to sleep. It was broad day when we awoke the second time. I became conscious that I was quite close to him. . . . I stirred.
>
> "Good morning," he said. "Did you sleep well?"
>
> "Oh I feel so much better!" I moved even closer to show I appreciated his conduct and was not afraid of him. We lay like this a while and then he said very quietly, "You arouse me greatly, Josephine. I hope you will forgive me. . . ." I did not move from him. After a little he said in a sort of gesture of apology and despair—"Oh, Josephine. . . !"
>
> He sensed my relaxation and put his hand gently upon my hip. "May I?" he asked hesitatingly. . . . He drew me to him. His arms felt like pliant steel. I stopped him, "No, I'm afraid. . . ." "Don't worry," he added tenderly, "I know how to protect you and I would not hurt you for anything, Josephine. . . ." His embrace was so strong, so virile, so tender. . . . Somehow, it was unlike all other embraces, it was like a benediction, a purification.

❖ ❖ ❖

By mid-September Josephine had been away from John Garth for six or seven months. A fleeting and largely meaningless affair with a diamond dealer was three months in the past, and she was deeply absorbed in an exhilarating intimacy with George Schuyler.

Basically, though, Josephine continued to feel an obligation to John under the implied terms of their temporary separation—an obligation not to be deceptive, at least.

While she viewed her involvement with Schuyler as a temporary one—because of seemingly insurmountable societal taboos, if for no other reason—the relationship was still too important to her to discuss it with a third party. No one, she thought, could possibly understand the subtleties of their affair. "It is better for Garth that my lover be George, than one of my own race," she rationalized in her diary, "for with George there would never be any question of a lasting relationship."

After much soul-searching, Josephine decided to tell Garth that they could be friends only, and that for now he should do as he pleased sexually, until they met again.

"This will be fair," she figured, "because then I will not be demanding more faithfulness from him than I am giving. He has always felt the tyranny of my possession which would not brook infidelity, now let him have his fling. Then when we go back together he will probably know enough to understand that love must be monogamous." Having mailed the letter to San Francisco, Josephine felt much relieved.

She returned home just as the phone rang. It was George telling her that H. L. Mencken had accepted his article "Our White Folks" for the December issue of the *American Mercury*. An irreverent, slashing appraisal of white superiority, "Our White Folks" so delighted Mencken that he was going to make it the lead article, a great distinction.

With the money from the article, George had rented a more spacious apartment on St. Nicholas Avenue and moved in immediately. He called to tell Jody about it and to invite her to share the first dinner there, the following night.

❖ ❖ ❖

Josephine's peace of mind about her relationship with John was short-lived. A few days after she had mailed the letter, Joseph, the doorman of her Village apartment, buzzed her at ten in the morning and said in an unusually agreeable voice that a gentleman from California was on his way upstairs. "He won't give his name, ma'am. He says he wants to surprise you!"

When she opened the door, there stood John Garth—a tall, plump, punk-faced, slightly balding man in a worn, tight, shiny black suit. She felt annoyance at his presence and distaste at his appearance.

"Come in," Josephine suggested in a hollow voice, and momentarily putting aside her annoyance, she let him kiss her.

"I thought I would surprise you," he offered, laughing with embarrassment, having sat down on her sofa.

Josephine said nothing. The silence was uncomfortable, and John looked down at his long fingers.

"Baber," he said after a moment, using his favorite nickname for her, "I felt you were slipping away from me and needed me. I had a dream Friday night. I saw you plainly in someone else's arms. A week ago I sold four hundred dollars worth of pictures and I thought the best way I could spend it was to come to see you—I guess I made a mistake."

"When did you leave?"

"Saturday."

"I sent you an airmail letter Monday."

"What . . ."—he looked apprehensive—"what was in it?"

"I wrote I wanted us just to be friends for the time being." There was a long pause while neither of them spoke. "I look pretty bad in this suit, I guess," he finally volunteered. "I ought to have shaved and cleaned up some but I thought it wouldn't matter." He grew silent, then added, "My bags are downstairs. I can clean up now."

The phone rang. Josephine reached for it without turning. It was George. He talked about the apartment. She did not guard her voice at all, plainly revealing an interest. "I can't come up tonight," she said. "I'll call you later about it. Someone is here." At his next words she laughed though she knew it was cruel to Garth. "Of course I do!"

John, who had been listening in frozen attention, suddenly rose. "I knew it," he sobbed, "my dream was right."

"Please don't cry," she said in great irritation.

"Who . . . who was that?"

"George Schuyler!" she announced defiantly, knowing full well John would recognize the name as he had often read Schuyler's articles himself.

"You . . . you . . . live with him?"

"Yes."

"You, living with a nigger!" he flung out, and his big frame shook convulsively. "What will your father say? What will Lena say? Your brothers ought to come up here and kill you both! And they will when they find out!" His voice was now shaking with rage. "Living with a nigger man, and I thought you were so proud and superior."

Josephine lost her temper. "I don't care what you or anyone thinks. I will have any man I choose at any time; I was true to you too long. It only made me miserable."

Wilted and disheveled, Garth slumped back down on the couch with his head between his hands. He sat like this in accusing silence for some time, obviously expecting Josephine to soften and retract her statements. He was used to her making devastating assertions in anger which in a calmer mood she would regret.

Instead, Jody launched into a litany of complaints about their life together in San Francisco: How for five years she had devoted herself to him, spending days in almost monastic seclusion, alone with her thoughts and her writing, living only for the moment when he would return, while he spent the day painting his nude models or whiling away the hours at the wharf with his friends. How she stretched her allowance to support both of them; to pay for two rents, to buy canvas and paints, for food. How she, an indulged "aristocrat" accustomed to luxury, had to forgo buying new clothes, wearing Lena's hand-me-downs. How he would accuse her of being irrational, in fact, mad; of being a millstone around his neck, claiming that he would be better off without her; and yet exert himself to keep her whenever she had tried to leave. That their life was devoid of basic understanding, of empathy, and that the only thing that had kept them together was their art.

Garth listened but his mind was on the present, not the past. He shouted, "All of that doesn't excuse the fact that you deceived me!"

Josephine tried to sort out in her own mind if she had purposefully deceived John and why. Groping for a reason, she found herself framing words that were assuming a life of their own. "I left San Francisco completely out of love with you; the shell of my feelings alone remained, the heart and blood had gone."

"Then why did you weep so when you left? Everyone was looking at you on the back of the observation car!"

"Because I felt it was the end of our love and the finality of the moment overcame me completely." She went over and sat down on the sofa opposite him. "Human beings are not very intelligent," she said. "This situation is unfortunate for the moment but it is better for us both that we separate."

"Yes, after you've taken the best years of my life!" he threw out.

"Don't talk like a woman!"

"If you had told me . . . been honest," he began again.

"You haven't told me what's happened to you in these five months either!"

He did not answer. He still sat slumped over, his head hunched between his shoulders, his hands half covering his face. It was an unbecoming picture. He looked like an overturned sack of wheat, she thought, and the thinness of his hair showed to full disadvantage.

"You have put on weight."

"Yes," he admitted, "I fasted on the train but it did no good, evidently."

"Well, no, not if you've been eating like a pig for five months. I thought you said when I left you would reduce. You've gained twenty pounds! You used to say my cooking and the absence of stimulating company made you fat. You've been free five months now."

They sat in silence again. Then he asked, "How long have you been Schuyler's mistress?"

His tone irritated her. "Oh, six weeks or so. Nor was he the first."

"No one else? Just three, huh? Well, that's pretty good! Schuyler's the lucky chap, eh? You and a nigger? I always thought you'd do something like this. You've always had a Negro complex. You were probably an easy mark for him."

"He's not that kind at all!"

"Oh, he's not sexual?"

"Yes, he is ultrasexual. But he's the *only* gentleman I've met in New York."

Garth laughed.

"You used to think he was terribly clever from his writings, yourself."

The confrontation was obviously heading for a fight with no holds barred. Josephine continued: "If some of your superior Nordics knew half of what he knows about love you would not have to be so afraid of black competition. He is a marvelous lover and possesses the most gigantic anatomy."

"I thought you described him as a short man."

"Only in height. He is a person of surprising contrasts." But Josephine did not stop there. In a calculated effort to hurt, she proceeded unabashedly to describe George's body and technique.

Her frankness and explicitness were deeply jarring to John. They had known and shared many passionate moments in their five years together, and John believed that their sex had been epicurean enough. He had occasionally joked about her greater sexual appetites and had called her his "Delilah"—"a woman will use up a man if she can," he had told her. But this confrontation with Josephine revealed to him a sex drive of unsurmised intensity, and he felt rage at the man who had unleashed it. Blind jealousy, hurt pride, puritanical disgust, and utter amazement chased each other in the vortex of his despair.

They sat in silence for what seemed a long time and then a strange calm, the quiet after the storm, began to take hold of John.

After a while he said that he would go down for his bags. She asked him to get some fruit at the grocery store, too, and he left.

When Garth had gone, Jody called George to tell him about her visitor.

"If you want to see me, get another girl and we'll come over and dance Friday night."

"What? Do you mean that you and Mr. Garth both will come over?"

"Yes."

"You must be out of your mind, sister," he said, but finally conceded.

When Garth returned Jody announced, "We'll go to Harlem Friday night and dance. I just called Schuyler and he invited us to come. He'll have another girl."

"Does he know you told me everything?"

"I told him I did."

"And he asked me over?"

"Yes. Of course he thought you might not want to come but I told him that you were sufficiently intelligent to accept things gracefully."

Garth thought for a while. "Well, I do want to meet him. I really didn't mean all those things I said, Baber. But will you promise me one thing, that you won't see Schuyler alone while I'm here? It'll only be two weeks."

She felt she owed it to him, though her reason doubted it and her emotions denied it. "All right, I promise."

❖ ❖ ❖

Josephine decided to make the best of a bad situation, spending the next several days with Garth, exploring New York. Together, she and "Belovedest" (as she had called John almost since they had met) sat on the grass in Central Park, the late afternoon sun warming them. At sunset they rode up and down Fifth Avenue on the top deck of a bus. Garth was impressed by the architecture, the sights, and faces of a city he barely knew.

They explored the Village and lower Manhattan, and one morning came across a small Franciscan church off Canal Street. They entered and, in the soft, subdued, candle-starred mellowness, she walked cautiously and reverently down the aisle. At the side of the altar stood two plaster saints, their small doll-faced heads bowed under heavy gilt crowns. "We stood before the circular stand of expiation lights," she would record in her diary, "where delicate pink candles burned in a rack of graduated rings, lifting pure chaste points of fire toward a disinterested infinity. And what a motley collection of souls those three dozen or so candles were bargaining for! People who would not have bowed or spoken in life, now in death united for the brief burning of a candle . . . their souls enjoying the only immortality there is and in the hodge-podge company of other dead sinners carelessly assembled by fate."

She took a new pink taper and held its wick to the fire of an unknown soul. Then, moving it in the ancient four-point sign of continuity, she said: "To our mothers: Ada Losier, Methodist minister's daughter, descendant of French Huguenots, wife of a suave handshaking doctor who hastened her to death with his pills and his impatience, smiling mother of Belovedest, the artist; and Lucy Norflette Duke, daughter of an aristocratic English southerner, wife of an Irish atheist trader, mother of Heba Jannath, the writer. Both ladies, delicate, modest, bookish, respectable, joined in one candle; they who never met in life, who never knew their favorite son and daughter had more than met—and side by side with who knows what rascal or broad, sharing the fine clear fire of an hour's immortality on the candelabra of a Catholic church in a New York Lower-East-Side slum." She remembered Shakespeare's line "Yet death we fear that makes these odds all even."

Then, her diary relates,

> a sexton appeared. . . . "Do you want something?" "No." A priest now entered from the aisle door. "Do you want confessional?" No, I said, thinking on the contrary that I should, had I known how to go about it. . . . "We were just looking at your church, it's a lovely church." The priest accompanied us outside. We wanted to go to Hester Street. "What's the best way?" "Oh, you would see nothing but Jews there," the priest disparaged. "You should go to Chinatown. . . ." "We'll go there too; we want to see everything." "Well, you had better hold on to the young lady," he told Garth, "the Chinks might steal her." "They don't do that anymore," I laughed. "No, they are different now. . . there used to be lotsa white slavery down there in Chinatown! White women used to have as many as twenty chinks a day, they'd all be laying around in one room." He knew because as a young boy he used to bring them in beefsteaks and pies and cans of coffee, "you know, for tips, run their errands. . . . Used to be a decent white woman wouldn't go with a Jew, either, twenty years ago. But now, they're crazy for the Jews! Just look at them today, and Chinese, too! Everything's changing like that. Pretty soon it'll be the niggers."

❖ ❖ ❖

On Friday, the day they were to go to Harlem, the weather had turned chilly. "Come get in bed with me, Baber," John invited when she got up to get a coat. "Come on."

Reluctantly she crawled into bed with him, and they talked for a while, her head on his arm. The next moment he was begging her to let him make love. She "took no enjoyment in his arms," and to her surprise, he did not consummate the act. "I want to keep my pep for tonight," he said, rising. "When we go to Harlem I want to be feeling strong and virile."

"Good God. What nonsense!" she thought and was filled with contempt. She recalled how he had often regretted his embraces for fear his work would suffer. Josephine's mind quickly turned to George, who made love to her two and three times an evening, danced all night and worked all day.

❖ ❖ ❖

When they reached Harlem that night, a light was on in the *Messenger* office, and as they entered, George and a girl rose to meet them. The men met and to Josephine's amusement both seemed straining to be worthy of the situation. George looked unusually black and male beside John.

Miss Jones was introduced. She was a tall, slender octoroon with gleaming close-cropped hair in tiny ebony waves that clung tightly to her well-shaped head. She wore long jade earrings and was dressed simply and tastefully. Her smile seemed to announce that she was ready for anything and that nothing shocked her. "Miss Jones is just out of college," George explained. "She has recently come to Harlem and is a graduate dentist. She intends to practice her nefarious art, one way or another, on poor Harlemites."

The quartet left the office for the Savoy. They were early, and the two bands that worked alternate sets were just warming up on the double bandstand. The number-one stand was for the regular house band and the other for visiting groups. They called it the "Battle of Music." Jody was eager for John to see Harlem as she did—a mixture of the elegant and the squalid. She pointed out the best-looking hostesses to him. George aided and abetted.

"There's a filly that would make a rabbit kiss a hound dog!" he announced, as a jet black Diana in red satin swayed past.

"Take this next little high-brown," George was going on, "now she's got a better underpinning but hasn't the dignity of the other. She'd be an ideal bed partner whereas the first one would probably have more brains. Though you can't ever tell till you try them!" He was himself again, she thought—sensuous, satirical, sardonic.

"What is your impression of Mr. Garth?" Jody asked later when she was dancing with Schuyler.

"Well"—George hesitated politely—"I don't like to say exactly on such short acquaintance."

"Don't be so darn polite."

"To tell the truth, I was surprised. He is pleasant, a nice fellow, but like a Rotarian. I can imagine perfectly his leading the luncheon songs at a noonday meeting of the YMCA!"

"Exactly," she agreed. "He's often done it. He's deplorably middle class."

"People can't all be aristocrats or rats."

"What do you think of George?" Jody later asked Garth when they were dancing.

"Well, he's smaller than I expected; he's not more than five feet six inches, is he?" John was on safe ground since he was over six feet.

"What difference does it make? He's well proportioned. You should see the muscles ripple in his shoulders. His body is like polished bronze."

"He certainly wouldn't take a beauty prize."

"Well, Arrow-collar ads don't interest me. He looks like a satyr and I like it."

"He does resemble one, with that broken beak nose and mischievous dimple and his coarse thick lips. I'd like to paint him as a satyr."

"His lips aren't coarse."

"Oh, well, voluptuous then."

"I don't want to dance anymore," she said brusquely.

After the Savoy they visited the Sugarcane Inn, which was deserted that night, and dropped into a seafood place for dinner. At 3 a.m. they separated. George and Mary Jane got in a taxi and John and Josephine walked over to the El.

"Well," she said, "How did you like Harlem? Of course this was a bad night. Thursday, Saturday, and Sunday are the gayest."

"There was enough color and animation to be interesting. But you can't expect me to gush over it. If I had a Negro mistress I might feel like you do, and anyway, being a northerner I can't get enthusiastic over people just because they have dark skins."

They sat in silence while the train rumbled through the "purple violet-starred morning," past innumerable deserted streets and endless barricades of black roofs.

"I couldn't help thinking how terrible it was to see you in such company, raised as you've been and dressed like the Ritz. It was positively painful," Garth finally burst out. "You ought to be hobnobbing with the most prominent people in New York. Haven't you used any of your letters of introduction? You're certainly a fool. You fall for the first man you meet. God! If I had your looks and charm and opportunities. And when your folks find it out, I hope I'm not around."

"Shut up!" Jody said rudely.

After a while he tried to make up by saying, "I ought to paint him as a satyr. Stripped to the waist with his profile turned. Yes, and you as the nymph, nude on a red cloak thrown over a ground of yellow flowers." He took out an envelope and began to sketch while Jody looked on.

"I'd do your hair red, just lighter than the cloak, a tone between the red cloak and the yellow flowers and then his black body leaning over, perhaps his hand lifting a fold of the cloak from your bare limbs."

"I am sure he could pose for it Sunday."

"You telephone him in the morning and I'll go out and get the paints and a board. I want to make you a screen, too, to hide your radiator—kind of fix your place up before I leave."

The next morning she called Schuyler as soon as John went out. "You're to come over and pose for the picture of a satyr!" she told him.

"What the devil's this?" he objected, and Jody told him all about the idea.

"Won't this be sort of running it into the ground," he protested, "painting me in the act of seducing you?"

"It's only poetic justice. He'll have the satisfaction of proclaiming our faithlessness in paint."

"It's damned improper!" George finally said.

❖ ❖ ❖

Nonetheless, the following Sunday George sat sideways on a stool, bared to the waist, his skin gleaming darkly against a blue satin background—one of Josephine's bathrobes draped over the screen to represent a sky. She would pose for her part later. Garth wore a dirty green smock and felt slippers. He used a board for a palette and concentrated ordinary house paints rather than expensive artists' colors. His brushes were stuck in wine bottles.

The session was occasionally punctuated by banter about the rise of nudity in America, and at least on the surface everything appeared pleasant enough. Josephine, who herself had briefly studied portrait painting, watched the progress of the work with great interest. For lunch she scrambled eggs for everyone.

On Monday, Josephine posed all morning as the nymph. Garth was painting rapidly and with total absorption. He had been taught the human figure the way the old masters learned it—in the dissecting rooms of hospitals. His father was a physician associated with the University of Chicago, and as a young boy John would frequently make accurate anatomical drawings for the elder Garth's lectures.

He honed his skills as a portrait painter and muralist at Yale (graduating with a bachelor's degree in 1916), traveled and painted in Europe on several grants, and taught academic subjects at an American school in Istanbul. He spent all his free time roaming the alleys of the ancient city, sketching.

From Turkey he vagabonded through the Far East to Hawaii, where again he taught and painted. Passing through San Francisco two years later, on his way home to Chicago, he fell in love with the city and decided to stay. The time was January 1919 and Garth was twenty-four years old. He met Josephine two years later, serendipitously, on a streetcar.

At some point while painting Jody as the seductive nymph, John commented that her figure was more pleasing than he had ever seen it; that she had the finest skin he had ever painted.

"You mean white skin—colored skin is always beautiful."

"Your lover had a good body," Garth conceded amiably. "He ought to be taller though." And this precipitated another violent argument between the two, culminating with Garth yelling, "I hope you are pregnant and have a black baby!"

"I hope I do, too!"

"You'll probably end by committing suicide!"

"Undoubtedly!"

❖ ❖ ❖

On October 8, John sailed away, "a white handkerchief on the prow of the New-Orleans bound ship, waving to a white handkerchief at the end of a pier."

The following night, Josephine sat barefoot, in her black lover's kitchen, wrapped in his robe, the stove on to keep warm, waiting for George to return from a dinner meeting. She felt quite uncertain about herself. She thought sadly of John, who had loved her well and came to tell her so with his last four hundred dollars. "For two weeks I have been aloof and critical and now that it is too late, I long to be his again," she would tell her diary. Suddenly Josephine disliked New York and remembered with yearning the ubiquitous flower stands and misted hills of San Francisco.

The fear that George might have impregnated her was on her mind, and she felt resentful—only to recall at the next moment how he had guided her back to herself whenever she felt lost. How he laughed at her moods, saying, "Who do you think you are! You get to work doing some writing." Or he would grow stern and turn his head away from her, showing his "African profile, brooding and proud. . . . Never was there such a profile," she wrote.

> It draws me, undoes me, makes me long to sacrifice for it. . . . I want to say "Devour me, Negro, devour me." Aloud I say, "I should like you to kill me, Schuyler. . . ."
>
> "Sweet Heart," he exclaims, looking shocked. Then a mischievous smile lights up his countenance, he leans over me with his long hands gesturing diabolically. He takes my white throat in his hands and the pretense leaves his face and a sensuous look of cruelty enters it as he sinks his fingers into my flesh. My lips meet his and I feel like a white rabbit caught in the coils of a glistening black snake. . . . Then I know that I love him. Oh God, how I love him as I've never loved before.

❖ ❖ ❖

Josephine continued her affair with George through the autumnal nights of 1927. Then, rather unexpectedly, her sister Lena arrived in New York. Their father had sent Lena to rescue Josephine from the "arty atmosphere" of the Village.

> We went to live on Park Avenue, where I was supposed to meet the "Best" People and forget my radical friends. I found the "Best" People merely better dressed and less intelligent; both groups were equally promiscuous and inebrious. When a "wild" party grew too intolerably dull I slipped out and taxied uptown to spend a quiet evening with [Schuyler, ending up] in some Harlem dance hall. . . .
>
> My sister was violently anti-Negro. On several occasions she [embarrassed me] by flaunting her prejudice in public. Like most Southerners, she had a complex on

> the subject and the least thing stirred it up. . . . Once she broke up a dinner party because one of the guests, a very blond and Nordic[-looking] movie star, expressed a regret that "owing to American prejudice she could not do a picture opposite a certain Negro actor," who, she said, was the handsomest man she had ever seen. Once, in Saks' Fifth Avenue . . . when the clerk waited on a well-dressed colored woman before she did on us. Again, when one of our friends, a dilettante artist, modeled a Negro girl. . . . Each time she went into one [of her diatribes] I was tempted to tell her I was in love with a Negro and watch her explode into restful eternity. She offered me a fearful example of what I might [have] become. Her egotism kept her upset and miserable. . . . She . . . never stopped thinking . . . how wonderful she was. I had enough characteristics in common with her to fear for my own soul if I continued in the same atmosphere. . . . I decided to marry [Schuyler] if he would have me and become a member of a race which was daily forced to be humble. No doubt, I wanted to marry [Schuyler] anyway. . . .
>
> One night, sitting in the Savoy between dances, I asked [George] to ask me to marry him. He gave a sharp look and said: "Are you sure you want me to?" I nodded, and he leaned over and very earnestly proposed. I accepted and the date was set for New Year's Eve, three weeks away.
>
> In the meantime [Ben], my sister's husband, had arrived from the South to take us back for the holidays. Compared to his . . . my sister's prejudice was mild. . . . A newspaper mention of a black and tan marriage greatly upset him the night of his arrival.
>
> "Why, niggers can't marry white people!" he announced from the sitting room where he was reading. "Why, it's against the law!"
>
> "Aren't there some states which permit intermarriage?" I asked casually from the bedroom.
>
> "Certainly not!" he answered. "Maybe Mexico does, but no *civilized* country permits such unnatural unions. . . !"
>
> He didn't get over his anger all evening, and next afternoon, as soon as he got in from the [Commodity] Exchange, he announced that he was right, niggers couldn't marry white people in America; he had asked the cotton brokers at the Exchange and they had said they couldn't!
>
> En route from New York to St. Louis, when we started into the [dining car, Ben] perceived a colored man eating at one of the tables. . . . "These ladies can't eat in here!" he [bellowed] to the whole dining car, and rushed us back to our drawing room where he had us served in private by a Negro.[4]

❖ ❖ ❖

Josephine began life with values similar to Lena's. Though a compassionate human being, she was steeped in the racial mores of her native South and was expected to accept them without question. It was her stay in California, in her late teens, that effected a profound change.

To have a "colored woman" seat herself next to her on the streetcar, to perceive a black man in the audience the first time she went to the theater, were shocking and disorienting experiences.

When a young colored man entered her literature class in school she was "dumbfounded." Josephine's first impulse was to "make some queenly gesture of disapproval," and the only thing that stopped her was "the fear of making a spectacle" of herself. She soon found out that the black student did quite well in class and in fact "recited brilliantly." She had feared that it was going to be "very painful watching a poor primitive mind struggle with Shakespeare and Goethe," and in a way she was relieved.[5]

Josephine became involved with left-wing activities, which were rampant then on the West Coast, marched in the red parades, sang the "Internationale" with great feeling, and gave away much of her allowance to political causes. Reading the *Liberator* from cover to cover, she saw an advertisement for the *Crisis*, the official journal of the NAACP. She had not known that there was such a thing as a Negro press; Josephine subscribed and discovered, to her amazement, that it had a fine literary style.

This was the height of the postwar lynching epidemic, and the *Crisis* reported these atrocities in gruesome detail. The thought that her "race and clan had for three centuries" exploited and oppressed the African American preyed on her mind. The savage sport of "nigger hunting," which she knew from personal experience existed, now gave her nightmares. Tossing and turning, she felt horror and pain for the victims.

"Then, one night, I had a very different kind of dream. Aniky, my favorite among our servants, introduced me to a strange Negro youth and we began to dance together. I felt an indescribable happiness and something seemed to say to me that now that we were dancing together there would be no more suffering and misunderstanding in the world."[6]

Josephine's decision to marry a black man, and in so doing, to fly in the face of a taboo accepted by a crushing majority of her 120 million countrymen, white and black, was the act of a truly lone iconoclast. Although women had won voting rights through the Nineteenth amendment at the beginning of the 1920s, the feminists were a movement of many, not one. They struggled in the supportive company of like-minded souls. Josephine, however, acted without the succor of her sororal compatriots. "The race barrier," she would write in retrospect almost a score of years later, "is America's last frontier and it requires all the courage and determination of a pioneer to enter into an interracial marriage."[7]

3

Interlude

Familiar smells emanated from the large kitchen, greeting Josephine through open windows as she came up the long driveway to the imposing house where she had been born and raised. The entire Cogdell clan was gathering to celebrate Christmas: her vigorous seventy-eight-year-old father (now a widower), five of her brothers and sisters, their spouses and their children. Three generations had come to Granbury.

Aniky, Rhoda, Jo, Mandy, Ivory, and other trusted black servants still alive would lavish affection on her, the last born. But the promised festivities only served to underscore Jody's feelings of conflict, which she fought silently. She could confide in no one about her impending new life.

Josephine's grandfather, Thomas Cogdell, had moved his family to Texas in 1848, just three years after it acquired statehood. Born in North Carolina, he was of Irish descent and a zealous Methodist. In 1842, he had married Pamelia Brown, an educated Alabama-born woman of Scotch Presbyterian stock. Both were strong-willed, independent persons who violently differed on religion and politics. Pamelia bore him ten children; Daniel C. Cogdell, Jody's father, was the fourth and their first child to be born in Texas (1849).

Thomas and Pamelia settled in the upper Texas prairie; for a long time the family endured considerable frontier hardships. This was before the "Great American Void" was discovered to be not only habitable but profitable; before the cowboy, the bronco, and the longhorn spawned a myth that would inspire torrents of written words, and millions of feet of film.[1]

Cattle were raised on the open range, and corralled only twice a year—for branding or to begin a thousand-mile trek to the feedlots of the midwestern cornbelt, over hill and prairie, over the storied trails, in an ungainly, almost continuous file of hoofs and horns.

The railroad age came to the area after the Civil War, in 1867, when Thomas McCoy, a twenty-six-year-old Illinois entrepreneur, persuaded the westward-bound Kansas Pacific to establish a cattle depot at Abilene, complete with livestock pens, saloons, and whorehouses.

Large-scale cotton growing did not move to Texas until the beginning of the twentieth century. By that time, some machinery had been introduced and the historical association of cotton with forty acres, a mule, and black stoop labor had diminished. Cotton came to West Texas without the masses of accompanying black men and women.

Josephine grew up in affluence. The Cogdell money had been made through the raising of cotton, cattle, and horses, through land holdings and mills. Her father also had founded the Granbury Telephone and Electric Light companies; and had been president of the town's First National Bank for forty years now.

D. C. Cogdell always spent money lavishly. He gave generously to charitable organizations; he indulged his children, especially his youngest. Jody had a string of tutors and made frequent trips to Fort Worth to attend ballet, theater, and concerts. His largesse was known throughout the county.[2] Although uneducated—he never went beyond the third grade—he maintained a respectable though eclectic library of books and magazines in his home. It was important to him to educate his children in values other than those of the frontier.

Though one of the founding families of Granbury, and raised there, the Cogdells—in the words of one of Jody's surviving nephews—"were not bound by conventions and . . . viewpoints of the small town. . . . They simply did not fit in."[3] The Cogdells still embodied the American pioneer: They were fiery, temperamental, uncompromising, ruthless, rapacious, arrogant, and strong-willed. They had a craving for danger and a desire for power. A "godless family . . . they worshipped one thing and one thing only: money."[4] Their strength was Herculean, their egos enormous, their pride overwhelming. They were known countywide for their hellish tempers.

A chronicle of one generation alone reads more like a gothic tale than a saga of southern gentility. Josephine's father had a black mistress for over forty years. Josephine's oldest brother, while still in his teens, sired a daughter by a black woman. He was shot in the groin by the woman's husband (whom he then proceeded to shoot at point-blank range) and was never able to produce any children after that. Her cousin Duke murdered a black man in cold blood (only six weeks prior to Josephine's visit home) and almost twenty-seven years later would murder his wife and critically wound his daughter. Another brother, Gaston, died an alcoholic. Jody's sister Lena allegedly had two husbands, her first murdered by a gambler while running for district attorney on the reform ticket. Yet another brother, Buster, was run over by a train; his arm was severed, then surgically reattached. For the rest of his life he required help in dressing, although he could play an instrument. He was reputed to be a musical genius who could not read a note but could play anything by ear. Josephine's mother, while being driven in the family convertible to a neighboring town by her daughter-in-law, was partially paralyzed after breaking her neck on the hood struts as the car went over a bump. Josephine's maternal grandmother was institutionalized as insane. Another sister, Zuma, was forced by D. C. to marry a "respectable" man whom, it was reported, she did not love. Her choice of the heart had been rejected by her father as too common to share the Cogdell name. The sister most like Josephine, Daisey, tragically lost a son, a football hero, while he was still in his teens. Jody not only ran away from home to marry a "no-good white" salesman, but she was about to marry much worse.[5]

Josephine captured the family ethos well when she wrote in her diary:

> As usual, there is a lawsuit hanging over the family. It is difficult to remember a year when some one of us were not being extricated from some scrape. This time it is for violating the 18th amendment [Prohibition], of course. . . . Thank Heavens it can never be said of us that we are law-abiding. . . . This continual shadow of the courts at least keeps the family elaborately polite to the officers of the law and important citizens of the vicinity. The sheriff must be bowed to in passing, the justice of the peace greeted with warm hand clasps, the deputies familiarly jested with. This compulsory courtesy is a bitter pill for the family. They smile and bow when they would prefer a long nose and a kick to encourage the distance between themselves and the populace.[6]

Josephine inherited the Cogdell independence and strength of ego, even their arrogance. Her models, after all, had been an all-powerful father who subjected a weak and apparently "savagely selfish mother to countless humiliations," and brothers who murdered, quarreled, raped, and used their money recklessly in order to achieve their goals.[7]

Jody had been a restless, active child, "liable to do most anything. She was histrionic and artistic, and prone to great mood swings. Always dancing, dressing up, horseback-riding, memorizing poems and plays or writing in her diary, she was continually in motion as if warding off something terrifying yet unknown."[8]

The family Negro servants, who protected and indulged her, influenced Josephine's development profoundly. As a small child she spent many hours with the Cogdell's six-foot ebony groom, Big Jim, who doubled as her nurse and who every morning took her astride his horse, trotting leisurely along the country roads with little Jody holding onto the saddle horn instead of trundling her around in a baby carriage, and who later taught her those very necessary skills of the Southwest: how to shoot, skin a possum, ride a bronco, and rope a cow.

Josephine spent a good deal of her early years with the family's cadre of black domestics, in part her own choice but also because as the youngest child by almost a decade, most of her siblings had already left.

Time distorts memory but Josephine's special fondness for these men and women never paled. "They were always doing something interesting," she would write in an article published many years later,

> . . . branding and dipping cattle, slaughtering hogs and sheep, shooting wild game or chasing coyotes out of the pastures, gathering pecans from the towering

> trees along the creeks. . . . I [rode] after cattle with the colored cowboys, [went] hunting and fishing with them and my nephews on moonlit nights, and on rainy evenings played blackjack or poker with them on the back porch or in the big fragrant kitchens of one of [the family's] several homes.[9]
>
> I preferred sitting in our large, airy kitchen with the honeysuckle and wisteria vines poking into the windows, amid pleasant odors of baking and frying, listening to the sardonic comments on life which were flung out in a running repartee between Jim, Linky the cook and Bup our long-legged black houseboy, to hearing the musty platitudes soberly pronounced by my mother and her friends in our marble and mahogany parlor. I particularly liked Monday, which was wash day, for then Linky would help Rhoda, who was washing under the big live oak in the back yard. The clean scent of the lye-soap boiling in the pot, the acrid odor of burning twigs under [it], the bed and table linen flapping on the line, dazzlingly white in the sunshine, and the two robust Negresses standing over their tubs of hot suds in the cool shade.[10]

Another cherished person in her childhood was Dolf, her sister Daisey's cook. He was Aniky's brother. "My cousin Joseph and I fished and hunted with him. We loved his gay rich voice and rollicking tales. On Sundays we played games and over-up and gambled for soda-pop."[11]

Now he was dead, and so was Big Jim. Both had died violently. Aniky's husband had shot Dolf in a fit of anger over a calf that belonged to Aniky's daughter by her first marriage. Jim had been murdered several years earlier. He had been sent to retrieve a horse from one of her father's white renters. But the man did not want to relinquish it, and when Jim insisted, the renter grabbed his gun and shot Jim in the back.

Josephine happened to be home visiting when both Dolf's and Big Jim's murderers were tried, and her diary reveals a striking dichotomy in her father's attitude: his paternalistic benevolence toward his black employees, and his uncompromising solidarity with the white man when the color line was crossed. One of her journal entries reads:

> The only witness to Jim's murder was another Negro, who was not allowed to testify at the trial. The day the case came up, all the poor whites in the county came to town carrying weapons and corn liquor and made a holiday of the farcical trial. The defendant was acquitted in five minutes by a jury of the same class. I asked my father why he did not testify. He knew that Jim was of good character and that the renter was a confirmed Negro hater. He replied that it would be bad for his business for him to take a Negro's side.[12]

The trial of Aniky's husband for the murder of Dolf was very different: The accused pleaded self-defense, claiming that Dolf had threatened to kill him first. Aniky testified on his behalf: "They didn't put anything over on her . . . believe me! She gave them back as good as they sent and with the righteousness of an Archangel. Aniky intends to free her handsome new husband, her third, and with Papa's help she'll do it too. Mama is very angry with Papa. . . . She says he helps the Negroes more than he does his own family. . . . Papa'll get him off tho' since Aniky wants it. Moreover he's a good mill hand."[13]

While Josephine shared many of the Cogdells' characteristics—their enormous egos, their fear of boredom that at times escalated into a craving for danger—there was also something different that made her stand apart from her self-indulged and hedonistic clan. Even as a child and adolescent, Josephine had compassion for the suffering of the peoples of the world. "She had dreams of flying from the nest and going far away. Although she had everything money could buy, she was not happy," her sister-in-law Susie May remembered, reading back into her memory across the span of more than half a century.[14] Josephine's nephew Gaston would recall, "She tried to have some kind of world view, something other than a parochial or provincial viewpoint. . . . She was groping. She had a fine mind."[15]

❖ ❖ ❖

All through the holiday season Josephine agonized. She continued to realize the tremendous risk she would be taking by marrying a Negro, the inescapable severing of the bonds with her family, the certain disinheritance and ostracism. Yet finally, her passion and admiration for George won out. Schuyler would be her emotional salvation.

The day after Christmas she told her father that she had been offered a job in New York and was returning east. Before her departure, she wired Schuyler to meet her in Philadelphia on December 31. They would be married there instead of New York, and, as a symbolic gesture, on the last day of the year.

During the long train ride north, serious doubts about her marriage began to assail her again. She reexamined what this union to a black man in America of the 1920s meant. She began to question the depth of their love, and now she feared to face him. She wrote and tore up a telegram stopping Schuyler from meeting her in Philadelphia.

Josephine marshaled anew all of her arguments: She would marry Schuyler because she wanted him to cure her of the white affectations which she feared would only make her as miserable as Lena and Papa and Buster were. She would marry Schuyler because "the white race, the Anglo-Saxon especially, is spiritually depleted and America must mate with the Negro to save herself." She would marry Schuyler because "marriage to him will mean an infallible insurance against boredom." She fell asleep to the soothing rhythm of the train wheels, realizing that it would be her "last pure white night."

But George was not at Philadelphia's North Station to greet her when she arrived. Confused and angry, she tried to reach him, but was unsuccessful. She took his failure to meet her as a warning from fate.

Josephine boarded the next train to New York. Pulling into Pennsylvania Station, she immediately called George at his office and found him hard at work. She was furious. George explained that he had never received any telegram from Dallas about meeting her in Philadelphia, that he had been so upset not knowing where she was. Appeased, Josephine told him to hurry down. "Let's try to get to the Municipal Building before it closes at noon," she said.

At a quarter to twelve Schuyler entered the waiting room at Penn Station.

Everyone gazed with astonishment when she ran to him, the redcaps as well as the passengers. She had wondered how he would impress her after their separation. When she saw him in the doorway, "tense, dignified, dapper, a sporting gentleman in every line, his face very aquiline and black between his soft turned-down grey felt hat and muffler of pale grey satin," she was thrilled.

He had a taxi waiting, but it was too late to make the license bureau, so they drove up to his home. "New York was good to see again," she thought. Schuyler kissed her all the way up Broadway. "I felt strange and self-conscious to be kissing his dark face and heavy lips. I looked at his black skin and suddenly panicked. Was it that Dixie had filled me with cowardice or was it that experimenting with a Negro lover was a vastly different matter than coming to his house in Harlem to live? 'I shall not marry him,' I thought, and then looking at his earnest face, was torn between fear and desire."

Not having heard from Josephine, George had made plans for New Year's Eve—to spend it with his friend Theophilus Lewis and Lewis's date, a woman from the cast of *Porgy and Bess.* But he told Jody he would cancel; he wanted to bring in the new year with his future wife. Josephine, trying to hide her uncertain state of mind, argued. But George insisted, and as soon as they got home he left to wire Lewis about canceling. Schuyler was barely out the door when Josephine grabbed her hat and purse, went to the corner delicatessen, and made a series of panicked telephone calls to the "respectable" men in her address book. But no one was home.

Over the next several days, Josephine had her things brought up from her Village studio and slowly unpacked. But she was continually ambivalent, even wondering if she might have lost some of her reason. One moment she was consumed by her all-encompassing love for George Schuyler, her admiration for his intellect; and the next moment she remembered that she "was an aristocrat, the daughter of one of the first families of Texas, a full white whose grandparents had been slave-holders, one grandparent the friend and fellow townsman of Jeff Davis who had with his own purse fitted a dozen soldiers with uniforms and horses and sent them with his son to fight for slavery. The other grandparent a Red River planter, owning a hundred negroes, a descendant of one of the oldest English families in America, ancestors on both sides fighters in the Revolutionary, Mexican and Civil Wars."

George was scheduled to leave on the seventh of January for two weeks of lecturing. He knew that Josephine would be alone in Harlem, with barely anyone to talk to, or to turn to. He worried about her state of mind, and about her safety. Two days before he was scheduled to leave, George asked Josephine to marry him.

"I love you," she said and secretly wept against his hair. Hair which she thought strange once and a little ridiculous. "Yes, Schuyler," she agreed. "Tomorrow. . . ."

4

Miscegenation

Sweetheart, hurry," Schuyler urged, after sitting fifteen minutes patiently on the bed. "The bureau closes at 4 p.m." He looked at his watch nervously. George had come home from his office at two, but Josephine was not quite ready; she was composing a rhapsody in green.

She had selected the green and brown silk. She tried jade beads but pearls were better. She put on the soft, dark green felt hat trimmed with pale gray and green violets, pulled on the brown and green suede gloves, and chose the green snakeskin bag. Her underwear was green and russet georgette. Green, she proclaimed, was the symbol of spring, youth, and fertility.

It was January 6, 1928, her wedding day. George had gone to his office early that morning. Lying in bed she had again spent hours agonizing over the step she was about to take. John Garth's paintings on the walls spoke to her, recalling another life. "Oh, let me be your friend, Belovedest, leave me alone now," she pleaded with his ghost. "Never once did you ask me to marry you. To be sure I wouldn't have done it but you should have asked me. You feared I'd be a burden to you. You said you needed only your work."

"And what about the others," his ghost interrupted, "the suave and rich men of your sister's influential circle. They'd put you on easy street."

"But they'd bore me to extinction. I don't want to play at life in a living room. The Cogdells were all miserable with their legal mates, their good women and men. We have no place in the anemic, conventional world. We are savages. We like our meat and our whiskey raw. I've always liked the night better than the day because the night is still primal and can never be tamed. Our kinship is with people who frankly, lustily fight, and love, and break the soil, and sing. It is in our blood, the love of black people. Yes, I *will* marry Schuyler. He is the only straight, honest-to-god real man I've ever met."

By 1 p.m. she had risen to shower and dress. It was almost three now and the tardy southern beauty was finally ready. She touched her ears with green perfume and they left.

George hailed a passing taxi and gave a subway station as the destination. "Oh, are we going by subway?" she asked, surprised and not a bit pleased.

"We won't have time to get there another way. We'll take an express down to the Municipal Building."

Josephine hated subways and hardly ever used them. The car was crowded and they had to stand near the door. People kept their eyes fastened on them. The men looked puzzled, the women astonished.

"The crackers are worried," Schuyler bent down to whisper in her ear.

Sensing her discomfort he said, "I know you don't like crowds. Neither do I. We'll be there in a few moments now." Then he added with assurance, "You and I are going to get some money some way, sweetheart, and live like we should!"

At their destination, they entered a large building with imposing portals, and took the elevator to the License Bureau on the second floor. "My head seemed to grow light," she later wrote, "and I lost all sense of sound. People and things seemed to float weightlessly and events crowded each other swiftly like tumbling cards."

A short gray-haired official handed them an application blank, and they sat down at a long oak desk. George studied the paper and wrote his name, his parents' names, his birthplace, and his occupation with great neatness and precision. At the question "Color?" he paused and looked at her, smiled, wrote BLACK. Then he passed the paper to Josephine.

"What shall I write?" she whispered urgently, pointing to the questions about her parents. She had visions of her family discovering her marriage, the pain it would cause her father, Schuyler getting shot by her indignant brothers. She wished she could expire at once, and not cause any more trouble in the world.

"Just use your married name as your maiden name. They'll never know. It won't affect the legality of it since it is your name." Josephine felt greatly relieved.

A mischievous idea occurred to her. She rose and sought the attention of the official.

"There's some colored blood in my family," she confided to him with great innocence. "Does that make me colored?"

"Oh, yes, yes!" said the old gentleman, and his relief was apparent.

When George saw her enter "colored," he smiled. "That'll throw everyone entirely off the track," he said.

They stood at a window, put up their right hands and swore something. The good-natured official mumbled some abracadabra and handed Schuyler the license. They were told to go up to the next floor if they wished to be married by the justice of the peace.

Lifted to the next floor with what seemed to her the most surprising ease and swiftness, they entered "a smaller room where a thickset man with an irritable impatient expression and a greasy bald head sat behind a desk. There were three men and a woman ahead of us to whom he was addressing curt questions.

They were foreigners. The other two men were witnesses and had to spell their names several times." Josephine, not wanting this unpleasant person to humiliate them with his brusqueness, said appealingly, "We have no witnesses." In an entirely different tone he answered, "Oh, that's all right. I'll fix you up." He wrote his name and sent them into another room.

"It was just four o'clock," she wrote in her diary that night.

> We were the last couple of the day to be married. Schuyler shook his head. "God, but they're hard-boiled around here!" A judge noticed us and beckoned us into his sanctum. Unlike the others, he had an amiable though skeptical face and didn't turn a hair at our difference in color. But I began to feel ridiculous, standing up before so many desks and officials, like a child petitioning permission to do something which no one seemed to have the slightest interest in granting or withholding. Their indifference was degrading.
>
> "Clasp hands," said the judge.
>
> Then in a daze I heard Schuyler saying, "I do." Next I was addressed. The official uttered an appalling catechism of my future intentions toward the man beside me. . . . I grew very sober and perhaps terrified. Simultaneously the outer edge of my mind decided not to say "I do" because it recalled how much comedy has been gotten off about it and when the judge finished his interrogation, I answered "Yes." The question had to be repeated. The next thing I heard distinctly was, "Now, kiss the bride." I turned and George planted a stiff kiss on my lips. Because I felt so sober I had an impulse to hide it, so I said lightly, "And is that all there is to it?"
>
> "It wasn't hard, was it," the judge returned amiably.
>
> "Well, I wish you happiness," he added with skeptical tolerance.
>
> We descended the marble stairs arm in arm. . . . As I stepped down and down, I thought—"Aniky, you won." And in the shadows Big Jim strode beside us with his death wound in his black back and now he looked content as if his debt had been repaid, and [Rhoda] nodded, and her ancient black mother who had said, "Niggahs'll nevah have no justice on this earth. No, lil missy, nevah!" Aniky nodded also and the distraught look on her gnarled chocolate face smoothed away, and she smiled. Gaston's mulatto daughter said, "Our blood is yours." Jo shook hands with me and Mandy took me against her soft yellow bosom, "Baby," she said "you done right." And all the cotton-pickers and millhands stood smiling at me, and all of them shouted joyously—"Things is changed! The ole man's baby has married a niggah!" . . . And the dream I had once, of Aniky bringing me to a youth as black as night and giving me to him and of our dancing together and of my feeling such a rapture and ecstasy in his arms as we swayed and pirouetted, came back to me in all its force.
>
> "It's a good thing George doesn't know how sentimental I am," I thought, full of embarrassment at these hallucinations. "Nothing can justify the pain his race has suffered at our hands. It is silly to harbor such illusions (in fact the negroes at home would probably disapprove of my action on common-sense grounds)."
>
> "We'll take a taxi home," said Schuyler, hailing a cab. In the taxi he put his arm around me. "My wife," he said. I gave him my lips and he kissed me tenderly. Outside the evening crowds clogged the intersections.
>
> "Do you know what I thought of as we came down the stairs?" he said. "The scene in *All God's Chillun Got Wings* where they come out of the church and the

people are gathered on either side with their hands pointing at them." He kissed me again, and I hid my face in his coat.

They looked out of the window at the smart shops and hurrying crowds. They passed the Village and she had no regret. At the Biltmore block Josephine thought of Lena and wondered if she knew. They entered Central Park. The lamps were lit, bright yellow buds against the late winter afternoon grayness, glowing like rows of tulips through the delicate brown tracery of bare tree-limbs. In the background on either side, tall buildings were ironed flat to the sky in somber stateliness. The lake lay far to the right like a plaque of frosted glass, and scribbles of tiny figures gyrated restlessly across its surface.

I felt like a bride being stolen . . . by an enemy prince and I understood why the Aztec girl had betrayed her people to Cortez because as woman fuses the color and characteristics of two races in her womb so her role in life is to stir up, confuse and mix; Mirana started a new half-cast civilization in Mexico as the Southern woman will eventually dilute the black and tint the white and create a legal mulatto civilization in the south of the United States.

They stopped by the office for George to look over his mail and to get the proofs of the magazine that had just come in from the printers. He wondered how he was going to write his editorials and make up the dummy before leaving the following day.

"Take it home and I'll help you," Josephine offered.

On the way, they stopped at a Greek market for a quart of Italian wine and at a fish store for two dozen oysters on the halfshell to take home.

For the first time she brought out her linen and silver and laid the table. She filled the decanters with wine and lit the green candles standing in their crystal candlesticks.

"To my wife," George toasted.

She turned out the lights and they sat in a golden glow while the tapers made nervous shadows on the walls around them. He touched his gold and silver oyster fork delicately with a long black forefinger and Jody read his mind.

"Yes," she laughed, "my mother left me her choicest silver and linen, little knowing that it would star in our wedding supper."

"Poetic justice," he mused.

"Sweetheart, we've got to get out of this country," he finally said. "Let's set a definite time to work to. Say we leave in six months. Go somewhere where we don't have to hide our love. Go where we can live in style. I'm fast losing what interest I had in the American scene and I am convinced we could be happier elsewhere in a more civilized country. Where people know how to live," he touched her arm tenderly, "where I can openly claim you as my wife."

Josephine told him about Mexico City and he liked the idea. "I can finish everything here in six months. Get that book of mine out, and if you'd do what I tell you, you would get your book published too."

"Can you write your editorials now?" she asked, aching to go to sleep.

He kissed her passionately, carried her over to the bedroom, laid her gently on the bed and covered her. "You go to sleep and I'll write my stuff." But George never made it back to his desk.

They woke at 2 a.m. and Jody reluctantly rose.

It was getting late. George sat down at the typewriter again, showed Josephine how to clip the magazine proofs, and went back to work on his editorials. When she had finished, he came over and as she handed him the numbered sections he pasted them into the dummy. They went to bed at six and slept till nine. George left for work while Josephine dressed and hurried to Grand Central Station to pick up his ticket for the five-o'clock train to Indianapolis. When Schuyler came home at two, she had luncheon ready.

George continued to worry about his wife's safety and her happiness. "I don't want you to be lonely," he said. "I want you to go anywhere and do anything you please, I do not wish you to be less of an individual because you have married me. That is our secret. I trust you absolutely. Do you know, Josephine," he said suddenly, "we stand absolutely alone? We can't count on anybody. The whole world is against us. The Negroes as well as the whites?"

"I'm not afraid," she said.

Just before he left, George took their marriage certificate out of his coat pocket and laid it reverently on the chiffonier. Then he confessed suddenly, "I wish I could frame it and hang it up. Of course I know we can't," he added quickly. "But I thought of it last night. You know I believe that marriage is sacred." He kissed Jody at the door.

"I watched him at the elevator, ringing the bell, holding his small pigskin locking suitcase, a small lean black gentleman in a rakish derby. 'That is my husband,' I thought, and I was thrilled and at the same time shocked and finally a warm wave of security filled me and I shut the door quietly, happily."

❖ ❖ ❖

"I have dropped completely out of sight," reads the last page of Josephine's diary. "No one in the white world but Mr. H. knows my whereabouts or will ever know. Lena and Zuma have both wired him asking for my address and if I needed money. I lie in bed with the window open toward the park and in the dark I hear someone . . . whistling that rollicking melancholy melody that as a child I would hear Ivory whistle, as he came up the railroad track at midnight . . . It has no name . . . it is just an expression of the passionate, triumphant acceptance of life."

II

Hybrid Vigor

5

All-American Newsreel

Got a story for you, Joe," the assistant editor said, raising his voice above the din of the City Room at the *New York Herald Tribune*. "There's that three-year-old colored kid up in Harlem. Some kind of a damn genius. Can read and write, they say, and spell big words." He added wryly, "Maybe you can talk to her about your Joyce or Proust—and learn something."

The cub reporter so addressed was Joseph Alsop, Jr., twenty-three-years old, right out of Harvard.[1] Alsop borrowed the file on George Schuyler from the paper's librarian, made the contact, and was invited to Philippa's birthday party, her third.

"The arrival of a visitor in the apartment's sitting room," his story would read,

> was the signal for [Philippa's] entrance from the dining room. She gave a last touch to her admirable silk turban, a final hitch to her pajamas . . . and walked in, offering a polite "How d'you do?" and a fascinating look from her enormous and liquid black eyes.
>
> After that it was time to snuggle up against Mrs. Schuyler, who also was in bright silk pajamas. From behind her mother's shoulder she vouchsafed the information that she was three years old that morning. There was a slight confusion when it developed that she believed she would be only two again on the morrow, but that was straightened out.
>
> "I want my globe," remarked Philippa, emerging from behind her mother . . . and casting herself on the floor in an attitude of which Cleopatra in her best days would not have been ashamed. The globe was produced. With a small, but unerring hand . . . she pointed out the continents and countries suggested to her. Asia, India, Africa, Australia were picked out, and each time she spelled the name, making a little song out of Australia.
>
> "That's Ceylon," added Philippa as a voluntary. "C-E-Y-L-O-N . . . ."
>
> This seemed to call for a certain emphasis, so she turned around twice gracefully and rapidly on the rug, ending flat on her stomach. From this difficult position she pointed out Liberia, "where George was"—(she calls both her parents by their first

names)—the Sudan, the Gold Coast, England, Iceland and Greenland, which troubled her by being yellow on the map. A long series of American places followed, ending with Central America "holding hands with North America."

". . . I want to do monkey."

At that she was up on the couch in one leap, from which point of vantage she continued to spell fluently Mesopotamia, hippopotamus, Hawaii, and a number of other exotics followed. Finally, the contents of the room were rapidly polished off, to the accompaniment of violent kicking with both legs and considerable smiling. Recitation time had arrived, and Philippa left her couch and mounted a chair. Her first choice was one of the verses that Mr. and Mrs. Schuyler have used in teaching her.

> The world is shaped like an orange
> It rolls around through space.
> It takes three hundred and sixty-five days
> To return to the same place.

"My Country Tis of Thee," Countee Cullen's "What is Africa to Me" ["Heritage"] and several other selections were half sung, half recited. . . .

The blackboard was fetched. Fish came first. So much time was devoted to fish that Mrs. Schuyler suggested . . . they would be tired staying on a dry blackboard so long, so some water was drawn in. Then the perch was rubbed out. "Now they're asleep," said Philippa, and began a house, which was replaced by the picture of a friend, the portrait of an ideal wife she had created for a bachelor friend of her family's, by birds and trees. Each subject was labeled in scrawling but legible print.

After a short bout of making one plus two equal three . . . she began turning round and round on one foot at top speed. Mrs. Schuyler took the hint, and started to explain her teaching method, which is to have no method. She teaches simply by showing Philippa that she herself is interested in [what] she is teaching, and Philippa at once picks it up. In this way, she has passed the eighth grade in her speller in nouns . . . and the fourth grade in verbs. . . .

Philippa, who had been amusing herself by picking out Marx's *Das Kapital* in the bookcase and spelling out the author's name and the book's title, strolled across to the huge tray of fruit on a table. It seemed almost the largest thing in the tiny sitting room. Philippa chose a ripe fig, and took a healthy bite.

"Say good-by, Philippa dear heart," said her mother.

"Good by. Come again," said Philippa and returned to her health-giving and protective fig. . . .

. . . unaware that she had made friends with a socially prominent young man—something that was very much on her mother's mind. The *Tribune*'s headline writer took it upon himself to title Alsop's article "Harlem's Youngest Philosopher Parades Talent on Third Birthday."[2]

Six months earlier, Lincoln Barnett had also written an article for the *New York Herald Tribune* on Philippa; it was entitled "Negro Girl, 2½, Recites Omar and Spells 5-Syllable Words."[3] At that time, Phil had a five-hundred-word reading-and-writing vocabulary. A reporter from *Pathé News* had also come to interview. Afterwards, he told Mrs. Schuyler that her daughter was the "most impressive child prodigy he had ever seen." Nonetheless, they did not film her, using such excuses as "the apartment was too small." In reality they did not quite succeed in hiding their disappointment that Philippa was "too white" to bill her as a Negro child, as they had intended to do. It would have been evident that one of her parents was white, and America's most mass-minded institution, the newsreel, was not ready for that. "It simply would not do to come out and show the product of a legal mixture as brilliant, beautiful, and happy," Josephine would write in the scrapbooks.[4]

Philippa had barely been into her second year when George had bought her a blackboard with some colored chalk and a complete set of wooden letters. As she teethed on her alphabet, she became fascinated with their shapes. She would grab a letter while looking questioningly up at George; he would pronounce its sound. Soon she had learned the alphabet forward and backward and how to duplicate on the blackboard the symbols she had just learned. At two, she was picking out words from headlines in the newspapers that lay strewn about the apartment and printing them on her board.

Both parents made learning into a game, and George tried to spend as much time as possible with his daughter when he was home. At the zoo, he would teach her how to spell the names of fishes and birds and animals as she rode on his shoulders. On the walks there, Philippa would run after the pigeons shouting, "Birdies red stockings." They would go by a garage and she would exclaim, "Auto sleep!" She saw her first steamship with her father on these walks and cried, "Boat bath!" Her ability to think laterally, even as a small child, seemed uncanny.

Philippa enjoyed other moments with her father: being bounced on his knee or riding on his foot in rhythm to a verse invented by George:

> Ride a horse to Granbury town
> One foot is white, the other is brown.
> Go in the morning, come back at night,
> One foot is brown, the other is white.

Phil needed little if any prodding in learning how to read and write. The uses of words intrigued her, and she constantly asked semantic questions: Why did Jody say "milkman" but not "cream man"? Why was there red pepper but not red salt? Could you say "go crooked to bed" instead of "go straight to bed"? Nor was she satisfied with simple explanations; hand in hand with this seemingly natural curiosity went a strong desire to understand, which distinguished her wordplaying from that usually encountered in children.

Josephine dutifully recorded the development of her daughter's language skills, noting when she progressed from such four-word sentences as "Jody break glass, bad," to longer, rather outlandish ones like, "Philippa urinate on Jody's clean white bed, bad! Jody wash."

Very early, George and Josephine noticed that their daughter possessed both a photographic memory and an ability for extraordinarily high levels of concentration. In 1934, when Philippa was first introduced to the typewriter and then the piano, she would spend hours at a time playing melodies and scales on one keyboard, then typing poems and short stories on the other.[5]

By the time she was four and a half, Philippa sometimes completed ten stories a week. (Despite her reputation as a master speller her typed pieces are full of misspellings—writer's frenzy, perhaps.) She often wrote fantasies, creating witches who spanked their children, put them into loaves of bread, and ate them all up; or tables and chairs who were kings and queens (respectively) and walked "softly down the street to eat a drink of whiskey," which was "verry [*sic*] good for them." Philippa killed off not a few of her characters in premature denouements; she once exclaimed after Jody recounted a tale with an unhappy ending, "That was fine. I like stories to end like that sometimes and not always 'and so they lived happily ever after' like fairy tales. That's not life, not real."[6]

Philippa's stories and poems, often sung to Jody, show a boundless imagination and a natural bent toward musical-sounding words. Characters were christened with coined names: Armarnia, Jolumbow, Salam, Ahuba, Lilchillobe, Wallaga, Varnetida, or Chilabow. The city they inhabited was called "Channa," the church they attended "Chilbensia," and the food they ate "thaga."[7]

One might ask just where little Philippa had been wandering when she invented these names. Certainly not in Harlem or Manhattan. Perhaps she had shrunk herself into the globe that her Aunt Louise (George's cousin) had given her that summer and discovered a place somewhere between Armenia and the Congo. The words she had recently learned to spell were no doubt a source of inspiration: *Zambo Anga*, *Manila*, *Ilo-Ilo*, *Russia*, *tuberculosis*, *chrysanthemum*, *zinnia*, *dahlia*.

Philippa made the news once again in 1935—this time for being the smallest and youngest American to spell the biggest and longest word in the English language,

pneumonoultramicroscopicsilicovolcanoconiosis

(a lung disease resulting from inhaling volcanic rock dust). Josephine happened upon the forty-five-letter whopper in her morning wanderings through the New York dailies. The *Herald Tribune* featured Philippa and her mouthful on July 29;[8] the article was syndicated. Ironically, the paper misspelled the word. Most of its editions had already been printed before someone discovered a missing "c" in the word, but only in time to catch the very last edition. One week later, Philippa got an "apology" (from the illustrious *Herald Tribune*), which was pasted into one of the scrapbooks next to the newspaper clipping with the missing "c."

A month later Philippa was highlighted on *Time* magazine's education page along with four other (white) whiz kids.[9] She was described simply as "daughter of a Manhattan Negro writer."

❖ ❖ ❖

Neighbors, friends, and acquaintances felt little compunction about offering the Schuylers advice on child rearing. George's champion H. L. Mencken—editor of the *Smart Set* and the *American Mercury*—alerted them, when Philippa was not yet two, to the "dire results" of such training in precociousness: "I begin to fear that your daughter is headed toward a literary career. If this turns out to be true, you must use harsh and even cruel measures to save her. Tell her that her father is a bootlegger or a clergyman. Maybe it will help to throw her off her dreadful interest in Beautiful Letters."[10]

George and Jody laughed at the warnings from the man who was virtually the dean of American letters in the twenties and thirties, if not its literary dictator, but they were less good-natured about the censorious criticisms from others: "I am quite alarmed about the progress of your child," wrote Louis G. Gregory, a friend from Atlanta. " . . . I would urge you to . . . consult some educational experts before allowing her training to go further. [The ones I have talked to] agree that the premature development of the mind of one so young may in later years do harm to both mind and body. . . . Also give her as much time as you can with other children. Too much adult society is not the best thing."[11]

As the critics grew more persistent, Josephine began to question her theories on child rearing. Had she done some irrevocable harm already by teaching Philippa so early? Publicly, Josephine defended her position: "Philippa constantly asked us questions and we answered them and she remembered our answers. . . . I couldn't be silent because she was very persistent. Once I tried it and she repeated the question 34 times!"[12] But in private, she elaborated in the scrapbooks.

> It seems to me strange that these educators claim reading and writing are not to be learned until 8 or 9, or only when the desire moves the child. What then if the desire never moves the child? Are we to raise a horde of unlettered artists—for yes, art it seems is permissible. . . . This seems to me a foul blow to art—is art then simpler than letters? Of course it isn't. . . . Now, I think, that letters are an art. That reading and writing are an art craft. That, furthermore, our civilization cannot be understood or assimilated without them. One can certainly get along splendidly in this age without being able to paint a picture, make a basket, or hammer a dish but one can only get a little way unless one can read and write.[13]

The scrapbooks that Josephine kept for her daughter resembled, at one moment, carefully annotated research notebooks, minutely recording in almost clinical language a subject's response to stimuli, and at the next moment, repositories for philosophical musings, or bulwarks against the day when a daughter might sit in judgment of a mother. From time to time there is a poignant—or maudlin—reminder of a mother's sacrifices.

Many entries in the scrapbooks were about the glories that were Philippa's. And Jody's desire to prove the success of her marriage, through her daughter, was clearly becoming an obsession:

> Everywhere your Daddy goes people ask about you, having read of you. When I take you to the Library on 145th Street the lady librarian makes a great deal of fuss over you. It is very nice having an accomplished daughter. I glow with pride. If only we can keep this up, darling, maybe you can be a great personality in the world. I hope so. . . .[14] If only I might make you sense the preciousness of life while you are young! If only you could see with the double glasses of youth and age. Then you might truly perceive life in the FOURTH dimension. It is a good thing to have a fine start in health and intellect but it is only a foundation. Upon it you may build a shack or a palace.[15]

❖ ❖ ❖

In June 1934, when Phil was not quite three, the Schuylers accepted an invitation by Columbia University's Child Development Institute to test Philippa's mental development and social adjustment. Using the Merrill-Palmer and Kuhlman-Binet tests, Columbia concluded that Philippa was "endowed with superior mental ability" and noted that "tasks involving ability at the upper limits of success were met with rather well-defined evasive techniques."[16] (This reaction is not uncommon to child prodigies, who tend to discover alternative systems for getting around things that do not come easily or quickly.) Yet both parents continued to shun the word *prodigy*, claiming that their extraordinary daughter was merely the result of hybrid genetics, proper nutrition, and intensive education.

When Josephine had Philippa's IQ tested about eighteen months later, in 1936, the results proved to her that her daughter had every tool available to succeed in life. In fact, Philippa was tested several times by different institutions that year, first on February 22, by Dr. Baker from Columbia. When he told Josephine the result, she reportedly asked, "Is that supposed to be good?" Later that afternoon, after they had arrived home and Philippa was out of the room, Josephine said to George, "Well . . . the worst has come to the worst. Your daughter's IQ is terrible." George looked shocked. "Yes," she said, "it's only 180."[17]

That same month, NYU tested Philippa as well. Her EQ was 200 and her IQ registered between 179 and 185. Josephine was adamant that her daughter's IQ should be a private matter. Although George may have felt similarly, he was bursting with too much pride to keep it a secret. On tour, he revealed it to journalist Hortense Young, who reported it on April 18, 1936, in the *Informer*. Jody was angry but it was too late and all of the subsequent media coverage invariably mentioned it.

❖ ❖ ❖

One month shy of 1935, after much palaver between husband and wife, Jody placed their three-year-and-three-month-old child in front of their apple green piano. In part this decision was made to deflect the continuing criticism of their teaching Philippa reading and writing "too early." The person who would guide Philippa through her first five years of music, her first nonparental mentor, was Arnetta Jones, a black graduate of the Juilliard School of Music.

From the beginning, Phil evinced the same prodigiousness in music as in letters. She approached the keyboard with exuberance. "If your music teacher gives you a piece in Middle C," wrote Josephine barely two months after her daughter's first lessons, "you learn it there, then the next thing we know you are playing it in the treble, in the bass and sometimes part of it in each."[18]

Nine months later Philippa debuted on a local New York radio station, WINS. And four months after that, on January 19, 1936, she gave her first performance before an audience, at the YMCA on West 135th Street. Twenty-six of Arnetta Jones's students—aged four to twenty—performed for their families and friends. Philippa was the youngest.

Until her YMCA performance, Phil had always insisted she would be a dancer, a writer, or an artist—all the things Jody herself had wanted to be. That night, however, as she was tucked into bed, she announced to her mother, "I'm going to be a musician when I grow up!"[19] And the next day she set out to prove it, composing two pieces—one a setting of the Irish ballad "Up the Airy Mountain," and the other composed from whole cloth, entitled "Oriental Dance of Little Egypt."[20]

Josephine entered Philippa into her first musical competition at age four years and ten months. It was the National Piano Playing Tournament, sponsored by the National Piano Teachers Guild (NPTG), and held at Aeolian Hall in New York. In addition to four required pieces, Phil played six of her own compositions. When the results were announced, she was one of seven winners out of 103. Reward for her superior performance was a gold seal certificate and a place on the NPTG Honor Roll. Philippa became the youngest person ever to be so honored. The judges were impressed by the facility with which she played Mozart's Minuet in G, and considered her "Nigerian Dance" to be "superior in every way."[21] Her other compositions were entitled "Rolling Home on My Roller Skates," "Pansy Bells," "The Butterfly," "The Wolf," and "Golden Fish in Silver Waters." Each had its own story, which Philippa related, while playing, in a singsong manner with traces of a lisp.

It was during the Aeolian Hall competition that, to even Jody's surprise, the judges discovered Philippa had perfect pitch. Irl Alison, then president of the NPTG, issued a prepared statement to the press saying he had found a "musical genius." *Time* magazine picked up the story and on June 22, its music page was entirely devoted to Philippa:

> Prodigious at more than music is this Harlem-born daughter of a white mother and a coal-black father. Mrs. Schuyler paints, writes for Negro newspapers. George Schuyler was a day laborer and a dishwasher before he became a novelist . . . a contributor to *American Mercury* and *Saturday Evening Post.* All three Schuylers subsist on raw vegetables, raw meat, a diet which Mrs. Schuyler claims is largely responsible for her daughter's precocity. . . . She is keen at mathematics, reads fourth-grade books, writes poetry, draws and paints, turns out neat letters on her father's typewriter.[22]

This was the second time in so many years that Philippa had hit the pages of *Time*.

What set Philippa apart from the "typical" child prodigy was her ability to compose. In fact, composing seemed to come as naturally to her as seeing and hearing. She would visit a toy shop with her mother and sit down to write "The Toy Maker's Ball" with the clacking of little mechanical figures beating steadily in the background. Watching WPA workers fix the street in front of her apartment, she would write "Men at Work," a whimsical piece. (As the foreman's attention is called elsewhere, the men slacken their pace but when they see him returning, they speed up again as if there were no tomorrow.) She would visit a shooting gallery and faithfully reproduce the din and the mechanical rhythm; she would go to the shores of Lake Ontario and write a composition in which you can almost hear the wind and the lapping of the water on the shore.[23]

These early pieces all show a seemingly effortless ability to synthesize story and music, and to fashion a work far beyond her chronological age. In fact, Phil composed music in much the same way she had written verse—as if she were telling a story. Even as an adult, Philippa responded to her musical muse first as a teller of tales and then, in some mysterious recess of her brain, transmogrified them into notes.

"The Wolf," for example, composed shortly after her fourth birthday, uses almost the entire keyboard. It spans five octaves—two below middle C and two above. The fourth measure has a dissonant chord (built on flat II) representing the wolf. Hungry, the wolf advances toward the innocent skipping lambs (dotted sixteenth notes). Fearful, they scamper away (eighth notes) but become more and more petrified as he approaches. Runnning faster and faster, they stumble one on top of the other (thirty-second notes cascading in descending scales). Finally the lambs are safe, all is quiet, and the wolf goes home, sad and still hungry (quarter-note chords).

Everything engaged Philippa's early compositional mind: a goldfish who thinks the sky is the sea and jumps out of its bowl onto the floor to its unwitting death; an onion, shunned because he smells so bad (dissonant chord clusters), whose only friend becomes a condescending potato (quarter-note block chords); and, perhaps most revealing, cockroaches. "The Cockroach Ballet," written at age four, was soon her favorite; she rarely missed an opportunity to tell someone the plot: The cockroaches are having a meal in someone's kitchen. Suddenly, footsteps are heard. The roaches flee swiftly into hiding, to wait while the humans feast. When the people have gone, they again creep out and dance gaily as they pick up the crumbs. But the humans return. There is a massacre; they are all crushed to death and the people march off in triumph. But one little roach who escaped destruction peeps out. He dances sadly until, suddenly, another little roach appears and together they dance, happy and confident that cockroaches will go on forever—unfortunately.

Philippa loved performing as much as composing. During 1936 she appeared in public three more times,[24] continuing to charm audiences with her diminutive size and to intrigue them with her gracious manner and musical abilities. George was away a good deal that year and heard her only once: on November 8, when she played at the St. James Presbyterian Church in Harlem under the auspices of the Federation of the National Association of Negro Musicians. He cried a little watching the way she got up, bowed, and played like a professional.

❖ ❖ ❖

Jody was now entering her four-year-old into every contest, every amateur hour, every musical event, she could find, even though she knew Philippa was too young for some of them. Aware of the pressures she put on her daughter, Josephine again turned to the scrapbooks to compose an itemized brief in defense of her actions:

> I realize, darling, that these contests are often stupid and I know many educators disapprove of them. But here is why I persist:
>
> 1. George and Jody have nothing to give you save opportunity. And because we have refused to be conventional in our way of life, opportunity will not come to us unsought. We must seek the best for you, go out and get it or it will pass us by. We, and especially you, are a challenge to the set notions of America on race. These prejudices, erected to justify a diabolical system of exploitation of man by his fellow man, will not easily give way. Only genius will break them down, and that you have. So I take you about as much for the education of America as for the education of Philippa.
>
> 2. Aside from all this, it is good training in social conduct for you. You learn to meet new people in a friendly, charming fashion, to do your part under all circumstances. . . .
>
> 4. Finally, I heartily wish my parents had given me the kind of help I'm giving you. As I look back, I see how I longed to be important, to be taken seriously, to be given a way of life, pointed a road that was interesting. Instead, I was given money, social position and treated like a baby. I loathed it.[25]

Although Phil acceded to her mother's demands, occasionally the strain was apparent—usually the result of the conditions under which she had to play. During a piano performance in front of a psychology class of 350 at New York University in April 1936, four-and-a-half-year-old Philippa lashed out.

They had no bench, only a stool, which she could not manage well. They crowded around her too closely. She played several pieces fairly well but when she began transposing "Little Pussy," after having done four transpositions perfectly, Philippa missed a note. She was obviously annoyed and asked her mother what the missing note was. Jody did not know; Philippa glared at her mother and said, "Well, you ought to know." She refused to go on until the missing note was found. "I never go on 'till I finish a thing. You *know* that, Jody!"[26]

Another incident was her ill-fated audition for Major Bowes' Original Amateur Hour, an insensitive, sensation-seeking boiler-room program where

Phil played badly. Later, in retelling the incident to the scrapbooks, Jody would add: "But the fact is, I do not care greatly for you to appear on the Amateur Hour. That is purely a mob situation and your ancestry being what it is and you being in addition a so-called prodigy, both things would greatly annoy the mob. There is nothing Americans generally hate and fear so much as a prodigy. He belies their . . . 'every man is born equal' idea. You can be inferior in America but never, never superior."[27]

At other times Philippa remained totally unfazed by the unexpected. On February 25, 1936 (her father's forty-first birthday, for which he was away on assignment), she appeared before a psychology class as part of the continuing battery of tests undertaken by Columbia and New York University. She was in high spirits: "One key stuck on the old battleship and you pulled it up without stopping," the scrapbooks read.

> Then you got up and bowed very low, which brought down the house, for you act like you are broken in the middle. Then we mounted a small lower platform where there was a blackboard and I put you through your paces. . . . When I asked you who was dictator of Germany you said "Hitler," but added "because he hits everybody!" "Hush," I whispered, but Dr. Baker was grinning and so were the front rows. . . . We ended up by you writing in script "I love you, George," which you spelled Gorge. I said, "is that right?" and you shouted G E O R G E and changed it in high humor. . . .[28]

Phil was in even higher spirits that night when they returned home: George was sitting, unexpectedly, at the kitchen table. He had so much wanted to be with his family on his birthday that he cut his assignment short. The table was set with flowers and candles, and George had prepared filet mignon and mushrooms for them. His grandmother's picture graced the table; it was "the happiest night of his life, to have his family all together. . . . You were terribly flattered," wrote Jody for her daughter, "to stay up till 11:30. And you went to bed with a rose in your hand. Next day George left for a two-month lecture tour, and again we are widows."[29]

6

Psychological Care of Infant and Child

The light that guided Josephine's footsteps in raising her daughter came from a man named John Broadus Watson. A disciple of Ivan Petrovich Pavlov and a precursor of B. F. Skinner, he was the guru of habit training and the Dr. Spock of his time.

A distinguished professor at Johns Hopkins University, Watson had founded a school of psychology called behaviorism. Rejecting the concepts of introspection and instinct, he described behavior almost exclusively in terms of responses to stimuli. In 1928, he published an influential book entitled *Psychological Care of Infant and Child.*[1]

Watson contended that the newborn comes into this world naked not only in body but also in mind. It has only three unlearned responses: fear (of falling or loud noises), love (when gently stroked), and rage (when restrained). All others are learned.

His advice was to treat children

> as though they were young adults. Dress them, bathe them with care and circumspection. Let your behavior always be objective and kindly firm. Never hug and kiss them, never let them sit in your lap. If you must, kiss them once on the forehead when they say good night. Shake hands with them in the morning. . . .[2] Mothers just don't know when they kiss their children and pick them up and rock them, caress them and jiggle them upon their knee, that they are slowly building up a human being totally unable to cope with the world it must later live in.[3]

Watson advocated a rigid schedule: wake-up time at 6:30 a.m.—sponge lightly; breakfast including orange juice; allow to romp until 8 a.m.; put on toilet for twenty minutes, without toys and with door closed (a dictum which Jody followed, coming in to wipe Philippa only after a bowel movement); and as soon as possible thereafter put the child out into the sun—naked in the summer; lunch at midday; one-hour nap after lunch; a tepid bath at 5:30 p.m.—"the object is to get the child clean not to entertain it";[4] a light meal after bath;

to bed at 7 p.m. (for the two- to five-year-old), with hands under the covers if a thumb-sucker, over the covers if a masturbator. Josephine adhered to this routine faithfully.

No equivocator, Dr. Watson largely discounted heredity and instinct, and believed that a child's character can be "spoiled by bad handling [within] a few days."[5] He assailed the competence of most parents to bring up baby and questioned "whether there should be individual homes for children—or even whether children should know their own parents. There are undoubtedly much more scientific ways of bringing up children which will probably mean finer and happier children."[6]

In an even more argumentative vein, Watson suggested that "No one today knows enough to raise a child. The world would be considerably better off if we were to stop having children for twenty years (except those reared for experimental purposes) and were then to start again with enough facts to do the job with some degree of skill and accuracy."[7] It can be safely assumed that Josephine thought herself above these criticisms.

Jody subscribed wholeheartedly to Watson's ideas that "practically the whole course of development of a child is due to the way [it is] raised."[8] She relished the behaviorist's belief that "If you start with a healthy body, the right number of fingers and toes, eyes, and a few elementary movements that are present at birth, you do not need anything else in the way of raw materials to make a man, be that man a genius, a cultured gentleman, a rowdy or a thug."[9] She happily accepted the premise that it was her superb training and special diet that had produced Philippa's talents and abilities.

Josephine had great difficulty, however, accepting the thesis that "almost nothing is given in heredity."[10] After all, she was dedicated to the notion that it was the hybrid vigor resulting from their miscegenation that had afforded Philippa the potential for high achievement.

Jody believed in instilling in her child a sense of self-reliance, in slanting her toward a rewarding life, and in determining her daughter's vocation. This was in direct opposition to the teachings of John Dewey and other educators during the first quarter of the twentieth century: that training should allow the child to develop from within and in its own time—a doctrine that, in Watson's words, "had made us lose our opportunity to implant and then to encourage a real eagerness for vocations at an early age."[11]

Watson's advice that "no child should get commendation and notice and petting every time it does something it ought to be doing anyway"[12] was strictly followed by Josephine. She also believed that children needed discipline. "Children want and need 'ORDER' above all, order in their little orbits," Jody would write in the scrapbooks just after Phil had turned three.

> One must learn first how to do things by rule before one can know how to deviate from the rule advantageously. The world is a strange and marvelous and terrible place for children; they crave a guide line to take them safely through the maze. . . . Children want and need to imitate someone. They need the continued inspiration of pleasant and agreeable adult companionship. After all, childhood is

simply the preparation for adulthood, it is not a state within itself. It is a stepping stone, a preparation for conscious living. . . .[13]

It would appear that Josephine was either ignorant of a child's developmental milestones, or, because of her rigid behaviorist bent, had a total disregard for them. She expected constant confirmation and love from her child, and her unrealistic expectations occasionally led to disappointment and even rage. At times, when Philippa was, in fact, behaving no differently from other children going through a similar stage, Josephine would get annoyed with her daughter for being "ornery, difficult and arrogant."

Perhaps projecting disapproval of her own upbringing, Jody recorded:

> You are stubborn and self-willed and want to do things your way. I try not to give in to you, for I would hate for you to grow up like that. Last month I spanked you half a dozen times hard. . . . You remember the week after if I spank you and talk about it quite rationally and say you will not be bad again. While I am whipping you, you often put your arms around me and say most plaintively, "Oh, Jody, don't you be bad to me. Oh Jody, please be nice to me." But at the same time you stubbornly resist doing whatever I have told you to do. . . . I simply must not let you grow up too mulish, it will ruin your whole life, and no matter what your talents, it will be like a huge mountain in your path. I must teach you to adjust yourself swiftly to new situations (you do this well though) and to forget anger and forgive quickly (this you don't do).[14]

When Philippa was four, Jody reported a telling conversation between mother and child. It was New Year's Day, 1936, and Josephine had suggested it was time to make some resolutions. After Jody explained what a resolution was, Philippa said,

> "Well, I think you ought to make a better rule about whipping me. I don't think you ought to whip me. When you whip me it hurts more than you know. . . ."
>
> "But darling, why do I do it? I don't like to. But I must teach you the rules. If I don't you will not be able to take your place later in society."
>
> "But teach me some other way. . . . I don't think it does me any good. . . . It doesn't make me want to follow the rules. It makes me think you can't love me as much as you say you do."
>
> "What shall I do then?"
>
> "Put up my play things till I am good again."
>
> "All right! That's exactly what I will do. . . . I promise you there will be no more whippings," and you hugged me.[15]

A week later when Philippa would not eat her lightly broiled heart, after George had gone to a lot of trouble to get it for her, Josephine became cross and said, "Very well, then, you won't have your bedtime story tonight."

Phil was not happy but after a while she remarked, "Well, anyway, I didn't get a whipping."

It is troubling to read of Josephine's frequent "beating," "whipping," and "slapping Philippa across the face." It was in stark contrast to George's view of corporal punishment. On July 2, 1934, there is a small entry in the scrapbooks in Jody's hand: "George had to spank you the other night, the first time he ever did it, and it almost killed him."

❖ ❖ ❖

George and Josephine began to treat their daughter, from a very early age, as an intellectual equal. When they talked, the child listened attentively and when she did not understand something, she would ask for an explanation. Over time, Philippa's natural curiosity waxed, not waned. In particular, their evening meals, when George was not away on assignment, became full-fledged forums on every topic imaginable—from the political situation in Bolivia, to the anatomy of the ear, to the meandering Brazos River in Jody's native Texas. The meals would go on for a long time. Later, when Phil was a little older, they would hold a dinnertime "quiz." The loser did the dishes.

The parents also followed Watson's advice to talk to children frankly about sex at the earliest possible age—even before they ask questions. What's more, they would "make it interesting," as the guru urged. "You are competing for credibility with 'secret' information learned in the streets and parks from older kids."[16]

As early as age three, Philippa had asked George when they were taking a shower together, "What's the matter with your all-wet?" Later she had said to Jody, "Men have mustaches on their faces. George has a mustache on his all-wet, too."[17] But she must have forgotten that, because two years later at age five, Philippa climbed up on the couch, pointed at the naked savior on the cross and asked, "What is that?"

"What do you mean, exactly?"

"*That*," and she indicated the penis.

"That," Jody explained, "is the organ of reproduction. Have you not noticed George? That is what makes a male a male, and different from a female."[18]

Jody also remembered that Phil, after seeing some boys urinate in the park, asked her what it was "that stuck out in front of them when they urinate." And having been provided with an explanation, very seriously said, "I should think it would be hard for them to sit down. Doesn't it get in the way?"[19]

During a test at one of the universities, Philippa had sketched the male member in her drawing of a man, labeling it "Zeb."

Two weeks later, just as she was going to bed, Phil asked out of the blue, "Jody, if a man . . . cut off his pipe organ, would he become a woman?"[20]

❖ ❖ ❖

Reading the early scrapbooks one is struck by Josephine's obsessive need to control her daughter. Her domination is evident on almost every page. At times Philippa was literally forced to "play for her supper." When Josephine discovered what she considered Philippa's "first lie" during Christmas week of 1935 (she had asked her to go to the bathroom just before dinner and found out that Phil only pretended to go), Jody sent her daughter to bed without a meal. Phil cried; Josephine replied, "Only if you go to the piano and play six pieces can you have your steak."

Mother and daughter began to develop a mutually manipulative relationship: Philippa became as demanding of her mother to "perform" by asking the same questions over and over again, as Jody was of her. And Phil became more savvy at playing one parent off the other—a perfectly normal behavioral pattern. What was less normal was Josephine's reaction. "You do shift loyalties," Jody wrote for her daughter's benefit. "Whoever has just taken you out, or been nice to you, you adore over all others and will even say 'You go on and read Jodie, I stay with George.' You decide only your favorite can have the honor of washing your face or assisting you in any way. But with a few smiles I soon win you back."[21]

But Jody was Philippa's whole world, and vice versa. Phil constantly sought her mother's approval. Jody's persistent demands for excellence did not create, on the whole, a rebellious child, but rather one who placed equally harsh demands on herself. Josephine in turn saw her daughter as an extension of herself, a way of recreating what she might have become. Jody found it increasingly difficult to recognize Philippa as a separate person and their relationship grew intensely symbiotic. One is reminded of the Italian author Ambrosio Donini's point that all slave and seigneurial societies throw up the figure of the *padrone*, who simultaneously protects and abuses, nourishes and punishes.[22] Often, it is a difficult relationship to end.

Josephine continued to instill in her daughter a desire for the best. As Philippa grew more proficient at the piano, George and Jody began paying her a penny for each piece played perfectly. In her fears of failing the rising expectations of her parents, Philippa created an alter ego when she was five. Her name was Rosewings, the fairy. Rosewings was always good, never bad, never failed at anything. Philippa was just the fairy's rather stupid shadow. It was her alter ego who each morning leapt out of bed, tiptoed to the piano, and woke everyone up with her morning concert.

George and Jody encouraged these childhood fantasies. They took to leaving little notes scattered on the music stand or the bench, to both Rosewings and Philippa. "Dear Fairy," read one. "Please do all the pieces this morning so you won't forget them. Do the Wild Rose '6' times. Thank you. Fairy Queen."[23] Or from George: "Dear Mistress Philippa: May I suggest that you play your Minuet with Variations once with the Music book? I noticed the last two or three mornings two or three wrong notes."[24] Usually the notes to Philippa pointed out her mistakes; those to Rosewings, how well she played.

Rosewings had had a precursor: Muffy-Muffy Andress, who came with Philippa in April 1935 when the Schuylers moved into a larger apartment at 320 Manhattan Avenue. Muffy-Muffy Andress was a little cutout doll with a string tied around her ankle.

Muffy was not only a friend; she was, again, a kind of doppelganger, who could accomplish all the things that Phil could not. Philippa wanted Muffy (and her three children), George, Jody, and herself to move to a city she had just created: Minnelos, poised high atop a mountain. There were no "bad" people or policemen because a fence prohibited them from entering. Communally owned fruit trees abounded, and anyone could pick what he or she needed. No noise but rain, no animals but dinosaurs long dead and turned to dust, no one to "holler" at Philippa. The town had "lots of bathtubs and houses of stone that could not grow old." When Josephine asked where it might be, Phil said, "Texas."[25]

Perhaps the most disturbing aspect of Philippa's early years is her almost complete isolation from the company of other children, at least until she was five years old. Without exception, child psychologists, regardless of their orientation or bias, strongly advocate the need for peer companionship. Even Watson recommended this.

But Phil had no playmates, black or white, of her own age, no peers with whom she could share toys, bicker and fight, and generally find the limits of acceptable behavior. No doubt a brother, sister, or steady companion would have helped. And it is this isolation more than the "premature" intensive teaching that educators should have feared.

Apparently George wanted another child, but Josephine would not agree, arguing that they could not afford it. To her friends she would occasionally claim that she nearly died delivering Philippa (which was apocryphal) and feared another pregnancy. It may be closer to the truth that Josephine felt another child might dilute her singleminded, if not obsessive, preoccupation with nurturing a genius.

The fact that Philippa spent almost all of her time with Jody is recalled by several friends, especially the Harlem Renaissance artist Elton Fax. He first met Phil when she was posing in the nude for Augusta Savage's art class (at age two and a half), and he remembers seeing her often in the park with her mother, but not in the company of other children.[26] But even if the Schuylers, and Jody in particular, had sought the company of other children, their "selectivity" would have made this a difficult task.

Philippa was not yet five when Jody enrolled her in dancing lessons at Doris Humphrey and Charles Weidman's Academy of Allied Arts on West Eighty-sixth Street, but classes were only a couple of times a month and hardly provided playmates. The Schuylers had also sought to enroll Phil in various prestigious schools, but there was always something wrong, or some other obstacle stood in their way. Jody inquired at several musical institutions such as the Curtis Institute in Philadelphia, but her daughter was too young. They applied to Walden, the "progressive" school in New York, where she was accepted, but New York University advised against it, insisting it was mostly for the "maladjusted" and might "warp" Philippa. They then tried the Lycée Français, but Philippa was rejected, in part because of her age (she was too young again), in part because they expected children to speak French fluently, and in part because they claimed a child of Philippa's ability would not fit in with the group. But perhaps the exclusive social nature of the school was not conducive to accepting a black child.

During the fall of 1936, New York University's School of Education, which continued to monitor Phil, decided to send a tutor, gratis, to Harlem two days a week. Thus a normal arena in which Philippa might have interacted with her peers was again denied.

7

Sergeant Jackson

It would be difficult to find two more dissimilar Americans than Josephine, the lily white heiress to pioneer bounty, and George, the black child of American segregation.

Superficially judged, George had a typically middle-class upbringing. He was raised in Syracuse, New York, in a rambling two-story house that sat serenely on a tree-shaded street with plank sidewalks. The interior was spacious—a carpeted parlor, a large sitting room where the family gathered nightly for Bible reading near the big coal stove with isinglass windows, an elegant dining room with an oval table that sat eight, a sewing room, an enormous kitchen that was mostly the domain of his martinet grandmother, and numerous bedrooms, both upstairs and down. His was the only black family on the street. The parlor held many books about Negro war heroes, poets, and writers.[1]

George's family was a proud one. If any of his ancestors had ever been slaves, it must have been before the Revolutionary War. A great-grandfather fought under General Philip Schuyler (who became his eponym). His maternal great-grandmother came from Madagascar, was bound for service in New Jersey, and married a German sea captain who had settled in America. The family looked down upon those who had been born in servitude; they neither cherished nor sang slave songs.[2]

His father,[3] who died when Schuyler was three, had been chef of a local hotel. He "was an aristocrat in the . . . community [who] affected baronial living, insisted on a good table, and dressed well."[4] George remembered him as a balding brown man with a stately mustache, who had traveled the seven seas as boss of many a ship's galley and told lively tales of his experiences. The women in the family talked about him long after he had passed away.

When his widowed mother remarried, the family moved to another house on the outskirts of Syracuse. George, who had learned to read and write before he was five, was now old enough to attend school, and he eagerly looked forward to his first day. But it was not what he expected. An Italian boy called him a nigger. The insult escalated into a fistfight, and George came home, his face tearstained and his shirt bloodsoaked.

His mother washed his face and gave him a fresh shirt. His stepfather drew him toward the mirror. "You are always to fight back," his mother said standing behind them. "We are as God made us; it is what's inside our heads and what we do that is counted."[5] It would be a moment George never forgot.

As Schuyler grew, he realized there was little if any opportunity for him in Syracuse. The colored people, he wrote, were social and economic pariahs.[6] The tiny black community there was so fragmented by class divisions that any group unity was out of the question. Many of the blacks were "not his kind"—they were frequently in trouble—and his mother had forbidden him to associate with such "riffraff." The world outside offered a young Negro little better than redcapping or working on the railroads as a cook or sleeping-car attendant.

George turned his eyes toward the only road he knew would allow him a way out—the United States Army. Where else could a poor black teenager have the opportunity to travel, to receive pay more or less the same as that of whites performing similar jobs, and continue his education at no cost? While regiments were still segregated, blacks could compete, as an example, with whites in athletics—something they could not do in civilian life. That the black community thought well of its soldiers was another enticement to join the service.[7]

George enlisted during the spring of 1912 when he was barely seventeen and had not quite finished high school. His mother, "the apostle of the possible," as George kindly referred to her, unhesitatingly lied about his age when she enlisted him.

In mid-July he was sworn in at Fort Slocum, in New Rochelle, New York, and assigned to the all-black Twenty-fifth Infantry.[8] After three weeks of intensive basic training, he was shipped with twenty other young Negroes to Fort Lawton, Washington, an old, comfortable post on a peninsula that jutted out into Puget Sound.

Nearby Seattle was a pleasant, cosmopolitan place to visit when off duty. Racial prejudice was moderate, and considering that a beer was ten cents, carfare five cents, and a burlesque show twenty cents, the fifteen-dollar-a-month pay could go a long way, if one did not gamble it all away the first night.

It was, therefore, with both excitement and regret that the men learned at year's end of their impending move to the Schofield Barracks in Hawaii. En route, in San Francisco, George and his buddies spent four wild nights in the dives on the Barbary Coast. The reform movement had not yet struck San Francisco, and the town was wide open. Pandemonium reigned nightly. In dance halls, combos of "black boys pounded out the jungle beat for boozy shufflers."[9] Or the soldiers gambled in neighborhood cellars. In one of them, cream-colored "Gold Teeth Mame" presided over the district's biggest crap game. She called the dice in a deep voice, a loaded .45 revolver parked near her right hand to reinforce her decisions. All this was heady stuff for a seventeen-year-old kid from upstate New York.

Schuyler found Hawaii to his liking. He saw many mixed couples and a veritable melting pot of races, making whites a minority. There were opportunities for long hikes; challenges to climb the volcanic mountains; occasions to read a book borrowed from the post library, or to swap stories with his buddies who came from all parts of the country. Some were educated; there was even a sprinkling of college graduates unable to find civilian employment. But for many, "education" was a rudimentary ability to read and write. So George soon found himself teaching English and geography to a group of grown men. And in no time at all, he had wangled a job clerking in the orderly room.

Opportunities for other diversions were also close at hand—Honolulu was only a short train ride away. There was some prejudice against blacks, but by and large the women of the Iwilei district, no less than the saloon keepers, appreciated the serviceman's readiness to spend a buck on booze and sexual pleasures.

Schuyler was a soldier's soldier who not only enjoyed a life of strict discipline and order but gambled, drank, cheated at cards, wenched, and hustled as well as any other enlisted man.

After his three-year stint was up, George tried civilian life for a few months. Pearl Harbor was being constructed and he worked briefly as a common laborer, outdoors, under the broiling sun. But this was not to his liking and he decided to become an entrepreneur. With the purchase of a secondhand Packard touring car, he and a buddy transported soldiers into town and spirited them back in time for reveille. Life was easy, and the job lucrative.

But George was again getting restless. He availed himself of a free passage to the States as an unemployed ex-soldier to see the Panama-Pacific International Exposition in San Francisco. When he began looking for a job, he found economic opportunities on the West Coast no greater than they had been in the East: only menial or servile occupations were available for the few thousand Negroes (in an ocean of whites) hanging on the fringes of western society. George reenlisted: The army, once more, seemed a better bet.

It was in his second enlistment, in a highly disciplined company run by an out-and-out martinet, that George was promoted to corporal. He also began writing. Schuyler composed eagerly read satirical skits for the *Service*, a weekly magazine; produced several pieces for the morning edition of the *Honolulu Commercial Advertiser;* and started a short-lived newspaper, the *Daily Dope*, which he tacked on to the bulletin board.

Soon after America declared war on the Central Powers, a separate Negro Officers' Training School was established at Fort Des Moines, Iowa. In what appeared to African Americans as a historic development, approximately twelve hundred young black men from many parts of the country, mostly with some college education, were selected as candidates. To train them, eighty noncommissioned officers from the Twenty-fifth Infantry were sent from Hawaii to Des Moines. George Schuyler was among them.

In three months, these "ninety-day wonders" were supposed to receive instruction for commissions as infantry, artillery, and corps-of-engineers officers. Along with them, some of the instructors were to be commissioned.

After the training had been unexpectedly extended into the fourth month, it became apparent that the government had no plans to send the black officers to the front, and, in fact, was not quite sure what to do with them. Nonetheless, 639 of the men were commissioned, from among both the candidates and the instructors. Schuyler was made a first lieutenant.

Life at the fort was not very strenuous. With plenty of time on his hands, George became a frequent visitor to Des Moines' night spots. One "bibulous evening," he met a girl named Jack Patterson. They danced the popular shimmy-sha-wabble, and George fell for her "like a battleship's anchor."[10] Jack Patterson was an "optional Negro"; that is, she could pass for either white or black. Years later, in his autobiography, George would describe her as having "long, wavy brown hair, blue-green eyes, a babyish face that would have enthralled a Hollywood casting director, a bosom menacing her chin, a slender waist, and limbs that would have enchanted Rodin."[11] Unfortunately, she was also a hopeless alcoholic—and insisted on driving her low-slung red roadster in all states of inebriation. Like Kipling's old soldier, George admitted, "I learned about women from 'er."[12]

After being commissioned, Schuyler promptly donned his new officer's uniform with silver bars and traveled on leave to Syracuse and Boston, where he was greeted warmly and lionized, before reporting to Camp Dix, New Jersey, to await further assignment. At twenty-two years old and with an army salary of $181 a month, saluted by the ranks wherever he went, George felt that life was good. From childhood he had wanted to be somebody, and now he was.

A dozen years later, he would write an essay for H. L. Mencken's *American Mercury* entitled "Black Warriors."[13] It is full of wondrous vignettes about George's army buddies and experiences. Told with compassion, the stories reveal Schuyler's powers of observation, an understanding of human nature, and a fine ear for "nigger talk" and its regional variations. Most memorable is his vignette on Sergeant Jackson:

> Everybody in the regiment expected Sergeant Jackson of the Machine Gun Company to be sent to the colored Reserve Officer's Training Camp when the news came in that there was to be one. He was a tall, smooth black fellow on his second enlistment and the idol of the regiment. He was originally from a small town in Mississippi, but he had been taken North early and had graduated from a polytechnic school in New York. He was an Expert Rifleman, a good pianist, short stop on the regimental baseball team, and a squad under his command had won the machine gun contest the year before.
>
> His name appeared on the first list drawn up and he was happy. He told me that night that he thought the war would bring a better understanding between the races. Because of the fact that we had both been turned down two years before when we had applied for permission to take an examination for commission in the

Philippines Scouts, he had thought we might not get a chance at the bars. Now he saw himself a captain or major.

On the long journey to the training camp he grew more optimistic and declared again and again that a new day was dawning for the lowly Aframerican. He was soon spouting as much nonsense about democracy as a Four-Minute Man. The older non-commissioned officers would smile grimly and say nothing when he held forth.

At the training camp Jackson was in his element. There were a thousand young Negro candidates for commission there, most of them college graduates and the majority of them believing that a new day was at hand. To hear them talking patriotism one would have imagined that the Germans had treated men of their color worse than the Americans, English and Belgians. It was exceedingly dangerous for any of the skeptical to suggest that there possibly might not be a love feast of blacks and whites after the world had been made safe for democracy. There was no more outspoken patriot at the camp than Sergeant Jackson.

Even the fact that the Negro candidates were being given only infantry instruction, although they were scheduled to be officers in a division composed of all arms, did not arouse Jackson's suspicions.

When the six hundred candidates received their commissions, he emerged with a captaincy. His joy knew no bounds and he strutted around the barracks like an American who had just been noticed by a titled Englishman. He expected to be assigned to the artillery or the engineers because of his knowledge of mathematics and horses.

At the big camp in the East to which he was sent, he found himself assigned to duty with a battalion without men—a paper organization, while former non-commissioned officers who were good material for infantry officers but poor mathematicians were assigned to the Negro artillery units with full complements of men. Jackson still could not see through it all. Finally he was assigned to an infantry company and his spirits rose. He said he was aching to get a crack at the Kaiser.

During the Christmas holidays he went on leave of absence to see his father and mother, who had returned to Mississippi to live. The Jim Crow laws did not yield to the war sentimentalism, so Jackson rode into his home town in the colored coach. As he stepped from it, he noticed a number of white soldiers in the station.

"Look at that nigger in captain's uniform," somebody yelled. Jackson hurried through the colored waiting-room with his suitcase. At the street door a crowd of soldiers and civilians met him.

"Don't think we're gonna salute you, nigger," they warned. Jackson tried to push his way through but they pushed him back.

"Where'd you get that uniform, darkey?" they asked. "Why don't you make us salute you?"

They had completely surrounded him by now and he glanced helplessly to the right and left. There was not a kind look on any of the faces circling him.

"Let me through, please?" he requested, with as much dignity as he could muster.

"Oh, so yah wanna git away, eh?" jeered the ringleader. "Well, wait'll we git some souvenirs."

He reached out and took the insignia off one of Jackson's shoulder straps. Another hand took the other. Somebody snatched his hat off. Deft fingers unfastened his Expert Rifleman badge. Willing hands pinioned his arms while equally willing fists pummeled him. They were all laughing.

Jackson suddenly broke loose with a wrench and fought his way through the mob. Down the street he ran with the pack after him. Mr. Sanders, a friend of his father, saved him by pulling him in his automobile and driving rapidly off. Later on by a circuitous route he took him home. "Better stay in th' house, Wilbur," the white man said as he drove off.

Jackson was a sight. His well-tailored uniform was dirty and torn. He was bruised and scratched and one leather leggin was missing. His mother had hardly finished attending his hurts when the telephone rang. Old Mr. Jackson answered it.

"Dave," came a kindly voice over the wire, "bettah tell youah boy tuh git outta town 'fore it gits too late. They're comin' after 'im."

In a suit of his father's clothes, with a slouch hat pulled over his ears, Jackson caught the next train out of town. Christmas Day and the day after he rode in a dirty, cramped coach, getting to safer soil.

The day after New Year's Day, when the officers on holiday leave had returned to duty, the comedian of the mess, a jovial second lieutenant, was reading aloud one of the patriotic blurbs from the Creel Press Bureau in Washington and a few of the cynics were smiling. Jackson came in and sat down. He ate in silence. Then suddenly, to everyone's surprise, he blurted out, "For Christ's sake will you stop reading that bunk!"

Captain Jackson's story is most likely based on an incident which occurred in George's own life at the end of his active army service.

One morning, probably in the summer of 1918, while awaiting his reassignment at Camp Dix, Lieutenant George Schuyler stopped at a bootblack stand at the Philadelphia train station to have his puttees shined. The bootblack, a Greek whose rudimentary knowledge of English showed he had been in America only a short time, refused in a very loud voice to serve "a nigger." This was the culmination of too many slights and insults over too long a time. "I'm a son-of-a-bitch if I'll serve this goddamn country any longer!" Schuyler muttered aloud, to no one and everyone.[14]

George calmly decided to desert. Packing his civilian clothes in a suitcase, he boarded a train for Chicago, and changed en route. A black officer was conspicuous enough and the change was noticed. Someone wired ahead to military headquarters in Chicago. On arrival, George was arrested. Fortunately, he had kept his uniform in his grip, and was an experienced enough soldier to convince the police that officers frequently changed into civilian clothes when not on duty, though it was strictly against regulations. Soon released, George caught the next available train to California. On reaching San Diego, he procured a job on a nearby ranch as a dishwasher, and stayed nearly three months. One day he heard that his old Hawaii regiment was coming through San Diego. Concerned that someone might recognize him, George decided to give himself up before the end of the three-month grace period, during which desertion is called AWOL. He was brought, handcuffed, back east and tried before a military court. He got five years. The sentence was reduced by President Wilson to one year, and ultimately he served only nine months, getting time off for good behavior. George was imprisoned in the ancient Castle Williams, on New York's Governor's Island.

As luck would have it, the warden was an officer under whom George had served in Hawaii. They struck up a friendship. The prisoners, it seemed, exercised self-government and lived under the honor system. Soon, George was their virtual head. When his time was up, the colonel hired him at $250 a month as a civilian in charge of the supply department. He occupied the position until 1920 when, in an economy drive, the job was eliminated.

George kept this period of his life a deep secret until he met Josephine and confided in her a month after they had started dating. Reputedly, the only other person of his acquaintance who knew about his prison stint was Solomon Harper, a black inventor, who, as Elton Fax put it rather vaguely, "took some secret about George's army years to his grave"[15] when he died in 1981.

Neither Schuyler's autobiography nor his oral history, recorded by Columbia University in 1960, mentions his desertion or imprisonment.

❖ ❖ ❖

In search of a new job, Schuyler took and passed with high marks a federal civil-service examination for first-class clerks. According to George's autobiography, he was the first Negro to pass. He received notice to apply to the U.S. Shipping Board in Hoboken, and was elated when told, after an interview, to go home to await assignment. However, a few days later a blunt notice arrived: "In view of the fact that you refused the position offered, your name has been removed from the list."[16]

George made ends meet through various menial jobs. After a hellish stint in a brass strip-mill, where he witnessed an arm being crushed in a rolling mill and an old man fall into an acid tank, Schuyler washed dishes in a restaurant at Ninety-sixth and Broadway. The hours were from 6 a.m. to 6 p.m., with one day off a week. A huge operation with a furious pace, the restaurant fed as many as two thousand for lunch, and hundreds for breakfast. During rush hour, dishes were stacked mountain-high, and George required an assistant just to scrape them. Another man washed the silver and glasses. The job was exhausting; he never felt fully rested. George's hands grew raw from constant immersion in hot soapy water, and it was a good place to develop flat feet.[17]

The job robbed George of personal time. He was perpetually tired and more and more disgusted with his circumstances and himself. He and Myrtle, a quadroon he had been living with, quarreled a lot. They finally parted, and Schuyler quit "the fetid world of steam, odors, dirty dishes and twelve-hour days." After a nine-year absence, he returned to Syracuse, to live with his cousins Lila and Mary Louise.

He again decided to become an entrepreneur. Advertising his availability for odd domestic jobs, George soon had a full schedule going from house to house. What's more, he now enjoyed some leisure to read. In November 1921 he joined the most active political group around town, the Socialist Party of

America. For them, he organized the Negro Community Forum, and wrote pamphlets. He also began to meet his intellectual equals.

Tired of the handyman role, with its uncertainties and hazards—which included unwanted advances from lonely housewives—he hired on as a hod carrier and building laborer. The job was strenuous, and occasionally, when a pouring had to be completed without interruption, the hours seemed interminable. But the pay was good, and he found the hodgepodge of a work force congenial. Yet Schuyler was again beginning to feel the stifling narrowness of the Syracuse scene, and when the building was completed he impulsively decided to return to New York.

He rented a clean and inexpensive room in the Phillis Wheatley Hotel on West 136th Street, named after the famed Negro poet of Revolutionary days and operated by the Universal Negro Improvement Association headed by the flamboyant Marcus Garvey.

Garvey's white-hating Back-to-Africa movement was riding high in December of 1922, following severe racial clashes that had occurred nationwide in 1919 and 1920. Out of curiosity Schuyler attended some of their meetings, but he was fundamentally opposed to the movement from the outset, and later would become its outspoken critic.

❖ ❖ ❖

Although life in Harlem was pleasant enough, and George was meeting interesting people, his money ran out after a month. Once again he began searching for a job, with little success in the depressed economy of 1922. For a while George worked as a stevedore, going to the docks in the morning on the El with a big hook hanging from his belt. He discovered, however, that life was cheaper on the Bowery, and he gravitated to the Lower East Side. One night as he stood in a sheltered doorway next to a cheap, brightly lit restaurant, a hobo came shambling down the street, leaning against the wind, smoking a big cigar, hands buried in the pockets of his shabby overcoat. "Friend," he said, "I know where you can get a feed and a flop until you get on your feet again. Wanna go?" It was a foolish question.

"We came to a big, darkened church on the corner of [Tenth Street and] Avenue A . . . stumbled down a dozen snow-caked steps to a basement . . . knocked and [were let in]," Schuyler would write in retrospect. "A motley crew . . . sat around a big . . . table. They were young and old, punks and weather-beaten veterans of highways, jails, almshouses, and hobo jungles." A pot of coffee and some food were on the stove. Three unmarried clergyman lived upstairs. The basement was the domain of Frank, the middle-aged janitor, a former hobo and gangster who refused to forget the world's dispossessed.

The rules were simple:

> "You take a day's work when you can get it, see?" [Frank] explained. . . . "When you get a couple of bucks, put one in the kitty. . . . That's how we're able to buy

> chow. When you ain't got nuthin' don't worry; some o'th' boys will have somethin'. Play fair with us and we'll give you an even break. Get me? . . ."
>
> For a couple of days the snowstorm afforded us all work. Twelve hours a day in the biting air . . . for ten [bucks] . . . and the kitty was overflowing.[18]

And there was always a buck to be made on Saturdays acting as the "Sabbath goy" for the Orthodox Jews, lighting their fires.

George would describe with some affection the denizens of this "lower depth": "the folks furthest down . . . from Sweeney, a big hard-faced . . . Irishman with . . . little green eyes who had once studied for the priesthood . . . to Robert, a young . . . lazy no-account dreamer . . . with pasty face . . . and red-rimmed eyes" who would talk for hours on the rights of men. Robert, though able-bodied, never put anything into the pot and ultimately was drummed out with a dollar from the kitty after the worst of the winter had ended. And Mabel and Cleopatra, two good-looking pals "in their early twenties, dissatisfied with the Creator for assigning them to the wrong sex."[19]

Schuyler spoke warmly about the basic lack of racial prejudice among the persons at the very bottom of the economic scale. This newfound existence held another attraction for George: he could spend hours in the reference room of the Forty-second Street Public Library, or roam through the spacious halls of the Museum of Natural History. He began to go regularly to the forum of Friends of Negro Freedom in Harlem, "where some of the sharpest minds in Harlem assembled to make irreverent comments on subjects sacred and profane."[20] It was there that he met A. Philip Randolph, organizer of the Brotherhood of Sleeping Car Porters and co-founder of the *Messenger*, which ultimately led him to his job with the magazine, and to Jody.[21]

8

George Schuyler, Investigative Reporter

In the early winter of 1930, at about the time that Philippa was conceived, George Palmer Putnam, publisher of the *New York Evening Post,* asked Schuyler to go to Liberia on a covert mission. The League of Nations had just released the disturbing report that a new form of slavery had sprung up in Africa. In Liberia, boys and young men were being sold to work on Spanish plantations on the Island of Fernando Po, a hellhole off the coast of Nigeria, a thousand miles east of Monrovia. The president and the highest Liberian officials were allegedly involved. Schuyler was to write a series of syndicated articles.

Black journalists working for white media were rare in those days, and a colored writer had never before served as a foreign correspondent for an important metropolitan paper.[1] Putnam asked to see some of Schuyler's published work, and invited him to his apartment near the Plaza Hotel. Arriving with a largish package, George was mistaken by the doorman for a messenger and directed to the delivery entrance. The error was soon corrected by an irate Mr. Putnam. Upstairs, George was graciously received by his host and his fiancée, the aviatrix Amelia Earhart, "looking exotic and relaxed in lovely oriental pajamas."[2]

Only Josephine, Putnam, and Arthur Spingarn at the NAACP knew the true purpose of the trip when George sailed in luxury on January 24, 1931, to Liverpool, continuing, on February 11, to Monrovia. On board he made the acquaintance of Charles E. Mitchell, the new United States minister traveling to Liberia to assume his duties. Mitchell, a respected black banker and educator from Charleston, West Virginia, was extremely helpful to George. Mitchell allowed his vice consul—who was eager to gain firsthand knowledge of the country's interior—to escort George on his trip into the hinterland.

With a crew of fourteen and a substantial load of provisions and trinkets, the two Americans sailed up the coast in a kru boat, propelled alternately by wind and ten oarsmen, to Robertsport near the Sierra Leone border. They then struck into the interior for more than a month of steady tracking through the jungle.[3]

By "dashing" village chiefs in remote settlements with salt, tobacco, safety razors, mirrors, and the mandatory bottle of gin, Schuyler was able to gather stark details relating to the slave trade.

George spent weeks in the jungle collecting information. By early April, with still a long way to go, he contracted malaria at the mountain town of Sublima. The trip now became an ordeal. When they could rest, George sat under the rough-barked trees, his body shaking violently with fever and cold. By the time he returned to Monrovia, he had lost thirty-five pounds. Nursed back to health by Mrs. Mitchell's staff, George sailed for the States in early May after a three-month sojourn on the west coast of Africa—up the Liberian coast to Sierra Leone and Senegal. Through most of this time he worried silently about his family. He had been totally out of touch with Josephine, who was pregnant with Philippa.

Despite his bout with malaria, George wrote six articles for the *Evening Post*, which appeared during the last days of June and first days of July 1931, together with some of his unusual photographs. The articles, syndicated to the *Buffalo Express*, the *Philadelphia Public Ledger*, and the *Washington Post*, brought Schuyler's career as a journalist to new heights.

By October he had also completed the manuscript for his novel *Slaves Today*. The plot is a faithful mosaic of actual happenings. All of the characters were taken from life, and only the names of the Americo-Liberians were changed. Interspersed with the stark record of man's inhumanity are sharp insights into the tribal society of Africa, its taboos, its dances and festivals, and lyrical descriptions of the jungle.

Schuyler's book and his articles intended to expose the cruelty and corruption in Liberia, the oldest black republic in the modern world, founded in 1822 by a private American philanthropic society as a refuge for liberated American slaves. The reaction from his fellow African Americans was that George Schuyler had been duped by white publishers and white newspapers. Seen in context, their reaction is not surprising. *Slaves Today* did little to reverse the image of the black. Furthermore, Schuyler's book brutally deromanticized the popular Back-to-Africa movement, which had painted an alluring picture of the black homeland.

"Slavery, in the form of forced labor with little or no compensation, exists under . . . various euphemisms today in practically all parts of Africa, the East Indies, and the South Seas," Schuyler wrote in his introduction to the book. "It is found as well in the colonies of Europe as in the Negro-ruled states of Abyssinia and Liberia. Regardless of the polite name that masks it, while bloody profits are ground out for white and black masters, it differs only in slight degree from slavery in the classical sense, except that the chattel slaves' lives were not held so cheaply."

Always the polemicist, George added, "If this novel can arouse enlightened world opinion against this brutalizing of the native population in a Negro republic, perhaps the conscience of civilized people will stop similar atrocities in native lands ruled by proud white nations that boast of their superior culture."[4]

❖ ❖ ❖

Even for such momentous occasions as his daughter's first birthday or second Christmas, George found himself, regrettably, far from home. In the winter of 1931–32, he was called on another "secret assignment"—this time by the NAACP's journal, the *Crisis*—to the Mississippi Delta, to discover the truth about labor conditions on a $325-million federal flood control project.

The NAACP had been receiving reports from the Delta about exploitation of workers bordering on peonage. George was joined in this investigation by young Roy Wilkins, then the assistant secretary of the association. Traveling Jim Crow and disguised as laborers seeking employment, they went from camp to camp along the levee. They quickly learned that conditions at some of the camps were indeed appalling. The men often worked twelve to sixteen hours a day, for which they were paid as little as one dollar. Housing was cold and drafty, and sanitation was poor at best.

The time of the year chosen for the investigation was not propitious; it was the dead of a cold winter and there was a freeze on hiring. Risks loomed large: Any Negro who seemed suspicious could easily wind up on a chain gang for loitering.

It was difficult if not impossible for George to keep in touch with his family. He wrote to Josephine every day, but moving so much over the frozen countryside he was unable to get any reply.

One night, in Vicksburg, Mississippi, staying alone in a rooming house, and comfortably propped up reading the current *American Mercury*, Schuyler was rudely interrupted by a loud pounding. Hoping it might be a Western Union message from Josephine, George threw open the door. There stood a detective and a uniformed policeman, both with drawn revolvers. "Throw up your hands, nigger, [and] get your things,"[5] the detective ordered. They hustled him off, handcuffed, to the police station, and while George was able to convince the duty officer that he was not one of the two Negro holdup men the police had been seeking, they locked him up anyway—first confiscating his notebook, some thirty dollars he had in the pockets of his overalls, and a Schaeffer Lifetime fountain pen.

His cellmates were a burglar and a road agent, the latter a disarmingly handsome man who openly confessed to a life of crime: robbery, murder, and rape.

As soon as the turnkey had gone, George bribed the trustee—with some money hidden in his shoe—to send a telegram to Walter White. Confident that this would bring swift action, he lay down on his hard cot and took it easy.

The next morning Schuyler was released. His notebooks were returned but there was no sign of the money or the pen. But, with the jail door open, George was not disposed to argue. Losing no time, he headed for the Negro YMCA, exchanged clothes, found someone to drive him to Jackson (forty-nine miles away), and caught the Illinois Central Express to Memphis, where he was met by Wilkins. Walter White had the wires burning that morning, preparatory to making Schuyler's imprisonment a cause célèbre. He had also informed Josephine. Four days later George returned to the Delta by another route to finish his investigation.

By mid-1935 George was again traveling in Mississippi, this time on assignment from the *Pittsburgh Courier*. With nearly a million blacks in Mississippi, the weekly had a circulation there of only two thousand, and George was sent to the state to drum up business.

He departed shortly after Philippa's fourth birthday (August 2, 1935) and remained until Thanksgiving, traveling the entire state by train, bus, and car, and securing a *Courier* agent in every county where there was none.

He also continued to write for the paper almost weekly, profiling southern life. George tried to present a balanced picture; to point out, whenever he could, successes of Negro initiative and entrepreneurship. Yet the conclusion that racism hampered progress and development in the region was inescapable.[6]

Schuyler undertook perhaps his most important roving assignments in the summer of 1937. The National Labor Relations Act of 1935 had established the right of workers to organize. The *Courier* wanted to determine whether the Negro could enter the productive stream of American industry and what opportunities, if any, and how soon, the union movement would offer him to that end. This was doubly important because of the escalating Negro migration to the North.

Crisscrossing the country, George logged over twenty thousand miles during those hot summer months and into the fall. It was an arduous job, and in addition to his investigative reporting and almost daily travel, he was writing the *Courier*'s editorials, his "Views and Reviews" column, and a one-column front-page summary of the general news.[7]

George wrote home almost daily, describing in an engaging narrative style some of his experiences. But more often his letters, written in tender and poetic terms, were about the pain of being separated from the family and his inability to watch his daughter grow. On one trip, after he had been gone four weeks, Schuyler mailed his daughter a rosebud that he had picked himself. He spent the extra money to send it special delivery so that it would arrive still fresh.

"Greetings on Mother's Day," he wrote on another occasion, "to one who with love, patience, intelligence and science wrought a miracle of development in her daughter in spite of economic obstacles and many annoyances from her husband who did somewhat less than his best but loves them both dearly and will soon be home."[8]

And on yet another occasion: "You and Philippa are growing together like Siamese twins. She is going to be a facsimile of you with the exception that she is having a much better start in life. She can become anything under your tutelage. She has already become a wonder. I often speculate on what she will become and what glory she will reflect upon us. It is a wonderful thing to look forward to. I just know she is going to be a marvelously beautiful and intelligent woman. We must do everything to preserve her, like a hothouse flower, for she is a rare and exotic breed. There are few beings like her in the world."[9]

9

The Prodigy Puppet

Philippa's reputation as a child prodigy began to grow. During the fall of 1937 Mother Stevens, director of the prestigious Pius X School of Liturgical Music at the College of the Sacred Heart (also known as the Convent Music School), read about the gifted child and decided to meet her. Since the school was on the grounds of Manhattanville College, then in Harlem, she invited the prodigy to come and play.

Mother Stevens, as Josephine wrote at the time, was not your "garden-variety nun. . . . She was a person of dynamic energy and intense earnestness but with a most unusual sense of humor. . . . She had a Chesterfieldian talent for paradox, a 16th century wit, an 18th century candor . . . and an unrivaled knowledge of early church music. Her devotion to God and music was inexhaustible. Though she must have been nearly seventy . . . she moved and spoke with the exuberance of youth."[1]

Philippa's recital at the Convent School included two Bach minuets, Schumann's "The Poor Orphan," and two of her own compositions: "Suite of the Seasons," in four movements, and her whimsical "Cockroach Ballet." After the performance, Mother Stevens complimented the child but asked why, when she might have written about angels, she chose cockroaches. "But Mother Stevens," Philippa came back, "I have seen many cockroaches but I have never seen any angels."[2] Later, Philippa and Jody attended a service in the church. All the candles were lit and sparkled like stars. Philippa was entranced. As they left, Mother Stevens placed a small cross on a thin silver chain around the child's neck and said, "The Virgin will now protect you."[3] The two seemed to understand each other instantly, and Mother Stevens invited Philippa to become a permanent member of her class. Two weeks later, Phil began her singing and composition classes at the Convent School.

That same year, 1937, and for the next six, Philippa garnered top prize from the New York Philharmonic's Young People's Society Concert Series at Carnegie Hall. This was not a performance award, but one based on the best program notes turned in by a child who attended the five concerts of each series. When Philippa first entered, she was six and actually too young to compete. But the committee was so taken by her thirty-six-page book with extensive illustrations that they awarded her a special prize for children under seven. Her "debut" at Carnegie Hall to accept the award was equally noteworthy: she stood, center stage, with an enormous stain on her white dress. So excited about going to Carnegie Hall, she had fallen into a mud puddle on the way down from Harlem.[4]

At about the same time, Deems Taylor was asked by CBS radio if he would go on *We, the People* to interview a remarkable pianist and composer named Philippa Schuyler.

> Now various painful experiences in the past had caused me to view child prodigies with dark suspicion [he would recall years later]. They almost invariably turned out to be little monsters who, by threats or promises, had managed to attain an extraordinary digital dexterity. They could play fast, Heaven knows, and they could play loud. But of what we roughly call "soul" there was little if any trace. And why should there be? Only by a miracle could a six-year-old possess the emotional and intellectual depth of a mature artist.
>
> But then I thought, "Oh well, if it will make CBS happy, I'll go along. Thank God she isn't a violinist. At least she'll play in tune, the piano permitting." So I went to the studio and the moppet and I gravely discussed this and that, and she played me some of her own works, and some by other composers. And at the close of the broadcast, as she and I faced the studio audience, a small hand tucked confidingly in my own, I suddenly realized something. "Why," I thought to myself, "this is no infant. This is a born musician." The miracle had happened.[5]

Philippa's performances continued to be extensive. For eight consecutive years, beginning with her Aeolian Hall performance in 1936, she placed on the National Piano Teachers Guild Honor Roll. When Philippa was eleven, the judges would declare that she had the most outstanding record of sustained superiority in piano in the city. She was also heard on WNBC and other radio stations, often premiering her latest composition. And she was a frequent guest on Madge Tucker's *Coast-to-Coast on a Bus* (formerly known as *The Children's Hour*), a Sunday morning show emceed by the legendary Milton Cross.

Philippa also began to travel, performing as far from home as Pittsburgh, Fort Wayne, Indiana, and Hartford, Connecticut. Her recitals included various of Bach's preludes and fugues (Book I), Mozart's Fantasia in D Minor, Chopin's Minute Waltz, Grieg's *Elfentanz*, Scarlatti's *Pastorale*, and as many as fifteen of her own compositions. Most of her concerts were sponsored by black organizations; those sponsors that were white were religiously affiliated, usually Catholic. Her reviews in the local and national press, both black and white, were outstanding.

For Philippa traveling on the trains was almost as exciting as performing. She first encountered the American railroad system one Saturday in May 1938, on her way to Hartford, to give a recital at a local YWCA. Jody and Philippa sat in the parlor car and later consumed a big porterhouse steak in the elegant diner (ordered uncooked, to the steward's consternation). "I want to do this more often," Philippa confided to her mother. "Why don't you let me go to Pittsburgh or Cleveland?"[6]

At first, Jody limited her daughter's appearances to weekends, so she would not miss too much schooling. But as Philippa became better known outside of New York, black organizations begged the Schuylers to present their daughter. Soon she was doing two-week tours.

The night before each tour, mother and daughter stayed up late, organizing and packing. Into Philippa's suitcases went ribbons and bows; her green jumper, new white-lace dress, and her rose organdy with the black-velvet sash; her white slippers, red slippers, green slippers, and black slippers; her wine-colored velvet dress with the matching beret; her Modern Library giant edition of Plutarch's *Parallel Lives*; three mystery novels; pens, pencils, and manuscript paper. Into Jody's luggage: colorful turbans, scarves, and hats; shoes, dresses, two coats, nylons, perfume, lipsticks, jewelry; for Philippa, carefully packed bottles of cod liver oil; several pieces of fruit; special bread made of figs and dates ground and mixed with egg yolks, melted butter, nuts, whole wheat flour, and honey; one hundred copies of Philippa's published music to sell at concerts (at thirty cents a copy, of which the Schuylers kept twenty cents); tickets; cash; and Philippa's elusive hairbrush—within easy reach in Jody's handbag. (Harlem lore has it that Josephine could never fix her daughter's hair; it always "stuck out." As one old-timer said, "We had to come over and show Josephine how to do it. White folks just don't know how.")[7] It was usually past midnight before they went to bed.

On the road, they always stayed at houses of strangers—recital promoters, friends, and acquaintances of George's, members of the NAACP—where Philippa was the center of attention. Although she loved the excitement, Phil needed at least ten hours of sleep a night, so she was often tired. And yet the opportunity to play her own compositions almost before the ink was dry compensated for the virtually constant pressure to perform. She was particularly proud of several of them which her parents had published during 1938—at their own expense.

Phil enjoyed the money from both the concerts and the sale of her sheet music. Although her honoraria began modestly (she was paid $31 for one of her first YWCA recitals in 1938), by 1940, during one of her Midwest tours, she received $175 per engagement, plus expenses. Most of her earnings supported her musical career, but George and Jody allocated her "tidy sums" from time to time, to spend however she wished. Philippa usually purchased gifts for her parents: one Christmas (1938), she bought George a very expensive radio. It cost $50 and she proudly selected it all by herself, at Macy's, the store that was to the Schuylers, as to many other New Yorkers, a "way of life" then.

Late in 1939, Philippa was photographed for *Look* magazine. It was a small but prominently placed entry at the bottom of a page. Posed rather stiffly at a grand piano, Phil smiles a bit ruefully in the picture, her face surrounded by a mop of corkscrew curls. The description of Philippa, though apt, is subliminally racist: "The Shirley Temple of American Negroes, Philippa Schuyler, precocious New York City Negro girl of 8, is probably America's best child pianist. Also an expert composer, she has won five prizes this year, given 14 recitals. Her father, George S. Schuyler, is a writer and her mother is a painter. They attribute Philippa's genius to a diet of raw foods and to the fact that her mother dieted for three years before Philippa was born."[8]

The piano at which Philippa was photographed is one Aunt Lena had sold Jody three years earlier, not knowing it was going to her "nigger" niece.

❖ ❖ ❖

After three and a half years of preparation, the 1939 World's Fair opened its doors on Sunday, April 30, on a reclaimed piece of land in the northeast corner of Queens once known as the Corona Dump. Twenty-eight men-of-war, the pride of the United States Navy, stood in from the sea at dawn to salute the event. As a preliminary, Mayor La Guardia dedicated the Bronx-Whitestone Bridge, the world's fourth-largest suspension bridge, providing direct parkway access to the fairgrounds from the north. Over one million people visited the fair on opening day.[9]

The political atmosphere that spring was tense. Fearing the outbreak of war, several European countries had shipped selected art treasures to their national pavilions for safekeeping. And indeed, four months later, Europe would be ablaze.

But for now, New York was looking forward to a century of progress. The theme of the fair was "The World of Tomorrow," its emblem two large white structures: the Trylon, a very slim triangular pyramid 728 feet tall, and the Perisphere, 180 feet in diameter.

The fair highlighted the Machine Age in all its glory. Young and old alike delighted in the technological marvels of the future, and many visitors saw television for the first time. On the lighter side were the usual handful of native village reconstructions; the Lifesaver Candy Parachute Jump with a giant steel tower and controlled drop chutes; and Billy Rose's famous "Aquacade," a show based on Hollywood's Busby Berkeley routines, but performed underwater.

Like other New York children of her generation, Philippa fell in love with the fair, and she never seemed to tire of wandering the grounds with Jody or George. By 1940 she had been there six times and had written a composition entitled "Impressions of the World's Fair." That year, for Phil's eighth birthday party, Josephine baked cakes in the shape of the Perisphere and Trylon; each weighed seventeen pounds.

But Philippa was not merely an onlooker. On July 25, 1939, the Women's Service League of Brooklyn selected thirteen New York "colored women" to present at the World's Fair, in recognition of their distinguished service to their race and sex. The League chose Philippa as one of the "Women of Tomorrow" along with such notables as Jessie Fauset Harris, Anne Brown, and Ethel Waters.

Exactly one month later Philippa appeared at the fair on an experimental national television hookup. Madge Tucker, of *Coast-to-Coast on a Bus*, had been asked to pick six young artists to participate in this historic event. She chose two singers, one mimic, two dancers, and Philippa.

Television was still very new in America. For the six little artists, several of whom (including the tap dancers) had never been seen but only heard by a wide audience, their appearance at the fair was a milestone; and their parents undoubtedly spent hours preparing them for their "world" debut. By comparison, Philippa was a seasoned trouper. She played two pieces: her own composition "The Circus" and "The Little White Donkey" by Jacques Ibert. Miss Tucker introduced her as "a remarkable talent."

In 1940 Philippa received yet another extraordinary award from the fair: she had a day named after her. On June 19, 1940—"Philippa Duke Schuyler Day"—she performed in the Little Theater of the Hall of Science and Education. The press, both black and white, was out in full force, and on July 1, she was spotlighted in *Time* magazine—for the third time in her young career.[10]

Although celebrated under the banner of "Negroness," Philippa's day at the fair placed her alongside some of the nation's best-known personalities. And, as one on-the-scene enthusiast reportedly quipped, the King *and* Queen of England had only rated *one* day between them.

Phil gave two concerts: one at 11 a.m., attended largely by children who had been excused from morning classes and were taking advantage of the ten-cent Wednesday entrance fee; and the other at 2 p.m., mostly for adults. For both performances, she wore a trailing empire gown of pale green lace with a gardenia fastened to the high bodice and five more in her hair.

"Philippa's white slippers carried her silently across the stage," Milton Bracker wrote in the *Times*, describing her morning concert. "She curtsied twice, sat down almost precariously near the edge of the bench, and began to play"—Bach, Daquin, Heller, Schumann, Lyadov, Ibert, and Rimsky-Korsakov. Then she took a rest, returned with a bow instead of a curtsy, and announced in a very small but entirely confident voice, "Now I shall play some of my own compositions."[11]

First came "The Goldfish," then "The Jolly Pig," followed by "The Cockroach Ballet." She continued with a couple of nature stories, her "Arabian Nights Suite," and finally "Manhattan Silhouettes"—a composite of her impressions of the circus, Spanish Harlem, WPA, and the World's Fair. After her final bow, another little girl walked up with a bunch of peonies bigger than Philippa.

Several reporters were waiting in the wings, and as her mother and her teacher began changing her shoes, getting them the wrong way the first time, Philippa granted a somewhat fidgety interview. When Bracker asked what her first composition was, she replied it was a nursery rhyme she had set to music, at about three years old—"Two little rabbits sitting in the sun, having their breakfast, fun, fun, fun."[12]

Phil left to change clothes and Josephine told of her daughter's astonishing school record, inexhaustible energy, and her passion for newspapers.

Did she really read the papers? the reporter asked when Philippa reappeared, cooler in a white summer dress. Yes, she did. Then perhaps she had an opinion on the war.

Philippa hesitated a second. "I hate to say it," she announced gravely, "but I think the Allies are cooked"[13]—undoubtedly reflecting political discussions around the Schuyler dinner table.

Mayor La Guardia—who had given Philippa an award earlier that year for her "perennial" prize at the Philharmonic Young People's Series—was also on hand, and presented her with yet another medal. La Guardia was so taken with Philippa that soon he became an occasional visitor to the Schuyler flat, stopping in simply to sit and chat with her.

There is no record of what the second-best-known American of his time (after FDR) and Phil talked about, but one thing is sure: They shared a passion for the comic strips as well as for music. In fact "the Major," as his friends and associates called him (using La Guardia's erstwhile military rank), deemed the funnies important enough to daily life and public health that he took to reading them over the airwaves during the newspaper delivery strike of 1945. He always did it with slam-bang gusto—occasionally adding a moral.[14] And it was to the mayor that Philippa would dedicate her second large orchestral work, "Rumpelstiltskin," in 1945.

When John Gunther interviewed La Guardia in 1940 for his book *Inside U.S.A.*, part of the exchange seemed to be emblematic of his relationship with Philippa:

> Gunther: What do you like best?
> La Guardia: Music
> Gunther: What do you believe in most?
> La Guardia: Children[15]

❖ ❖ ❖

Several months before Philippa's day at the fair, the Schuylers had moved (for the third and last time) into a more spacious apartment on 270 Convent Avenue, at the southwest corner of 141st Street. It was located on the western fringe of Harlem. The building, a Gothic edifice of slate gray stone catercornered from the elegant St. Luke's Episcopal Church, was two blocks north of Philippa's Convent School, a series of brown and gray buildings reached through an ornately carved arch of striking white marble. Situated on the top of a hill on a wide, quiet street, the apartments were tenanted by both blacks and whites.

The Schuylers lived on the ninth floor. Apartment 9C had three ample bedrooms and an airy living room with a bay window. Philippa had the largest of the three bedrooms. It accommodated her grand piano, a bed, a blue-and-white dresser, a cabinet, and some chairs. The crucifix sans loincloth had migrated to the living room, alongside some profane nudes bearing Josephine's signature. But the paintings somehow were less striking than the kitchen—there was no stove in sight. Josephine had it taken out when they moved in, since they ate only raw food, and instead had painted a tropical mural on the walls.

Philippa's days continued to be carefully regimented and choreographed. She now attended the Convent School for only two hours a day. At least a third of her waking hours were spent at the piano, practicing and composing. Yet she still made time to read mystery novels, Flaubert, political columns, and the comic strips; and to do crossword puzzles, design clothes for her dolls, exercise on her gym set, listen to swing (her latest passion), and learn all the jitterbug steps from George whenever he was home.

In August 1940, when Philippa had just turned nine, Joseph Mitchell of the *New Yorker* came one evening to the Schuyler's apartment to interview her for their series "A Reporter at Large."[16] He happened upon some clutter in Philippa's room which painted a rather surrealistic picture of her varied interests:

"On top of Philippa's piano there was a Modern Library giant edition of Plutarch, a peach kernel, a mystery novel called *The Corpse with the Floating Foot*, a copy of the New York *Post* opened to the comic strip page, a teacup half full of raw green peas, a train made of adhesive-tape spools and cardboard, a Stravinski sonata, a pack of playing cards, a photograph of Lily Pons clipped from a magazine, and an uninflated balloon. . . ."

Like Deems Taylor and Fiorello La Guardia before him, Joseph Mitchell was instantly taken with the unusual child. He admitted, however, to some discomfort at the very beginning of the interview when he was suddenly left alone with her, Jody having run off to the kitchen to put the homemade peach ice cream into the icebox. The experienced journalist was at a loss for small talk with a gifted nine year old.

"Do you mind if I smoke in here?" he asked.

"Of course not. I'll go get you an ashtray."

When she returned, Mitchell asked if she had been reading the Plutarch on the piano.

"Yes. I've read most of it. I got it to read on trains."

"Don't you find it rather dry?"

"Not at all. I like biography. I particularly like the sections called the comparisons. Best of all I like Theseus and Romulus, and Solon and Poplicola."

"What are some other books you like?"

Philippa laughed. "Lately I have been reading a Sherlock Holmes Omnibus and some mystery books by Ellery Queen."

"What book do you like best of all?" enquired the Reporter at Large.

"Oh, that's almost impossible to answer. You can't just pick out one book and say you like it better than all others. I bet you can't."

"I certainly can," Mitchell said, no longer bothered by their age difference.

"What book?"

"Mark Twain's *Life on the Mississippi*," he said.

"Oh, I like Mark Twain," Philippa was clapping her hands excitedly. "I like him very much. . . . I guess you're right. I *can* say that there's one book I like best of all. That's the *Arabian Nights*. George has an eight-volume set. It's an unexpurgated edition. I read it first when I was three, and at least four times since.

"I based my longest composition on it. I called it 'Arabian Nights Suite.' Oh, the stories in that book are absolutely wonderful!" She laughed, "Goodness!" she said. "I didn't mean to get so "—she paused and appeared to be searching for a word—"impassioned."

Mitchell was smitten, and for years to come he and his wife and the Schuylers remained friends.

When Jody reappeared, the conversation turned to one of Phil's favorite subjects—the funnies. A friendly argument ensued over which comic strip was the best. Mitchell sided with the *News*'s "Moon Mullins," while Philippa went for the *Post*'s "Dixie Dugan."

"The *Post* has the best funnies," Philippa was insisting. "I like 'Dixie Dugan,' 'Superman,' 'Tarzan,' 'Abbie an' Slats,' and 'The Mountain Boys,' and they are all in the *Post*."

At nine-thirty, it was Philippa's bedtime. She had recently bought a riddle book, and had been surreptitiously slipping in questions all evening from that book, most of which the reporter could not answer. "May I ask another riddle before I go to bed?" she now pleaded.

"Just one," George said.

"All right. What's smaller than a flea's mouth?"

"I know that one," Mrs. Schuyler said.

"So do I," said Mr. Schuyler.

"All right. All right," Philippa said. "Wait till tomorrow. I'll ask you some you couldn't guess in fifteen years."

Mitchell spent the remainder of the evening talking with George and Josephine, and looking at the scrapbooks, which were produced from their hiding place. He left late, and pausing for a moment at the door, he began, "That riddle about what's smaller than a flea's mouth—"

"That's an old, old nursery riddle," Jody interrupted. "I guess it's the only one I know. The answer is, 'What goes in it.' I'm very sorry she got hold of that . . . book. Tomorrow at breakfast she'll . . . ask us two dozen and we probably won't know a single answer."[17]

❖ ❖ ❖

Not only in Harlem, but in black communities throughout America, Philippa had become a "role model." Mothers would take their children to her concerts whenever they could, or gather them around the radio after church on Sunday mornings to listen to one of her broadcasts. A Harlem local, now a doctor, remembers being dragged to Phil's concerts while her mother demanded, "Now, why can't you be like *her?*" Another African American—now a professor of classical music at a prestigious university—remembered both Phil and her father. Growing up poor in Virginia, he had little opportunity or occasion to listen to classical music. But his mother, a domestic, knew he loved it, so she borrowed money to get her son a ticket to hear Phil perform at a local black technical college. "I sat in the audience," he reminisced, "and saw her father come out. He was so well dressed and so well spoken. Black as coal, too. He introduced his daughter and then she came out, her hair pulled back in a ponytail, and she started to play. I remember it as if it were yesterday. I was so inspired. From that point on, my life changed."[18] But perhaps the venerable sociologist Hylan Lewis described her influences most succinctly: "Do you know how many blacks took piano lessons *because* of Philippa?"[19]

Philippa's visibility was due not only to Josephine's relentless PR activity but in a large measure to George and his connections with the press, both black and white. Without George's long arm, *Time*, *Look*, the *Herald Tribune*, and others, might have turned a blind eye to the young "mulatto" prodigy.

Hylan Lewis, with typically sharp insight, once called Phil a "prodigy puppet," adding somewhat wryly, "—and she had two very good puppeteers."[20]

10

Godowski, Et Al.

When Philippa began her singing and composition classes in the Convent School during the summer of 1937, she also embarked on her first regular interaction with children her own age. Although she was the first and only black to enter the school, she was accorded a warm reception from the outset.

Several afternoons a week from January through May, 1938, she also attended Durlach and Emerson, a progressive school for the gifted, affiliated with New York University. In the mornings she continued her music lessons at the Convent. When Philippa turned seven, however, the law required her to enroll in school full time. Engaging a tutor, which was the Schuylers' preference, would be financially prohibitive, and special dispensation would be needed from the board of education.

Their choices were to continue at Durlach full time or to send Phil to the Convent's Annunciation School, on the same grounds as her music school. They asked their daughter which she preferred. "The Convent," Philippa replied. "If I just wanted to play games I could go to the park. But I want to learn something every day."[1]

Philippa entered the Annunciation School in September. She was again the first black child to do so. The Schuylers placed her in the fourth grade, even though she qualified for seventh, as Jody believed that days spent with physically larger children would have a damaging psychological effect. In recognition of her advanced academic standing, though, the nuns arranged for her to attend school half days, with the rest of the time off for music.

"She is very popular at school," Jody recorded at the end of Philippa's second term. "She is very good at all games, likes jumping rope and playing ball and climbing."[2] Earlier, the progressive school had written Jody: "She is quite a favorite in the group partly because of her innate sweetness and partly because of others sensing her capability and inner security." The teachers found her quick, ready to accept changes in routine, and considerate of everyone, young and old. "She deals with people remarkably well."[3]

By now, Josephine felt sufficiently relaxed about her child's upbringing to embark on one of her infrequent treks to Texas. Jody's sister Daisey (with whom she had little contact, if any) had written a letter to "Heba Jannath" at her Harlem address. Its contents, no doubt, caused some mixed emotions, but also hastened Jody's trip to Texas that summer:

"Dad seemed to be thinking about you Josephine and was wondering if you really knew about his illness. Somehow . . . in . . . the twilight of his life you were in his thoughts. God only knows how many times he has needed you to inspire him on his journey to the end. Somehow you really mean more to Dad than anyone else."[4]

The person in whose care Jody left her daughter during her seventeen-day absence was a white woman by the name of Edna Porter, who had been a close friend of Helen Keller and Anne Sullivan. A poet, an activist, an actress, and an inveterate chronicler, Edna kept a running diary of Philippa's daily activities that summer. And if one was ever tempted to doubt the accuracy of Josephine's scrapbooks—to suspect that she exaggerated Philippa's precocity or, in her parental exuberance, changed Philippa's language to make it sound more "adult"—such doubts can be put to rest by reading Edna Porter:

> Monday at Morningside Park . . . Billy came running and asked, "Can a boy and a girl in the bushes make a baby?" I said yes, who is doing it. He said "P is up there with three boys." I looked and so she was, they were running like wild deer. . . .
>
> Once I heard P say to Billy—"You had better go to the bathroom because you may want to go and not know it. My mother often tells me to go when I don't know I want to, and I go and find I wanted to and didn't know it." I followed this conversation to the end. She let him go to the bathroom alone.
>
> Tuesday. . . . P disappeared. . . . She'd been all the way to 124th St. and I said, you get a bad mark on your report card for that. She had such a look! It was as tho she thot, "Jody is away, Jody is in Texas, what's the good if I can't take advantage of a situation like that?"
>
> Thursday. . . . Great at Radio City. She was so at home there and they were all so charming with her. A man outside asked if she was my daughter and I told him yes.
>
> Monday. . . . When one thinks of the music part alone that Philippa has accomplished I'm sure she is entitled to run up Pike's Peak. Only I don't want to be waiting for her return! . . . She "killed off" Billy during the weekend in some way. He put his foot in a toy bathing pool she had built. . . . I asked her a few questions and the last speech was: "Well, he just turned out to be not the person he had represented himself to be in the first place."
>
> Tuesday. . . . It rained a lot today and spoiled the afternoon but P made up dances, hot as it was and pretended to be a child bride and dressed herself like one. It is all over too quickly, Jody dear. It was lovely![5]

❖ ❖ ❖

Phil's musical education, on the other hand, was much more problematic.

Although Philippa continued to study with Arnetta Jones, Jody constantly sought confirmation that her daughter was in the best possible hands. When

Phil was barely six, New York University sent her to the Cecilia Music School for an audition. There, frightened by the awesome presence of two women (described in the scrapbook as "hidebound Europeans"), who hovered over her as she played, Philippa "loudly banged out" her pieces. "You have a very bad teacher," yelled the director. "Your teacher is ruining you." (To which Philippa flared up with childlike loyalty: "My teacher is *not* bad!")[6]

The following week, Josephine, beside herself with worry, took Phil to the director of the Juilliard Preparatory Division, who assured Jody not only that her daughter was truly gifted and far ahead of her peers pianistically, but also that her playing indicated her current teacher was more than adequate. This did little to quell Josephine's fears, however. And in early July, she wrangled an invitation for Philippa to play for the legendary Leopold Godowski.

The visit was chaotic and inconclusive. It was interrupted by the arrival of Godowski's lawyer. Philippa refused to stop in the middle of a piece she was playing, and Josephine scolded her. Jody was doubly annoyed that the maestro, whom she described as a "darling, sweet faced benign gentleman and all wrapped up because of his arthritis,"[7] did not appear more enthusiastic.

Over the next sixteen years, Philippa would study with more than a dozen piano instructors, sometimes several in one year. Ultimately, this coming and going adversely affected her musical as well as emotional development. Jody fired most of Philippa's mentors, insisting her daughter was not receiving the best possible training. In truth, however, her decisions were invariably based on a small event or a whim; a disagreement about child development, payment, or concert scheduling; or sometimes even jealousy. George never interfered.

Early on, concert scheduling became a particular bone of contention between Josephine and her daughter's instructors. Some insisted that Philippa stop performing altogether in order to concentrate on building a technique. Jody strongly disagreed: Weren't the outstanding reviews proof enough? Two instructors resigned over this impasse, and Jody turned down at least two prestigious scholarships fearing the rigid schedule demanded by scholarship programs would interfere with the performances.

The disagreement was fundamental. Most of the income from Philippa's recitals defrayed costs of her musical education, which the Schuylers otherwise would have been unable to afford. Moreover, Josephine was incapable of relinquishing control over Philippa, forgoing the reflected glory of her daughter's accomplishments, or giving up the idea that her brilliant interracial child, now constantly in the public eye, could be instrumental in breaking the American race barrier.

Philippa remained with Miss Jones for two more years, until almost age eight. Arnetta Jones cared a great deal for her star pupil and Philippa reciprocated. Under Jones's tutelage, Philippa continued receiving excellent reviews.

But there were inchoate warnings. She now played with a grave face, as one reporter commented, and at the end of each selection looked up for approbation from her mother, always hovering in the wings. Her practicing at home grew shoddy, and it is difficult to know if Philippa was passing through a rebellious phase, not uncommon among child prodigies, or beginning to fear that she was not living up to Jody's expectations.

Jody was deeply worried, particularly the week preceding each performance; but she and Miss Jones were at a loss. Jody turned to her husband for help, and George began a vigorous campaign in his letters home, writing two, sometimes three times a day.

Philippa remained unmoved. In an uncharacteristic last resort, Josephine canceled all pending engagements and ceased accepting new ones. In George's handwriting, now almost nonexistent in the scrapbooks, is a note to his unhappy daughter explaining why she was being punished:

> You insist on Jodie accepting engagements for you like recitals very much, especially the money. . . . You like going on trains to recitals, too but you do not always want to practice your pieces with the accuracy necessary, although you are willing to practice if you are not too supervised. But Miss Arnetta Jones . . . insists on accuracy, and Jodie must see that you do it. This has of late made you defiant toward Miss Jones and Jodie, so that after the Philadelphia concert which you insisted on taking, Jodie would make no more engagements until you are old enough to understand the nature of obligations.[8]

Jody's tactics worked, and Phil was soon back at the keyboard practicing.

But the reprieve was short lived. Confident after a series of now resumed recitals, Philippa would return home thinking it senseless to practice pieces she thought she had already mastered. Again Josephine become perturbed, canceling an appearance. And to make it even more effective, this time she insisted that her daughter write her *own* letters of explanation. But Jody never mailed them. Instead, she pasted them into the scrapbooks where, no doubt, Philippa was shocked to discover them years later.

❖ ❖ ❖

On June 21, 1939, Jody fired Miss Jones. Philippa was shattered. She composed a sad, melancholic piece for her teacher entitled "Farewell" (appropriately in a minor key) and was left without an instructor for the summer. Although Jody had accepted a partial scholarship from the Convent School for Philippa to study once a week with Josef Hofmann's assistant, William Harms, the lessons were not scheduled to begin until late October, almost four months away. Meanwhile, precious time was being lost.

Jody was desperate. For some time she had had her eye on a young piano instructor, Miss Pauline Apanowitz, who taught at the Chatham Square Music School. Josephine had called her several times prior to firing Miss Jones, but Miss Apanowitz was too busy to take on another student. Jody also insisted that her daughter's teachers come to their apartment in Harlem, and Pauline was not "teaching out" at the time.

One night, Jody went to a recital where the pretty and multitalented Miss Apanowitz danced. From that moment on, Jody was set on Pauline. She called the next morning.

Almost half a century later,[9] Pauline Apanowitz (now Mrs. Styler) remembered that day well: "She called me saying, I'm not taking no for an answer. Jody . . . fell for [my dancing]. I think if I had played the most difficult piano composition, I wouldn't have impressed her [as much]!"

To appease the persistent woman, Pauline decided to make the trip up to Harlem and be interviewed, although she had no intention of taking the position.

"I entered this apartment," Mrs. Styler reminisced, "and [Jody] talked to me with the door closed. Behind the door, Philippa was [obviously] eavesdropping. [When we had finished, Jody] said, 'Philippa, you may come in.' And in this doll walked. She was the most enchanting thing you have ever seen in your life. She curtsied right down to the floor, extended her hand and said, 'How are you?' My heart sank. And that was it."

Pauline made the long trip from Brighton Beach to Harlem every day except Saturdays and Sundays, in every imaginable weather.

> But it was worth it. She was just the most delicious thing. . . . I was teaching there only a few days when Philippa . . . said "Do you mind if I stick a pin in your arm?" I said. "What for!" So she said, "I want you to be my blood sister." "I'll do that, but without any blood," I said. And every time I left, she'd cling to me and tell everybody in the elevator she was my sister.
>
> Shortly after that, she asked me if I had met George. "No," I said. "Well, he's a man of few words, but they're all important. You will like him. . . ." You can imagine where that came from! Philippa adored him. . . . One day . . . as soon as I came into the room, Philippa dropped down on the floor and remained motionless. . . . I was ready to run out and get Jody when Phil came to and said, "George showed me how to fall without getting hurt."
>
> And George showed her how to play poker . . . told her about happenings all over the world. . . . So did Jody. . . . But Philippa had such a yearning for knowledge. She was insatiable. . . . It was a real joy . . . not a chore to teach her. Musically, she was like a sponge. The minute you told her something, she understood. . . . Her memory was phenomenal; her hand was great. She had this wonderfully strong, supple hand, which could do anything.

When Harms came on board in the fall of 1939, the Schuylers continued with Pauline. She still came every day, and coached Philippa for her weekly lessons at the Convent. Philippa also continued with Mother Stevens, and she concertized a great deal. By the spring, another mentor had been added, the flamboyant Antonia Brico, from whom she took piano lessons, score reading, and conducting.

Although she still performed well and received outstanding reviews, Philippa's practicing became sloppy again. Both Brico and Apanowitz continued to feel that the root of the problem lay in concertizing. They both suggested to Jody that she take her daughter off the circuit. Pauline advised limiting performances to one month a year and Antonia urged discontinuing them altogether for the next two years. But Jody was adamant.

First Miss Brico resigned. "Philippa is extraordinarily gifted," she wrote to

Jody, "but she has not yet learned the value of pianistic discipline. As long as she plays in concert and receives praises from public and press she is bound to consider herself a better pianist than she actually is. . . . It is difficult for her to comprehend that she has to be corrected on pieces that she has been performing. . . . Too many highly gifted children disappear into oblivion because they play too many concerts during the formative years."[10]

Pauline projects an even sharper picture of the deleterious effect concertizing was having on Philippa's development as a musician.

"This particular day," she recalls,

> I was teaching Philippa a lesson and it was just dreadful. I was taking things apart—not angry, just [patiently correcting her]. Suddenly, Philippa looked at me and . . . said, "You don't love me anymore!"
>
> "Why do you say that?"
>
> "Because you think my playing isn't any good."
>
> "I'm just correcting the things that aren't good."
>
> "But you don't love me anymore. My public thinks I'm wonderful. And you think I'm not."
>
> "On the contrary. I think you're wonderful and I know that you can do even more than your public knows you can do."
>
> "No, I don't believe it. You don't love me. . . . I'm going to the roof. I'm going to kill myself."
>
> Now, when I heard that [continued Pauline], you know, all the stories about the piano teachers who were doing such terrible things to kids like Ruth Slenczynska's father and the Ricci kids —I kept seeing my name up in headlines: PIANO TEACHER DRIVES PRODIGY TO SUICIDE. I rushed out to find Jody:
>
> "Philippa ran to the roof! Get her!" I said.
>
> "Oh, come on. Calm down. She's just being dramatic."
>
> "I'm going up there!"
>
> "Forget it. Don't worry. She'll come down."
>
> And Jody didn't move. . . .[11]

Pauline resigned then and there. The following day, she received a tear-stained letter from Philippa begging her to come back. Miss Apanowitz "continued to love Phil, to see her, to care greatly for her, to follow her career," but like Miss Brico she refused to be her teacher under the circumstances.[12]

Philippa's music education remained in a state of disarray. She again was left without a teacher, and for a while, it seemed as if Antonia Brico's predictions might come true.

Jody, finally, secured an unusual man as her next mentor—Paul Wittgenstein. An Austrian by birth, he had made his piano debut in Vienna to great acclaim in 1913. Early in World War I, he was wounded, lost his right arm, and became a prisoner of war in Siberia. After repatriation in 1916, he devoted himself to playing with the left hand, acquiring an amazing virtuosity which enabled him to perform compositions formidable even for a two-handed pianist. He was a great success in Europe and such luminaries as Strauss, Ravel, Britten, and Prokofiev wrote works for the left hand, for him. In 1938, Wittgenstein fled Austria and settled in New York, where he taught privately as well as at Philippa's Convent School. It was there that he had heard her play.

Herr Wittgenstein took a few scholarship students each year and consented to include Philippa. He was a stern sort, and at first she "wept a little at his loud voice. Then he said, 'Darling, you must not mind if your teacher shouts a little. He can't help it!' Then when you are ready to leave he kisses you."[13]

Ultimately, he was not the proper teacher for Philippa, although she remained with him for almost four years. Her playing deteriorated during that time—which the critics seemed oblivious to—and it wasn't until 1945, when she started studying with Herman Wasserman, that she began playing up to her potential. Nonetheless, Wittgenstein had given her two invaluable tools: a phenomenal left-hand technique, and a grace and ease with pedaling which he had probably developed to compensate partially for the loss of a limb.

As busy as Philippa was with music, and as much school as she had missed due to her touring and her daily schedule, she managed to skip sixth grade and to complete seventh, in June 1942, taking the English prize and graduating with highest honors.[14] That fall, at age eleven, Philippa entered a parochial high school, Father Young S. J. Memorial High, again attending only two hours a day.

Relatively happy in high school, Philippa developed interests in lipstick, clothes, movie stars, and boys—although she continued to insist that George was still her only boyfriend. She was even featured in the teen journal *Calling All Girls*, where, she lamented, "eleven was a mediocre age," that "nothing ever happens to an eleven year old," and that she hoped her teens would be "more glamorous."[15]

It was about this time that Jody began to toy with the idea of redirecting her daughter's career from concert artist to composer, an idea that was fueled by Philippa's induction, in 1942, into the National Association for American Composers and Conductors. At age eleven Philippa became its youngest member. Jody wrote the distinguished composer Carlos Chavez, who lived in Mexico at the time, asking if he would take their daughter as a student. They were prepared to spend six or seven months in Mexico, but as it turned out Chavez was not accepting any new pupils. They then approached Otto Cesana who lived in New York. After a few lessons, Cesana noted that "Philippa has a tremendous natural talent . . . and does naturally what many modern composers try to do. [I] intend to let her develop along her own lines in order to preserve her wonderful originality."[16]

11

Black and Conservative

Wartime themes dominated Philippa's eleventh birthday. She entertained black soldiers and their dates at the Harlem Defense Recreation League, playing Chopin, Grieg, and a dozen of her own compositions. Jody's birthday cakes were the finest ever: an armored truck made of walnuts and raisins, with wheels cut from whole oranges; a fearsome-looking tank bristling with guns fashioned from ground dates and almonds, and manned by gingerbread soldiers; and an aircraft carrier (escorted by several submarines) molded out of cashew nuts and figs, with tiny planes covering its flight deck. The edible sculpture weighed forty pounds.

The party was well attended. Blacks had been drafted in substantial numbers since 1940, and others had enlisted in various branches of the military. Although the armed services would remain segregated until 1948, there were now opportunities for qualified blacks to become commissioned officers in the army, or to learn a useful trade in the navy, which had not been the case in World War I. In a dramatic contrast to George's experiences, the War Department announced that blacks would be trained as pilots in the Army Air Force[1] and admitted into the Marine Corps, thereby smashing an exclusion policy as old as the Corps itself. Black women also enlisted, in the WACS, and by war's end they numbered four thousand.

In addition, the home fires also burned stronger for the African American, with many more employment opportunities in the defense industry. But there were still constant and ubiquitous reminders of racism. The American Red Cross, for example, refused to accept blood from blacks, and even after protests changed this policy, the donations were separated according to race. These policies were especially insulting because the plasma discoveries of a black physician, Charles R. Drew, had made the creation of blood banks possible. And in New York, the socially prominent British War Relief, in a singular act of insensitivity, chose to show *The Birth of a Nation*—perhaps one of the most racist motion pictures ever made—for a major fund-raiser.[2]

As a result of rising awareness, many blacks adopted the "Double-V" campaign: Victory at Home, Victory Abroad. They were going to eradicate Jim Crow as well as Hitler.

Unlike his wife (and daughter), who supported America's efforts in the war, Schuyler did not. He had no use for the Double-V campaign and advised his countrymen, in no uncertain terms, *not* to enter the European war. To underscore his position, George joined in 1940 the America First Committee, an isolationist group advocating nonparticipation in the war.[3] His columns from this period hounded FDR for being a warmonger, and chastised the U.S. government for its wartime discrimination. He exhorted his *Courier* readers to look elsewhere for essays on patriotism. A loyal American, George "refused to be a flag waver because of the 'blather' about fighting for democracy, freedom, and the American way."[4] Jim Crow still dominated the land south of the Mason-Dixon Line, and although a fratricidal war was almost a century in the past, and slavery had been abolished, the black man's equality had made little progress. (*Brown v. Board of Education* was still a decade away [1954], and the Civil Rights Act would come another ten years after that.) George contended that America must first put its own house in order; then the African American could fight for "the American way."

The Schuylers had always shielded Philippa from overt racism, but it had not been easy. To counterbalance a world outside of racial prejudice, they tried to foster a feeling of personal security through a close family life. And indeed, Philippa's relationship to the black part of her heritage had yet to become complex and problematic. For now it was positive and relatively untrammeled.

As a child, she adored her mahogany-colored father, who lovingly had instilled in her the fascination with the written word. She was devoted to her cinnamon-hued Aunt Louise, "an Andalusian-looking quadroon," from whom she had inherited her haunting dark eyes. The shelves of their living room were lined with books on the accomplishments of the Negro. Exquisite African artifacts were scattered throughout the apartment. Cultured black visitors from overseas, distinguished men and women, and students frequently found their way to the Schuylers' home. And to top it all, Philippa was the darling of and role model for young black Americans.

When her school texts declared that the southern slaves had been happy and docile, or when her teachers insisted that Africans were savage, uncultured, and wild, Philippa would attribute this to general ignorance. She knew from the books in her father's library that there had been five hundred slave revolts, that a quarter million Negroes had served in the Union Army, and that Abraham Lincoln himself had said that the war could not have been won without them.

By contrast, and much to her chagrin, Jody was unable to provide Philippa any contact with her own white family. On the rare occasion when Josephine went home, it was by herself. And even then she could not share with her folks the pride in the achievements of her child, for the Cogdells barely knew that Philippa existed. Josephine's marriage had never been talked about "except in whispers, if that. Even if they had known about Philippa's extraordinary accomplishments they could not have taken any pride in it, for the simple reason that to have done so would have been to acknowledge the unthinkable. . . . The family could not bring themselves to accept the fact that their baby had shamed them and brought the stain and loss of caste and status forever."[5]

Philippa felt this estrangement, for she idealized, even yearned for her mother's native Texas, creating innumerable stories about a "perfect home" there. It was not without pathos that for her daughter's fourth birthday, Josephine had created a magnificent edible sculpture of the Cogdell's ancestral home, built by Philippa's great-grandfather after the Mexican War, and where her grandfather, D. C. Cogdell, was born. (It was a story-and-a-half log house constructed of ground cashew nuts mixed with honey, cream, butter, eggs, spices, and lemon juice. Roof, chimney, and stoops were fashioned of raisins, ground and molded to form. In the yard was a well with little buckets, and finally, to be shockingly realistic for a cake, there was a conveniently placed outhouse. Whipped cream dripped like snow from the roofs and lay on the window ledges, the windows shining with subdued light through thinly sliced oranges. The ground was strewn with coconut and cinnamon footsteps.)

Yet, in the final analysis, Philippa's primary identification was largely with the white world, even when it reached her via the black intelligentsia. Her precocity was fed on notions and conceits of the white milieu; and her passion for classical music would essentially reflect the same bias.

❖ ❖ ❖

On August 1, 1943, the eve of Philippa's twelfth birthday, Harlem boiled over: six persons died, several hundred were injured, and approximately two million dollars' worth of property was damaged. Like the Harlem riot of 1935, this too was sparked by a clash involving the police and spread by inflammatory rumors. A white policeman had superficially wounded a black youth on leave from the army. As the rumors swept through Harlem, the incident went from "wounding" to "killing." A race riot erupted virtually in Philippa's backyard.

The Harlem riot came on the heels of many other racial explosions that long, hot summer. The distresses of ghetto living, the high rents, the slow pace of integration, the hypocrisy of fighting abroad for something the Negro did not have at home, the treatment of blacks in industry and the military, and the feelings of first hope and then despair since the Great Depression, all collided. The "American Dilemma" had unleashed its fury.

The contradictions which characterize human struggle for self-definition are perhaps more dramatic for African Americans than for any other group in this country. W. E. B. Du Bois described this phenomenon as "double consciousness . . . this sense of always looking at one's self through the eyes of others, of measuring one's soul by the tape of a world that looks on in amused contempt and pity. One ever feels his two-ness—an American, a Negro; two souls, two thoughts, two unreconciled strivings; two warring ideals in one dark body, whose dogged strength alone keeps it from being torn asunder."[6]

George Schuyler was a prime example of Du Bois's double consciousness.

Caught, as an adult, between two worlds, neither of which he considered fully legitimate, denied equal opportunity, often repudiated, Schuyler had decided early that the pen was mightier than the sword. And as he became a professional writer, he would scourge and excoriate when he banged out his essays—his fist gloved in satire.

Thus, in his article "Our White Folks," which H. L. Mencken made the lead of the *American Mercury*'s December 1927 issue, Schuyler irreverently slashes white superiority. "The Aframerican, being more tolerant than the Caucasian, is ready to admit that all white people are not the same," George writes, "and it is not unusual to read or hear a warning from a Negro orator or editor against condemning all crackers as prejudiced asses, although agreeing that such a description fits a majority of them. . . . In this respect, I venture to say [the 'Ethiop'] rises several notches higher than the generality of ofays, to whom, even in this day and time, all coons look alike."[7]

Four years later, George would publish a science-fiction novel, one of the earliest in this country, called *Black No More*. In this sardonic Swiftian satire of racial prejudice, a young physician invents a process for instantly "whitening" Negroes, thereby disrupting the American social and economic fabric.[8] The primary theme of the book is the divisiveness and foolishness of American obsession with skin color. The work underscores Schuyler's conviction that simply "getting white," as do virtually all the Negroes in his novel, will not solve the race's problems.

The book spares neither blacks nor whites. It parodies W. E. B. Du Bois, James Weldon Johnson, the NAACP, the KKK, Herbert Hoover, southern aristocrats; it even savagely spoofs a lynching. What Schuyler espouses is not always readily distinguished from what he condemns.[9]

In a way, George was a direct disciple of Booker T. Washington. He believed in education, self-help, and a certain looking inward. And he firmly held that despite all the obstacles so obviously placed in the way of blacks in America, the United States was still the country which offered them the greatest opportunity.

He believed that much of the Negroes' difficulties and problems could be greatly ameliorated by their own efforts in cooperation with willing whites, to a mutual advantage. An avid student of the Cooperative movement in Europe, especially in Britain, George organized the Young Negro's Cooperative League in 1930. Its purpose was to place groups in each black neighborhood that would study the principles and practices of consumer cooperation. George envisioned black-owned stores that would deal directly with the white wholesalers, avoiding the obstacles that, in his opinion, had tripped so many interracial efforts in the past. Unfortunately, Schuyler lacked the time and the financial support to make the idea a reality. Finally, the Negro merchants objected to it.

Considering George's basic philosophy, it is not surprising that he had no use for the Garveyites, who romanticized the idea of a return to an African homeland, or the American communists, whose latest gambit was to advocate the creation of a separate "black belt state."[10] George, who prided himself on not pandering to the latest fad to promise salvation for the blacks, began early to view with deep suspicion the idea that communism was a collective way of solving the Negro problem in America. In particular, he felt that the communists were manipulating black frustration to create chaos and to sensationalize racial troubles in order to polarize America. For George, their assertions of brotherhood with black America were hypocritical; the Negro was merely a pawn in their grand design.[11]

Schuyler became particularly vociferous when an arm of the Communist Party, after a fight with NAACP lawyers, took over the defense of the Scottsboro and Herndon cases[12] and made an international issue of them. George contended that for the Party, the cases were merely a propaganda vehicle; the Party showed scant concern for the human beings involved.

He capsulized his two-pronged attack on both the Back-to-Africa movement and what he called the "Separate State Hokum" when he wrote:

> Whenever the Aframerican, flailed unduly by poverty, prejudice and proscription, yammers aloud for succor, there is a wild stampede of hungry psychological shamens, loaded down with weird and colorful nostrums, clamoring to assuage his hurts.
>
> Thus we have been periodically afflicted with the Back-to-Africa dervishes from the early days of the Republic to the advent of infantile paralysis of Garveyism; the high-pressure Group Economy salesmen who view segregation through rose-tinted spectacles, and the wistful witchmen who see Zion amid the snows of Alaska or the swamps of the Amazon.
>
> The beleaguered Brother has grabbed at one or the other of these nostrums as the rowels of adversity have bitten deep, but his saving sense of humor and fundamental cynicism have, after cursory examination, restrained all except the lunatic fringe from swallowing such crackpot proposals. The shamens rattle their shells and toss their gri-gri bags for a season, charm a moron minority with their bombastic amphigories, enjoy a grateful change of diet from neckbones to fillet mignon and then, when the disease has run its course and dues grow scarcer than dinosaurs, they hock their wardrobes and hold off their landladies until they glimpse another glorious vision of cash.[13]

Reading such powerful prose, one is not surprised that Mencken ranked Schuyler one of just a handful of people who he believed enriched language. Shortly after her fifteenth birthday Mencken would write to Philippa, "Of all my journalistic friends, and I must have first and last at least a thousand, [your father] is one for whom I have the most respect."[14]

❖ ❖ ❖

George's legacy to his daughter was a complicated one. If Josephine ruled the psychological roost, George guarded the political one.

Politics were always discussed around the Schuyler household, and by the time Philippa entered her teens George had already completed his labyrinthine journey from moderate left to conservative right. Ultimately he became an ultraconservative, prompting one Harlem friend to describe him as further right than Barry Goldwater. The author John Henrik Clarke put George's politics in deeper perspective: "I used to tell people that George got up in the morning, waited to see which way the world was turning, then struck out in the opposite direction."[15]

Philippa, who admired her father in many ways, became the direct inheritor of her father's extreme and often out-of-step thinking. Rather than reject his politics, she adopted and tailored them to her own evolving needs, never radically altering them.

It was a legacy that would not always stand her in good stead.

12

Manhattan Nocturne

As Philippa entered adolescence, she began developing into a tall, slim beauty. At five feet four inches, she towered over her mother and would soon be taller than George (who was barely five feet six). Her skin, the color of cinnamon, glowed. And her jewel-like black eyes—which had captivated so many—now had a slightly coquettish cast.

Philippa's transition into her teens was also marked by a renewed isolation from her peers. She had been accepted by New York's unique High School of Music and Art, but at the last minute Jody decided against it. Instead, she withdrew Philippa from school altogether. With the help of Mayor La Guardia, she and George obtained official permission to let their daughter continue her education through correspondence, provided that a tutor came to the Schuyler home once or twice a week.

There was another important change. The idea of redirecting her daughter's career from performing pianist to composer had finally gelled in her mother's mind. It was a decision not easily taken, and Jody was caught on the horns of a dilemma. On the one hand, she was weary of traveling, and fed up to the teeth with Philippa's "difficult" behavior on the road. On the other, the Schuylers needed the money that performing produced.

"Concert tours are not much fun, I tell you," Josephine wrote her husband in 1944. "I am constantly worried. But the money comes in so handy that I always feel forced to do it." She added rather frankly, "Without it, we would not be able to live in the style we like."[1]

Perhaps engaging in wishful thinking, Josephine had convinced herself that Philippa's earning power as a composer could in time equal and eventually exceed that of a concert artist. She also knew that if Philippa continued on the circuit, she would sooner or later need an impresario or an agent, and Josephine loathed the idea of someone sharing in her daughter's management. Nor is it entirely unlikely that in some dark recess of her mind, Jody—always so demanding (and critical) of her daughter—had begun questioning whether Phil had the talent to become a world-class performer.

Phil, who still implicitly trusted her mother's judgment, warmed to the idea. "My dream," she told a *Herald Tribune* interviewer on her thirteenth birthday, in obvious imitation of her mother, "is to become an outstanding American composer."[2]

To this end, Philippa's concerts were drastically reduced. And to broaden her musicianship, she was given violin lessons with the black virtuoso Clarence Cameron White and, later, conducting classes with Dean Dixon, another outstanding black musician. She continued her twice-weekly composition lessons with Otto Cesana, and piano with Wittgenstein at the Convent. By the fall, she owned a violin, a cello, a viola, a clarinet, and a flute—all gifts from various friends and selected relatives interested in supporting her burgeoning new career.

Thus, studying, practicing, and composing, Philippa spent most of her time in the apartment—and never far from her mother's view.

❖ ❖ ❖

In early 1944, the once-proposed but later abandoned trip to Mexico (to study with Carlos Chavez) was revived. Jody decided on a seven-month "compositional sabbatical" for them: Phil would write her first piece for orchestra, and Josephine would rest.

Circumstances for their sabbatical seemed propitious. George was home to look after the apartment and the cats;[3] Jody and Phil could live in Mexico inexpensively; Philippa had taken enough composition lessons to tackle a larger work; and although Jody did not speak Spanish, her daughter did.

Traveling by rail, Phil gave concerts along the way. In Texas, they took a circuitous path, consciously avoiding Fort Worth. But in Austin, Josephine left her daughter with a friend while she made a solo three-day trip home to see the Granbury estate. Her one letter to George during this trip confirms that her marriage and subsequent birth of her child were still kept from the family. It was a nostalgic trip for Jody, who had not been home for many years.

> I left Austin just as the sun was lifting on the horizon. . . . It was my first Texas Sunrise in many years—yet so short a time ago it seemed I had been sitting on my horse watching the sun rise in glory out of the East, over the rolling green prairies, bringing with it all the mystery and beauty of life. There is something here about the dry, exhilarating air, the sweeping sky, the low far-reaching plains and prairies carpeted with brilliant patchwork, there is something eternal and magnificent about it. . . . One feels unlimited, one can reach into the sky and pull down the stars. . . . There is nothing one cannot do if one desires—but one desires only that which is great, unusual, wonderful, magnificent. Only that which is worthy of the Gods. Well, I am still moved by these pagan, barbarian feelings.[4]

Jody had almost forgotten the grandeur of the seventeen-room mansion and the wonderful objects still preserved—linen, silver, carpets, and furniture. She found "surprisingly good" the landscapes and portraits she painted during her "other life."

Yearning to acquire at least a few of these heirlooms for her daughter, she asked George to send her sister Lena, now sole guardian of the sprawling estate, some money: "I want you to send Lena $50 in the next 2 months—$25 dollars at a time. . . . Just put my address on the envelope as I told her I would have a friend send it to her. And don't put *Geo. Schuyler* on the stuff to Hetty [one of the Negro servants still alive, who wanted some copies of the *Pittsburgh Courier*]. Just your business address but not your *name*."[5]

❖ ❖ ❖

Arriving in Mexico City sometime in April, they roomed for a few days in a musty, midtown hotel near the magnificent Instituto Nacionale de Bellas Artes before moving to a more reasonably priced flat.

This was wartime Mexico and the narrow streets were clogged with fine American cars driven by wealthy Mexicans in spotless tropical suits and jeeps packed with soldiers in uniforms of a dozen nations. Weaving in and out of the traffic were barefoot, serape-draped Indians in huge hats. On the corners, amiable street vendors made fresh orange juice while ragged urchins sold lottery tickets. In the meat stalls of the open-air market hung carcasses of lamb, carbito, beef, and fowl, aswarm with flies. More appealing to Jody and Phil were tables piled high with fruit and melons.

In search of a piano, they telephoned a Hungarian lady, "Madame B," recommended by a New York friend. She was a refugee—her husband had been French, and she had escaped from Europe with her son. Her house appeared to be a moldering stucco dwelling, but inside was a French-style flat with exquisite Louis XVI furniture and, best of all, a fine rosewood Bechstein, which she said Phil could use.

"Mme. B. was attractive, red-haired, vaguely in her thirties. Her son, Stepan, about 19, had black hair, sharp, severe features, and an . . . indrawn expression," Phil wrote sixteen years later.[6] Nonetheless, she liked the young man for he treated her like an adult.

Philippa began to come daily. Often Stepan was alone, except for the grim, sour-looking maid who was always hovering.

One day the maid was the only one there. After an hour's practice, Phil started toward the bathroom. But the maid pulled her back, saying, "Oh, no. Not zat one. Zat ess kaput. Today you go ze other. Come here."[7]

She led Philippa into a room, magnificently furnished, all black and silver, with silver lamé draperies, and an enormous bed with black silk brocade bedspread and satin canopy. At the foot of the bed lay a pair of gray pajamas, and a sheer green silk nightgown of chiffon and lace.

She pointed to the bed and said: "Zat ees where zay sleep. Zey sleep togezzer. Eet ees a sin. . . . God weel punish sem."

Indeed, this was the only bedroom in the flat, except for a small room off the kitchen for the maid.

Philippa left, and never returned.

A week later, Mme. B. visited Josephine and recounted a long story. Her son, she said, had been killed before her eyes in an air raid in central Europe. This boy had been his best friend. They resembled each other, and she had given him her son's passport. They had escaped together, and had come to Mexico hoping to get into the States but had been unsuccessful. Now they were waiting, just waiting. They had fallen in love, and since they had nothing left but each other, they had seen no reason to hesitate.

Neither Josephine nor Philippa quite believed her story.

Mexico turned out to be much more expensive than Jody had anticipated. Food was also a problem. In part from general inclination, they both refused to eat meat, and the sight at the stalls had certainly not encouraged them. Milk was often rancid. Philippa would not eat hot meals, which also limited their choices. And much of the Mexican standard fare was fried. Their diet was quickly reduced to fruits, avocados, selected vegetables, and freshly squeezed orange juice.

Philippa had hoped to compose something characteristic of the country. She and her mother studied Diego Rivera's murals, went to museums, saw native Mexican dancers, and attended rehearsals of the city's opera company and symphony. To "inspire" her, they went to a bullfight shortly after they arrived. But Jody wept as six defiant bulls were killed. Philippa threw up.

It is not at all surprising, then, that her first orchestral work, written in their small apartment in the heart of Mexico City, expressed Philippa's intense longing and nostalgia for home. She called it *Manhattan Nocturne.*

❖ ❖ ❖

Phil and Jody stayed in Mexico City only six weeks—far short of the planned seven months. They returned to New York in time for her thirteenth birthday. In a break from tradition, Philippa acted as hostess. "Slender, charming, as eager as a child, and as poised as a duchess," wrote reporter India McIntosh from the *Herald Tribune.* "Her modesty is one of the girl's most amazing characteristics. Getting her to talk about herself is a man-sized job, for she adroitly turns the conversation into interesting little side lanes which soon wander far afield. Her accomplishments embrace numerous branches of learning, but she speaks of them objectively, as if they had no relation to her."[8]

The highlight of the evening was Phil's playing, for the now traditional assembly of guests and reporters, of a piano reduction of *Manhattan Nocturne.* Nora Holt wrote the first review of the work in the *Amsterdam News*:

> It was a creation of a mind young in theoretic pronouncement but mature in conception. . . . Her melodies were fresh and spontaneous, with a feeling of joy and longing, mixed with pain and anguish, Manhattan night moods. . . . When she played the work on the piano, the real talent of this remarkable child was revealed. . . . Most of her compositions, and she has written more than 100 published and unpublished, are images and descriptions of what she has seen and heard, but her latest work, *Manhattan Nocturne*, is indicative of her capacity to speak of

> life, which will undoubtedly ripen as her technique and experimental facilities broaden.[9]

Eight months later, on April 7, 1945, Philippa's orchestrated *Manhattan Nocturne* was performed under the baton of Rudolph Ganz at the New York Philharmonic's Young People's Concert and taped by WQXR for radio broadcast. It seemed to mark a turning point for Philippa. At the rehearsal, the musicians, many of whom had seen Philippa come on stage to accept her awards over the past decade, gave her a standing ovation.

She remembered the day with mixed emotions.

> I . . . felt anxious, at thirteen, when I heard the New York Philharmonic . . . give the first performance of my symphonic tone poem. . . . "They're playing it too slow" I thought, in anguish . . . at the rehearsal, and though this was remedied at the performance, I lived years of self-torture listening to each note. One can never enjoy hearing others play a work one has written, for one goes through, in one's mind, as the music spins out, all the torments one underwent during the composition, the uncertainties, confusion, anger, months of revision, hundreds of wee morning hours spent laboriously copying out score and parts.[10]

George sat in the front row, feeling as if he were "sitting next to the gods." Josephine's response was equally sublime. "My heart almost stopped beating as the strains of 'Manhattan Nocturne' filled Carnegie Hall," she wrote. "I thought the spheres must have played your music when you were being born and I wept."[11]

The reviews were glowing. The *New York Times* said that her music, "written in traditional harmony, shows more than expert workmanship in the handling of form and orchestration; it is truly poetic, the expression of genuine feeling, a gentle, soft beauty and imagination."[12]

Phil had already begun work on another orchestral composition. Her new piece, a playful scherzo (the third movement of a planned four-movement work entitled *The Fairy Tale Symphony*) called "Rumpelstiltskin," she dedicated to Fiorello La Guardia. Her dream of becoming an outstanding American composer was drawing closer to reality.

13

Scrapbooks

The full awareness of being a manipulated human being, a "puppet," dawned on Philippa rather unexpectedly. Shortly after her thirteenth birthday, she was shown the scrapbooks, all fourteen of them, which her parents had lovingly and secretly kept.

They contained not only countless clippings and letters—every line written about her—but also page after page of notes, philosophical and clinical, tracing their daughter's development. The books unfolded the story of a brave genetic and behavorial experiment, designed to create an extraordinary child; the parents' insistence that the offspring's superiority was the result of nutrition and training; a mother's tireless efforts to promote, and a daughter's to succeed. All this was revealed to her in almost one sitting, as she pored over the books. And finally they underscored a mother's sadly mistaken belief that her brilliant, racially mixed child would topple America's race barriers.

Some of it was of course known to Philippa. It is most unlikely that she, a voracious reader, would have been unaware of the many reports and reviews that studded the newspapers and magazines strewn about the house: the *Pittsburgh Courier*, the *Amsterdam News*, the *New Yorker*, the *New York Times*, and *Herald Tribune*, *Look*, *Newsweek*, *Time*. But now they seemed to acquire a different meaning, to exist in a changed landscape. For what she had not realized was the extent to which her existence had been manipulated. Suddenly she felt horribly and irrevocably displaced. Her life seemed to be owned by everyone but herself.

Her musical career also acquired a different meaning. "Before then," she would write many years later, "I had played for enjoyment, without realization of the weighty importance of each concert. As soon as I knew, I felt a sense of responsibility about each appearance, which precluded enjoyment, and opened the door to anxieties . . . and apprehensions."[1]

Her sleep became fitful, punctuated with a recurrent nightmare. Philippa dreamed over and over of committing suicide. In these night terrors, newspaper men and journals would tell the world of her private thoughts, of her desire to kill herself. Telegrams began to flood the Schuyler apartment, begging her not to do it. They expressed dismay and confusion. Why would anyone as lucky and famous as she, they asked, even contemplate such an act? And she would awake, in the darkness, crying.

Philippa went into a kind of musical overdrive, practicing obsessively. Withdrawing into her music, she and Josephine quarreled often, something even the reporters began to sense. The rift was particularly hard on Phil, as Jody had not just been a mother; she was her best and only friend, her confidante, counselor, teacher, spiritual adviser, promoter, and manager. Aside from George, who had been away so much of the time, and her various music teachers, who always left too soon, her mother was the only person Phil had been allowed to get close to. In intimate matters, Jody was virtually all she knew.

Phil might have turned to her father, who was home much more now, but Josephine, who considered her daughter old enough to understand such matters as she had recently "become a woman," began to hint at George's philanderings. Although this confused Philippa, and would ultimately create an even stronger bond between her and Jody, she did not feel comfortable with her father anymore. A distance was created, a distance which she would never really bridge.

Phil, too, might have turned to the nuns at the convent, but Jody had switched teachers (she was studying with still another teacher, Herman Wasserman) and her formal contact with the school was now negligible. Philippa began to feel profoundly lonely; her music became her only accomplishment and source of self-worth, a friend, and a place to hide.

It had been a difficult time for all the Schuylers. George was experiencing problems at the *Courier*. His unpopular stand on so many issues, which had earned him in many quarters the epithet of a "vicious pen prostitute" or an Uncle Tom, began to conflict seriously with the paper's editorial policy. Always allowed to speak his mind, George found his series on George Hewitt, a defector from commumism, "abruptly terminated."

Josephine also suffered an upsetting year. On February 2, 1945, she received a telegram from her sister Lena, then living with their father in Granbury. Addressed to "Mrs. Heba Jannath"—the pen name Jody had not used in years—and sent to her Harlem address, it read: "Papa has cerebral hemorage [*sic*] and pneumonia cannot last through the day. Can you come?" He died two days later, at age ninety-five. The obituary listed his last born as "Miss Josephine Cogdell, of New York."[2]

14

Rumpelstiltskin

The year 1946 would provide a sharp contrast to the previous one. Not only did Philippa debut in the unusual double role of composer and performer, but she also garnered several national composition contests and had some of her works performed throughout America. All this helped extirpate the aftermath of the scrapbooks: Philippa seemed, at least superficially, to rebound from her depression.

Her double debut occurred on Saturday, July 13, at Lewisohn Stadium, an open-air arena belonging to New York's City College. The day was sunny and mostly clear. Occasionally a few clouds appeared, but they flew over the city quickly. At a quarter past seven the temperature was already sixty-nine degrees, and by late afternoon it reached into the low eighties. But it seemed much cooler. There was almost no humidity and a northerly wind caressed New York. The day augured well.

At ten o'clock, Philippa and her mother walked the few blocks from their home to the stadium. It would be her only rehearsal with the New York Philharmonic for her performance that evening, playing Saint-Saëns's Piano Concerto in G Minor. Not only was she going to be the first black pianist ever to solo there, but her recently orchestrated "Rumpelstiltskin" was also on the program.[1] As a composer, Philippa would appear alongside Mozart, Paul Creston, Robert Schumann, Rimsky-Korsakov, and Saint-Saëns. Tickets for the concert were still available—at 30 cents, 60 cents, $1.20, and, for those with deeper pockets desiring reserved seating, $1.80.

When mother and daughter arrived at the back entrance of the stadium a worried stage manager met them. No piano had been delivered from Steinway Hall, he reported. The Stadium Society had failed to order one or to tell the Schuylers to do so. It was impossible to secure a piano now, he added; Steinway Hall was closed on Saturdays.

There was a battered, out-of-tune upright in the wings, used by City College students for rallies, which the manager now rolled out for the rehearsal. Thor

Johnson, the conductor, and members of the orchestra entered into the spirit, and cheerfully went through the concerto with Philippa. She, too, tried to be confident.

Meanwhile, Josephine was making frantic telephone calls. Herman Wasserman located Mr. Theodore Steinway just fifteen minutes before his departure for Europe; he graciously agreed to order his watchman at the store to have Arthur Rubinstein's favorite instrument ready for the truck. But, he apologized, his contract with the union did not provide for weekend transportation.

Now an even more frantic Josephine suddenly recalled that Mother Morgan, Philippa's first harmony teacher at the Convent, had a family in the trucking business. She contacted Mother Morgan, who promised to do her best. The Schuylers retreated home, nervous wrecks.

Philippa slowly and apprehensively dressed for her debut. The piano—would it arrive? Had Mother Morgan been able to find a mover? She felt that her Lewisohn Stadium debut, both as a performer and composer, would be the most important night of her life. For a short moment she sat motionless on her bed. Knowing her daughter's penchant to dawdle, even on better days, Jody rapped on the door, asking her to hurry.

Phil got up and finished dressing. She wore all white—white shoes, white stockings, white beads around her neck. Six white gardenias swept her hair up, letting it fall down her back gracefully. Her dress had a slight décolletage, was short but full skirted, and was studded with sequins.

At 7 p.m., mother and daughter again walked to the stadium. And this time George accompanied them, holding his little girl's hand proudly, assuring her the Steinway would be there—but it was not. It arrived only half an hour before the concert, too late for Philippa to try it. The in-house tuner did a quick run through, and the piano was wheeled out, center stage. Over twelve thousand people filled the horseshoe stadium.

After intermission, with the orchestra returned and in place, Thor Johnson raised his baton to the puckish introduction of Phil's "Rumpelstiltskin." She listened with apprehension to every note. But his sensitive rendering put her at ease. The applause was enthusiastic.

Next it was time for the Saint-Saëns concerto. Philippa stood in the wings terrified by the capacity crowd, before walking to the piano. But as soon as she played the rapidly cascading opening arpeggios, her nerves calmed. Philippa performed each movement with rhythmic incisiveness and surety, bringing special fire and bravura to the closing Presto. Almost before the concerto had ended, the audience rose in a storm of applause. Nor would they allow her to leave the stage until she had given two encores.

In the greenroom, Jean Tennyson,[2] Gladys Swarthout, Carl Van Vechten, Deems Taylor, and many others embraced her. Later, scores of young people walked her home. There was a party for her but she was too stunned to enjoy it.

The reviews next morning were glowing.

"She revealed herself as a pianist, without regard to age, of extraordinary natural talent . . . and she disclosed imagination to be found only in artists of a high level," wrote the *Times*. Her composition, the reviewer continued, had "authentic musical insight and imagination . . . expertly written, with a broad melodic core that has genuine charm."[3]

Music critic Robert Hague described her piano debut as "remarkably persuasive, often brilliant";[4] and the *New Yorker* referred to her as "uncommonly gifted."[5] Thor Johnson wrote to her: "So many wonderful remarks were made concerning your exceptional playing and your genuine compositions . . . that I continue to feel especially fortunate in having had the opportunity of appearing with you on that memorable evening."[6]

Three weeks earlier, Philippa had given the same program in a "pre-debut" with Arthur Fiedler in Boston on an evening designated "Colored American Night at the Pops." Fiedler too had written the young performer encouragingly: "I enjoyed hearing you play with us in Boston and I hope we will have another opportunity. You must work on some other concerto so that you will enlarge your repertoire."[7]

Though the Boston audience had rewarded her with boisterous applause, insisting on an encore, the reviews were decidedly lukewarm—as if underscoring both the vicissitudes of musical criticism and the pitfalls of having been a prodigy. "So much has been written about her from the time she was a very small child," began the reviewer for the *Boston Herald*, "that it was hard not to expect another Mozart. What we actually heard was a talented young woman whose ability as a composer is far greater than that as a pianist.

"Unfortunately," the review continued, "it was the pianist we heard first. . . . Miss Schuyler, who charmed everyone with her stage presence and modesty, performed studiously and seriously, but her interpretation was almost entirely devoid of inspiration or brilliance. She has obviously been well-schooled, but her coloration was monotonous and her technique uneven."[8]

But on the whole, 1946 was an extraordinary time for Philippa. Not only had she scored a success at Lewisohn Stadium, but earlier that year she had also won both top prizes in the National Composer's Competition, sponsored by the Grinnell Foundation of Detroit. The competition was open to anyone under eighteen. Philippa first entered *Manhattan Nocturne*, but when she had not heard from them, fearing they had dismissed it, she sent them "Rumpelstiltskin." Even to Phil's surprise, they awarded her both first ($100) and second ($50) prize.

First prize also included a performance by the Detroit Symphony. On March 13, they played her *Manhattan Nocturne* to an audience of seven thousand schoolchildren. "Wearing a grown-up black dress and a black mantilla, Philippa sat with queenly poise in a box of Detroit's gold spangled Masonic Temple."[9]

Only a month earlier, Rudolph Ganz had conducted her Nocturne with the San Francisco Symphony Orchestra and the Chicago Symphony. By 1946, Philippa's first orchestral work had seen four American performances by four prestigious orchestras. It is no wonder that *Time* called Philippa the "brightest young composer in the U.S.,"[10] in no small measure vindicating Josephine's courageous step of directing her daughter's musical activity toward composition.

Philippa continued in her double role of performer and composer. On February 8, 1947, "Rumpelstiltskin" was performed in Carnegie Hall, in a Young People's Concert. On April 24, as part of the eighteenth annual Festival of Music and Art at Fisk University, Nashville, she played a piano transcription of her *Fairy Tale Symphony*, which included "Rumpelstiltskin." At not quite sixteen, her composition shared the roster with works of some of America's most outstanding living composers: Virgil Thomson, Marc Blitzstein, William Grant Still, Aaron Copland, and William Schuman.

"I am always astonished by Philippa," reported Carl Van Vechten, who was there for the duration of the festival.

> She comes out on the stage or platform as would a very young girl. The moment she begins to play, she becomes a giant of power. She plays with the greatest clarity. The thing that impresses one is that "something" inside her—that deep musical feeling, her genius, and her interpretations. She is a very beautiful girl and invariably wins people with her charm and personality. . . .
>
> That Thursday night when Philippa played, not only every inch of the concert hall proper was crowded, but also the entire stage. Students, white and colored, followed her around on the campus, and Philippa seemed to be having a wonderful time. Everywhere she turned, Philippa was hounded by autograph seekers.[11]

The supercritical Virgil Thomson was also there and was as equally if not more taken with her:

> Though thoroughly accustomed to platform appearances, and with every other circumstance lending itself to personal exploitation, she makes no personal play whatsoever. She plays music, not Philippa Schuyler, even when she performs her own compositions. And she gets inside any piece with conviction. . . . The sincerity of her musical approach and the infallibility of her musical instincts [are prodigious]. . . . Her emotional understanding of longer works is . . . immature. But her understanding of music as a language and the confidence with which she speaks are complete. Miss Schuyler is a musical personality of the first water. . . . Her symphony, composed several years ago, is the work of a gifted child. It is harmonically plain, melodically broad, brilliant but not original as figuration. Like her piano playing, it shows a thoroughly musical nature and a real gift for expression, for saying things with music. It is in every way as interesting as the symphonies Mozart wrote at the same age, thirteen.[12]

A few months later, Phil finished orchestrating her highly touted *Fairy Tale Symphony* and plans were afoot to have it performed in America, and, with the help of Rudolph Dunbar, in Europe. In September, she was invited to participate in a "Salute to Africa Day" as part of "United Nations Week" in Tarrytown, New York, where Phil performed (in piano reduction) two more movements from her *Fairy Tale Symphony:* "Rip Van Winkle" and "The Headless Horseman of Sleepy Hollow"[13]—both appropriate tributes to an old Tarrytowner, Washington Irving, creator of these fantastic tales.

❖ ❖ ❖

Success seemed to engender a physical response in Philippa: She grew more strikingly beautiful almost daily. Philippa's face had always been expressive even as a child. Her wideset almond-shaped eyes, fringed by long, naturally curled eyelashes, and her amber-colored complexion, gave her a warm and alluring aura. At fifteen, this warmth was rapidly transforming her into an intriguing beauty as the deepening of her inner self melded with her striking features. Her black eyes took on an even more probing quality; her mouth grew fuller, her skin richer.

And Josephine, the tireless promoter, was not about to hide her daughter's light under a bushel. Profiles about Phil now appeared regularly, not only in black publications such as *Ebony* and *Sepia*, but also in popular white ones, like *Calling All Girls* and *Seventeen*, as well as *Newsweek* and *Time*.[14]

Philippa seemed more relaxed now, even occasionally flirting with a male reporter: "I like dresses with full skirts," she told a feature writer for *PM Magazine*.

> She got up and twirled happily around, fluffing out her skirt, "fitted bodices, things with feminine lines. And I like evening dresses that have decolletage here," she demonstrated, "and long skirts. I don't have any long ones now. When I get to be 16, I guess I'll wear long skirts and wear my hair up, too. I understand that I'm too young now to wear long skirts, because if I put on a long skirt I would look 18—at least I hope so." She smiled gaily, and dimples appeared at the corners of her mouth.[15]

Philippa was invited to become a junior advisory editor for *Calling All Girls*, sharing the masthead with Elizabeth Taylor and Peggy Ann Garner, no small feat for a fifteen-year-old mulatto from Harlem. Frances Ulmann, the magazine's editor, wrote Jody, saying in passing, "I am sure Philippa will be glad to know that the chief comment I got from the men around the office was, 'Isn't Philippa pretty!'"[16]

For a brief moment, at least, it appeared that the gifted teenager might be heading toward a social life. "She is going to the opera at City Center next week with a young man, all alone," Jody confided to Phil's godfather, Carl Van Vechten. "One of those who came to her birthday party. They have been out to dinner and a show also recently! She has a friend here from Bridgeport, Conn., who won the recent Pepsi Cola contest there for scholarship and is going to Barnard. So she is all excited with girl and boy friends who share her intellectual interests." Beginning to sense the loss of her little girl, to whom she had devoted so much of her life, Jody added with a tinge of regret, "She will probably grow up fast now."[17]

❖ ❖ ❖

Philippa was riding the crest of a wave. Recognition for her compositions and performances seemed to be coming her way in an unbroken surf. Yet there was a darker side. The Schuylers were experiencing financial difficulties. In part this was due to the rising costs of Philippa's specialized training: three-hour piano lessons, composition, theory, orchestration, conducting, violin, and so on. (In 1947, one piano lesson was a whopping twenty dollars, which, considering that she took lessons several times a week, was extremely expensive.)

But the expense of Philippa's education was not the only reason for strained finances. Jody tended to live beyond their means, opting for a more flamboyant style than her husband might have chosen, and George's income had suffered because of his ideological difficulties with the *Courier* publisher. Nonetheless, George insisted on contributing part of Philippa's earnings to charity or black causes.

Another area of contention was Josephine's belief that she, as Philippa's tireless manager, was entitled to a substantial share of their daughter's income. George did not agree. In one undated letter to his wife—after forwarding a little extra money—he wrote, "Now perhaps you can pay Phil back some of the money you owe her."[18]

Jody began to complain to the press about the sacrifices she and her husband had made: "Mr. Schuyler and I have literally sunk all our money into her development," Jody told the *Baltimore Afro-American* in 1947. "We have, as a matter of fact, used all our money above bare living costs, to give her advantages in music. We have no car, we take no extra vacations—we couldn't and give Philippa the things she must have. . . . Of course, she could have had a millionaire to sponsor her education, but there were too many strings attached to such an arrangement."[19]

As financial concerns for the Schuylers mounted, so did the urgency for devising a profitable strategy. In the spring of 1948, Josephine decided to arrange a nationwide concert tour. Mother and daughter were on the road for three months, stopping in more than twenty cities from New York to Los Angeles. Phil performed predominantly for black organizations—churches, sororities, interracial benefits, the NAACP.

Jody had dreaded this tour, and not without cause. At sixteen and a half, Philippa was no easier on her mother than she had been at twelve and a half. "Philippa is most difficult," she wrote George from Texas.

> She sits all day at the piano, won't eat. Says she is too busy, or it makes her tired or sick to eat. She plays the scales over and over. This Virgil Thomson piece which got the best review had scarcely been practiced *at all* and Fidelman [Joseph Fidelman was yet another new piano teacher] heard it only *once*. But when I told her that this morning, she said I was just against Mr. Fidelman. She is completely devoted to him, cares for nothing other—all because he talks sentiment and looks grave. But we cannot afford to suggest that she be careful and not go overboard. She will put that down as jealousy on our parts. So I will say nothing but I am not going to make *any more* engagements or go on any more tours for several years, ever again. She is not congenial, does not act gracious to people until I force her to and cannot even keep track of her hairbrush. When this tour is over, I am through! You don't know how much I do to make them successful. . . . I will put her on a plane . . . for

> Austin and go on to Ft. Worth. . . . I am really low in spirit. There is no satisfaction in this for me when she is so cold and demanding, so irritable and exacting.[20]

But Philippa's behavior was much more than teenage revolt. She had encountered and confronted on this trip America's racist traditions on a much more personal plane than ever before. Most disturbing was being snubbed by her mother's family.

Not for a moment was the race issue out of Philippa's life. In Nashville, some of the area reporters confided that it was unsettling to see the blond mother and her black child together accepting the applause—a living tableau of miscegenation. Even the well-meaning *Austin Statesman*, which blared "Half Texan, Half Yankee; Half White, Half Negro, Piano Prodigy is Wholly Musician," filled Philippa with disbelief.[21]

John Rosenfield, while writing enthusiastically in the *Dallas Morning News* about her playing, pointed out that "the event drew an audience of 1,000, mostly Negroes, to the St. Paul Methodist Church. Had any effort been made to sell tickets to the Whites the patronage would have been larger, as interest has been engendered nationally in the girl who has shown such precocity at piano playing, composing and solving the conundrums of higher mathematics."[22]

Jody enclosed Rosenfield's review in her letter to George. But her account of the racial makeup of the audience differed subtly from Rosenfield's: "[The church] was almost filled. They did not go to the [Arts] Auditorium. There it would have been segregated. Here it wasn't. There were about 25 or more W's [whites]."[23]

This had been the first time that Philippa played in the vicinity of Granbury, but the only Cogdell "family members" to come were a few of the black servants, including old Jo. Josephine had insisted that her daughter send an invitation to cousin Buster (Jody's nephew), enclosing Virgil Thomson's glowing review. Phil naively expected that he—and maybe other Cogdells—would come. But Buster could not (or would not) attend.

Several months later, however, Buster started to correspond with his first cousin. Jody encouraged it. Although meant as a kind overture, his first letter to Philippa more likely reinforced her feelings of always being, and remaining, an outsider. "You do not know my brother Budge (Gaston)," he wrote. "He is now in Abilene, Texas, going to school studying for the ministry—a fulfillment of one of my mother's dreams, by the way. I had occasion to talk to him some time ago and he asked me point-blank some questions about you. In the course of our conversation I learned that he is far more liberal than I had supposed; he evinced a tremendous interest in you and expressed the desire to correspond."[24]

Coming face to face with her southern white roots for the first time had been a jarring experience. Sensing the difficulties of getting white America to listen to her music, Philippa was deeply frustrated and angry. To make matters worse, she could not share her feelings with her mother without turning them into accusations.

But the Schuylers had almost succeeded, had almost accomplished the impossible. Josephine's machinations and perseverance and George's connections had taken their daughter's genius and charm and solidly established her as a prodigy. Black audiences had received her with unbounded enthusiasm and pride. But now Philippa's continued success in America, the establishment of a long and distinguished career for her, critically depended on attracting whites as well as blacks. She was no longer a child prodigy, and the few mixed audiences she had played for, at the World's Fair, the Boston Pops, Lewisohn Stadium, Carnegie Hall, even Tarrytown, would become harder and harder to find.

By the western end of this grueling three-month trip, Philippa was both physically and emotionally exhausted. For almost the first time, she refused to present a charming front to the public at large. Her Los Angeles debut was followed by an elaborately planned gala in her honor. Philippa made a brief half-hour appearance, signed programs, and then abruptly retired to her bedroom. Weariness and anger had finally won out over graciousness.

A decade and a half later, searching for a palpable turning point in her life, Philippa wrote: "I was born and grew up . . . without any consciousness of America's race prejudice . . . [but] I became intellectually aware of it when I . . . entered the world of economic competition as a full-fledged adult. Then I encountered vicious barriers of prejudice in the field of employment because I was the off-spring of what America calls a 'mixed marriage.' It was a ruthless shock to me that, at first, made the walls of my self-confidence crumble. It horrified, humiliated me. But instead of breaking under the strain, I adjusted to it. I left."[25]

15

Farewell

Philippa's world was not the only one that had tumbled. George and Josephine's marriage had reached an impasse. It was not that they were physically or intellectually incompatible or had failed to find a shared purpose in life. On the contrary, Jody and George cherished each other's company, enjoyed each other's creativity, and found a strong common focus in the rearing of an uncommon child.

They did not quarrel much over money. Nor did husband and wife fight over the way their daughter was being raised, for George invariably deferred to Josephine's judgment even when he disagreed.

The interracial nature of their matrimony, though seriously frowned upon by most of America, did not seem to subject their own marriage to an overwhelming societal strain. From the outset their union was well accepted in Harlem. Not that unpleasant incidents were rare. More than once Jody returned home, her face tear-streaked as a result of an abusive remark made by some stranger about her marriage or her "nigger daughter."

At the core of their marital problems lay irreconcilable differences in temperament, background, and life-style. Josephine dogmatically adhered to ideas about pure and natural foods and believed in devoting her life and love to one man (at a time), while George was a dedicated epicure and an unreconstructed womanizer. Knowing George's nature, Jody would become desperate when she suspected that her husband had betrayed or "abandoned" her in any way.

Nothing in George's life since the Bible-reading days on his mother's living-room carpet had prepared him for fidelity or moderation. After all, he had been raised and initiated into manhood on a varied and rich diet, and his bent for fried foods, liquor, and cigars increased through the years. Never insisting on partaking of the forbidden fruits in his own home, George sought his culinary pleasures elsewhere. When in New York, he frequently enjoyed lavish meals with cronies such as Elton Fax or Solomon Harper.

Hardly anything could have illustrated this dichotomy better than a letter George wrote Jody in the fall of 1940, from Washington, D.C.

"Dearest Jodie:" he began.

> After a rare T-bone steak, lettuce with French dressing and figs with cream topped off with a fat cigar, I am in an expansive mood, as befits a man of parts. From this eminence of bourgeois satisfaction I glance down upon mere proletarians with cavalier indulgence. The class struggle and the fate of the Finns fail to elicit more than a moment of disinterested contemplation. Nor am I worried by the difficulties of the Chinese, the Poles, the Javanese or the Ethiopians. With stomach distended with good food, I rock along through the night thinking of you and Philippa—all I am really interested in."[1]

The letter was preserved because Josephine pasted it into the scrapbooks, as "incriminating evidence," with the notation "Confessions of satisfaction by his Lordship."

Women had adored George as long as anyone could remember. This was not easily explained since he was rather short and not handsome in the conventional sense. Perhaps it was his rare combination of sardonic wit and romantic charm which he could custom-blend in any proportions that lured females, and his manly prowess that infatuated them.

Well aware of Schuyler's attraction for women (and vice versa), Jody had questioned his ability to be faithful almost from the moment they met. In a curious lapse into racism, she believed her whiteness (in addition to her beauty) would be a bulwark against future infidelity once they became husband and wife.

When George began seeing Josephine in 1927, he was involved with several women (all African Americans), one of whom believed she and Schuyler were to be married. Although George insisted he had never proposed to "Miss Arkansas" (as Jody rather scathingly referred to her), he apparently gave her some money toward the "purchase of a coat" on the eve of canceling their putative engagement. Miss A. then threatened to sue George for breach of promise and George found himself confiding in Jody (probably against his better judgment). Josephine had recently moved in with Schuyler, who feared Miss Arkansas might take it into her head to harm Jody bodily.

Josephine never quite forgot the woman, nor did she entirely believe George had never proposed. She continued to suspect her husband of philandering—a suspicion which may not have been unfounded; Harlem lore has it that "George never lost interest in the girls at the Cotton Club."

This recurring mistrust tainted their marriage from the beginning. When George was in Liberia, and Jody pregnant with Philippa, she had discovered some photographs inscribed by her husband to another woman. Believing he had been unfaithful, Jody sent him a cable threatening to leave.

Schuyler was beside himself. He could not get in touch with Jody except by telegram from Monrovia, which he visited only at the beginning and end of his trip. He had no way of ascertaining if Josephine and their yet-to-be-born child would be in New York when he returned to America. The malaria he would contract later on that trip seriously contributed to his state of anxiety. Three desperate telegrams from George survive, sent at the very beginning of his Liberian ordeal: two on the same day only three hours apart—at 3:26 p.m. and 6:14 p.m. He insists she has tragically misconstrued the photographs and pleads with Jody not to leave him.[2]

Josephine did not abandon Harlem or her husband. There were few places in America that would have welcomed a pregnant single woman about to give birth to a "half-caste." But the marriage was strained. Several months after Philippa was born, Jody began insisting that her husband stay closer to home, give up his lecture tours, and concentrate on writing. She wanted to keep a watchful eye on his wandering one. George, of course, balked; out-of-town lectures and extensive travel as an investigative reporter provided a major share of their income.

Josephine suspected—and again with reason—that on the long and lonely nights on the road, in the empty hotel rooms, after the racial abuses he no doubt faced, yearning for some momentary companionship, George found himself spending his evenings with casually encountered females. But in the sober light of day, with the aftertaste of a peccadillo in his mouth, George would pen Josephine passionate letters, proclaiming his undying love. "As time goes on the conviction grows deeper and firmer that you are the one companion in the world for me," he wrote in one.

> We meet on *so many* planes. With most persons one is fortunate to be able to meet well on even *one* plane. I can feel thoroughly at home only with you. . . . You are the only person I've ever known of whose company I've never tired, and you know I exhaust my interest in most people very quickly. They are, on the whole, so transparent, so inadequate, so insufficient. . . . I lay down no claim to courage but I would willingly die to prevent you from being dishonored or debased. . . . I have never spoken more truthfully when I say that atheist though I be, there is only one god I worship, and that is my Josephine.[3]

But George was absent much of the time, especially during the late 1940s when he seemed to be on constant assignment. In the winter of 1947–48 Schuyler had been asked by the president of the *Courier* to visit all the state capitals and report on the freedom of accommodations, work, and schooling for the American Negro. It was an exhausting schedule and frequently he was unable to find lodging.

At that time, the vast area from the Missouri River to the Pacific Coast was a veritable no-man's-land for the black American traveler. In fact when Joe Louis and his entourage went to Salt Lake City for an exhibition, they had to fly back to Los Angeles that same night because no local hotel would accept them. At least separate though far-from-equal accommodations were available throughout the South.[4]

Ironically the one assignment close to home, to study in depth the socioeconomic conditions in Harlem, occurred during the spring of 1948 when Jody and Phil were on tour. George, who had been around a good deal by this time, considered Harlem the most maligned community in the United States and his articles tried to point out why Harlem was a haven for the African American.

Having finished this job, George was immediately sent by his newspaper on a Latin American tour, to explore "Racial Democracy in Latin America." He left during June 1948 and stayed for six weeks, visiting ten countries.

Returning to New York, Schuyler plunged into a series of articles on "What's Good about the South?" This involved visiting fourteen southern states and in each, interviewing one urban and one rural black family.

George's assignments for the following year, 1949, were equally taxing and again took him away from home frequently. During the spring and most of the summer he was in Brazil, the West Indies, the Guianas, Puerto Rico, the Lesser Antilles, Jamaica, Haiti, and the Dominican Republic as the *Courier*'s special correspondent, investigating color discrimination.

Frequent absences from home accelerated the impairment of the Schuyler marriage. By 1948 they had become estranged. George and Josephine were now companions and had ceased to be husband and wife.[5] They tried to enjoy each other when they could. Fewer and fewer questions were asked.

Deep down, a resentment smoldered in Josephine's heart, and she began to confide in her daughter about George's longtime infidelity. Philippa had heard rumblings before from Jody, but now her mother seemed to abandon any pretense. With her mother's crisis coming on the heels of her own, Phil felt even more uprooted and crushed. She had always adored and admired George. He was in many ways a perfect father to her. Now he had become the main cause of her mother's unhappiness and frequent depressions. As she had in earlier years, Philippa withdrew. She practiced ten to twelve hours a day, took lessons (Philippa was studying with yet another teacher, Mr. Levin), wrote no new compositions, saw no friends, and worried a lot.

So did George, about his daughter. Apparently unaware of this new cause for Philippa's emotional turmoil, he wrote Jody that summer: "Ten hours a day is entirely too much for anybody to work at music or anything else, regardless of what Levin . . . says. This is especially so where a person does *nothing else.* Philippa is ruining her girlhood with too serious application and other fool notions. She needs to have more optimism, hopefulness and buoyancy, and stop so much damn worrying about the future. It is especially silly at 18."[6] Adding, perhaps revealing his own agenda, "Success is not a career or money, it is inner peace and smiling contentment and a minimum of grandiose illusions."[7]

Nothing seemed to help Philippa. Even a proposal of marriage in the spring of 1949 from a handsome African prince by the name of Dr. Nwafor Orizu, who would later become acting president of Nigeria, did not boost her flagging spirits.

The final blow fell that spring, when her father was away on assignment in the West Indies. In an unusually embracing gesture, George had written his daughter and wife, inviting them to meet him in Cuba for the Fourth of July.

Jody was surprised by the invitation and happily accepted, hoping for a wonderful reunion.

"Our time is short here," she wrote June 2, 1949, "and we must not be foolish and act as if it were not. Phil will soon be eighteen and soon perhaps leave us; if death does not break up the trio before then. So let us seize the moment and live happily. I shall be glad to spend a little money and have us more united. Thank you for saying that our fortune was the unity of us three."[8]

However, two weeks later she was suggesting that Philippa should make the trip without her. She closed her letter to George with the same intimations of death she had in her previous one: "Sorry I can't come, too. But I fear to spend the money, around four hundred extra. After all, if you and Philippa see the islands that is all that is necessary. It will help you both. Nothing can help me anymore. I am lost."[9]

Josephine had always been somewhat manic-depressive, her mood swings often triggered by some invisible landscape that made little sense to George or her friends. She also had a disturbing death wish—even as a child; John Garth had been witness to her frequent threats of suicide when he had lived with her in California. But George refused to allow his wife the luxury of martyrdom. He wrote back, insisting she come; he still planned to meet them both. He had conceived of this meeting as a much-needed affirmation of the family's love and commitment to each other. He was determined that Jody be a part of this effort, for without her the family would be a shell, this affirmation a sham.

Finally Jody decided to join him in the islands and, in a reversal, without her daughter—perhaps with some encouragement from Philippa, who may have relished the prospect of being alone for the first time in her life. In any case, Jody left for Camagüey and for George, accompanied by two fancy hats, a suitcase brimming with brand-new clothes, and parting compliments from her friends that she looked ten years younger.

When she arrived in Cuba, George was not there. Jody was not immediately alarmed since he had warned her about the difficulties in travel between the Caribbean islands. As a precaution, they had made alternative plans to meet in Haiti, which was closer to the islands George was covering. And so, on June 27, Josephine flew to Port-au-Prince, relieved to find there a telegram waiting for her: George was en route. She cabled him at his next destination, Antigua, saying she had arrived safely. Jody now settled down in the hotel, but hardly went out to explore Port-au-Prince for fear she might miss a message from George.

Three days later, on the first of July, the hotel informed her that George had never received the cable she had sent to Antigua. Feeling alone and frightened on this strange exotic island, she concluded her husband had betrayed her once again. In a bruised haze, Josephine packed her bags and caught the next plane back to New York City.

George arrived in Haiti just twenty-four hours later, having been delayed by all the factors he had feared—bad weather, the Caribbean irreverence for clocks and time schedules, poorly managed connections. He had wired Jody and Philippa (whom he still believed to have accompanied his wife) in Port-au-Prince several times, telling them to sit tight—he would be there as soon as was humanly possible. None of his cables had arrived. He was heartbroken.

"Darling Jody," he wrote practically the moment he landed.

> How distressed I was to come here this noon and to find you gone. I could have cried right there in the air station when I learned that you had gone after I had tried so hard to reach here against all sorts of obstacles. . . .
>
> Jody, this must not get us down. We shall lose nothing by it except the time we would have had here, and that can be made up somewhere else, some other time. As for the money, I'll do my best to make that good, too. I repeat that we shall not lose by it.
>
> Indeed, we shall gain. Although I could not reach you, I am *closer* to you now than ever before and shall ever remain so. I have been terribly lonesome at this work, more so than ever with no one really to talk to. I would give anything just to talk with you and Philippa now. It was pathetic in retrospect, the way I sat and planned on boat and plane how we would dine and talk, and see things together.[10]

The following day, he wrote another letter to Jody: "I did not know until today that you were alone here, thinking all the time that Philippa was with you. I am now even sorrier that plans miscarried."[11]

And again three days later: "Maybe you don't want to hear from me after all that has transpired, but it is past midnight and I am very lonesome and have to have somebody to talk to. . . . I am terribly sorry about what happened. I did my best and underwent real hardships to make this come off but apparently it was not to be. It is hard to go on but the worst is that one has to."[12]

Back in New York, Josephine was severely depressed. She was considering suicide:

> I had a most miserable four hours waiting for the plane to New York. I prayed for it to fall in the ocean on the way back. I wanted so desperately to die. I did not want to see anyone again. I had told everyone I was to meet you and they had all been so interested. And I had sent postcards from all the places signing them by us both so they wouldn't know you had never come. . . . And I would have committed suicide if I had not feared for Philippa; but then on the return I found she was pretty good and so I felt better about her but not about you. I have been so terribly miserable for the last two weeks I did not think I could remain whole.
>
> Please come back safely. And please do not treat me like this again, I am too emotional. . . . Philippa was so very upset, too, for she had wanted us to have a good time. It depressed her so that it went wrong. We both felt we would rather die. . . . And she was already very afraid about her career so it seemed terrible. We are both very emotional. And we count so much on you. If you fail us we are lost. For if you fail me I am lost. And if I fail her she is lost. . . . I felt you had failed me for some strange reason. I felt you had long ago stopped loving me. That you had long ago become very hard boiled and callous and did not care about anything much; that you lived on another plane different from ours; that you cared nothing for us but felt you had to pretend to us for you felt that you could not entirely let us down. But that you did not love us or understand us at all. That you

> lived a life away from us and only wanted to be away from us and that we were doomed. That it was only a question of time when we would have to die by suicide. First me, then Philippa, for she will imitate anything I do. . . . That is the way it is for some reason, because she has no other influence but me; and if my influence fails her, she fails; like a certain kind of bird that lives just in one kind of tree and if that tree is chopped down, it flies and flies until it dies of exhaustion because there is no other tree. And that is true of a certain kind of woodpecker in Florida. You see I cannot live without you—and you fail me when you leave me and meet others. That I can't accept. . . . Though I try to I can't. It kills me. And I've only lived on because I knew Philippa would die if I died. But now, I could not even live for that; I was completely lost. Lost. Lost. The world has become very strange to me. Very strange. Everything was out of alignment, distorted as in a strange mirror. And that mirror—for all I knew—was the right mirror.
>
> I hoped, too, I would be murdered on my nightly walks. I have been going out later and later and wandering far afield, talking to strange people, and hoping never to come back, Never.[13]

Again and again, George tried to explain in detail why he had been unable to arrive as scheduled in Haiti. But it was all in vain. Jody accused him of duplicity and trickery. Appalled and hurt, George finally wrote back curtly: "I never heard of a more ridiculous thing that I brought you intentionally on a wild goose chase. In the first place, it was I who suggested the trip and was under the impression that BOTH of you were making it. . . . I would not do a thing like that to anybody in the world, and certainly not to those whom I love better than myself! You must think I am a fiend! I have gone through hell as a result of attempting to make this rendezvous, and it is bitterly ironical to have it regarded as a trick."[14]

It was only after George began admitting how an embarrassing lack of money had played a part in their miscarried reunion that Josephine found a reason to forgive her husband. She read all his letters over, scouring them for any mention of lack of funds: "As late as Wednesday the 22nd I thought I had a chance of reaching Haiti at least only a day or two late because I was expecting additional money from the *Pittsburgh Courier*. That, however, did not arrive until Monday, the 29th,"[15] which was transmitted in Guadeloupian francs, a currency which proved virtually unacceptable on other islands.

"To add to my woes, the third day after I left Martinique, the $750 arrived I was expecting for a month—and I was on a ship to Trinidad. Now I'm trying to get it transferred immediately because I have almost nothing but money I cannot change. . . . I had to borrow cab fare from the airport to here."[16]

And rather poignantly on July 7: "I have an opportunity to go far into Central Haiti near the Dominican border where a new town is being built for the Haitian refugees from Trujillo's terror. It is a wonderful opportunity but I am thinking seriously of excusing myself when Mr. Aubert and the others come with the Embassy station wagon because I simply do not have enough money to buy myself meals en route, and it will be a day long trip. And I'm just tired of being embarrassed."[17]

Suddenly Josephine began to perceive not only the real cause of her husband's tragic failure to meet her on Haiti, but also one of the main causes of their estrangement. "Why did you not let me send you money there?" she now wrote repeatedly to George.

> If you had intimated all along you were in need, or hadn't enough to meet, I would have understood and helped. This would not have happened if you had confided in me your circumstances. Isn't it better to let me know your embarrassment than outsiders? You always figure the other way around. But a family is a unit and should help each other in every way. In fact, that has been the rock on which we have stumbled so long. You refuse to let me be a part of your life—except for a corner of it that you choose. But don't you see how that doesn't work? Surely, it is better that you let a few of your defenses down to me, confide in me rather than in others or get yourself in these fixes. Don't you know that I love you and would do anything for you? You have only to say you are in need or unhappy when I long to help you. Why do you keep me at such distance—yet expect me then to trust you blindly or cooperate sometimes and not always? I don't understand when I am to cooperate and when not. . . . But my dear George . . . you cannot really offend me except by being cold and callous—never by being mistaken or helpless or in need. . . . Please let us not have any more misunderstanding. . . . I love you.[18]

❖ ❖ ❖

To add to Josephine's problems, relationships between herself and her daughter had reached an all-time low. At age eighteen, Philippa was depressed over her lack of concerts, and beginning to doubt her ability to make it to the top as a pianist. Nor had she completed a composition in almost two years. Her self-confidence had deserted her.

Philippa began to project her mood: "A slender fragile looking girl . . . as taut as a violin bow," was one reporter's impression.[19]

No innocence is quite as provoking as that which is being lost, and Philippa's face began to reflect this. A photograph from this time in the *Amsterdam News* captures her on the cusp between open hopefulness and strained wariness. The picture shows a pretty teenager, dressed a bit more sophisticatedly than her years, wearing pearls, her hair up. Her smiling eyes, accustomed to the flash of a camera, are clearly proud of the accomplishments that have brought her to the photographer's attention. Yet behind the dutiful smile is a new skepticism, a scrutiny that has concluded that the world's promise of reciprocation she was raised on is a precariously poised thing at best, and a sham at worst. Perhaps it is this struggle that makes her most beguiling—a quality that would persist, impelling the writer Ishmael Reed to call her "hauntingly beautiful" as she grew older.

Philippa continued to be withdrawn and edgy with her mother. Predictably, Josephine was bearing the brunt of her mood swings. Both mother and daughter, being past masters of manipulation, went at it hammer and tongs. And like a refrain, Jody volubly complained to George (who again was away on assignment):

> Philippa flung off something which sounded like she thought I had done something wrong. So I said "What have I done? Tell me and I will be glad to change. What is the matter?" She said "Oh, you wouldn't understand." "Well, try and see." "No, you won't understand!"
>
> It really is an awful burden on me to have to manage anyone so erratic, irrational and inconsiderate . . . and yet she is constantly indifferent to my feelings, never does one thing I ask except after I have been made to wait and beg. . . . Now, this can't go on. If she finds me so impossible, so unpleasant to have around to wait on her, maybe she better get someone else. But I'm afraid no manager would put up with this.[20]

George's only response was to write his daughter encouraging letters: "You are a great pianist, Philippa, and don't ever think you aren't," one of them read. "Even more, you are a great girl. Now remember what I told you—*enjoy* this trip [to the Carribean]. You can work AND play, you know, and in some ways the latter is as important as the former."[21]

But Jody was still Philippa's only confidante, no matter how much they quarreled, and it was to her mother that Phil turned after a performance at the United Nations. "Philippa . . . said that those people last night expected so much of her that she was . . . worried," Jody reported. "I said that . . . they already believe in her, so half the battle is won. If only she wouldn't get hysterical she could do it. But she says she can't compose and be a pianist too. So I say," Jody concluded like a master puppeteer, "if she will finish her concerto I will not ask her to do any more compositions. . . ."[22]

Such was the general mood in the Schuyler household as the year 1950 began.

16

South of the Border

The year 1950 opened a new chapter in Philippa's life: She began her extensive professional travels abroad. Through George's contacts, a series of concerts in the Caribbean was arranged for late February through mid-March—in Haiti, Cuba, Puerto Rico, and the Virgin Islands.

Mother and daughter flew to Havana at the end of February and checked into the Hotel Plaza, near everything one might wish to see or avoid. Philippa played at the University of Havana and for a broadly mixed audience at Club Atenas, the swank social club that Cuba's upper-class non-whites had established. In early March they landed on wind-lashed St. Thomas, where they lodged at an enormous old mansion that had seen pirates and slaves. Philippa played for Morris de Castro, the governor of the island, and flew to San Juan, Puerto Rico, where they stayed at the brand-new Caribe Hilton. But the highlight of the tour occurred in Haiti, where Phil gave two concerts in honor of President Dumarsais Estimé.

While the day-to-day relations between mother and daughter continued to be disastrous and Josephine complained to George, the tour accomplished one important thing: it renewed Philippa's faith in her ability as a performer—the reviews were uniformly excellent.

In December of that year, Philippa was invited back to Haiti to play at the inauguration of their new president, Paul Magloire. She composed a piano concerto, *Rhapsody of Youth*, for the occasion; was awarded the Medal of Honor and Merit by the president; and was made a Chevalier de Haiti. But perhaps even more significant, Philippa went on this trip alone. For the first time, Josephine was not there to watch over her every step, and the statuesque young woman turned quite a few heads.

The sights, sounds, and smells of Haiti filled her alternately with delight and horror: the sadistic cockfight; the voluptuous scarlet flowers splashed over the crumbling villas; merengue dances that lasted until dawn; the collection of jumbled objects on the *pé*, the altar of the small voodoo sanctuary, with its bloodstains from the previous night's ritual, and a huge black cross clad in top hat and frock coat.[1]

She flew north in a tiny plane over the lush Antibonite Valley to visit the decaying remains of Sans Souci, the once ludicrously splendid palace of Henri Christophe—the freed full-blooded Negro slave who, along with Toussaint L'Ouverture, had helped liberate Haiti from France only to become its tyrannical ruler. She climbed on horseback to the ruined Citadelle Laferrière, formidable fortress on top of a mountain surrounded by precipitous cliffs, where Christophe waged a savage but inconclusive struggle with Alexandre Pétion, the champion of mulatto supremacy. She heard the tale of Henri Christophe marrying his two coal-black daughters to white men, decreeing that all Haitians are black, regardless of the color of their skin.

Thus began Philippa's love affair with the Americas south of the border, an affair that was to continue for several years.

❖ ❖ ❖

Between 1951 and 1955, Philippa visited and performed in virtually every South and Central American country, and numerous Caribbean islands in between. She was sponsored by a variety of agents. Some fleeced her. Others, such as George Westermann—impresario for such artists as Marian Anderson, Paul Robeson, Ella Fitzgerald, and Hazel Scott—were honest and supportive. Her most colorful agent was the Catalan impresario Francesco Navarro. A short, thin, nervous man whose black-rimmed glasses dramatically framed his pale blue eyes, he talked in a mixture of six languages—about his awesome Spanish ancestry, his honesty, the hundreds of women who had been in love with him. "Marquesas! Condesas! Duquesas! Princessas! Een long lines zey wait for me!"

Navarro had all the earmarks of a prodigious liar and a selfish scoundrel. Yet there was something that drew Philippa to him—and indeed he would later present her in numerous South American cities as well as in a tour that led all the way to Spain.

Philippa was constantly amazed over the business codes south of the border. Often recitals were arranged at the last minute. If she made an appointment for 2 p.m. the person was likely to call at 4:30 to say that he was on his way—and appear an hour later. Cuba was the only place she had ever been where even the radio programs did not start on time.

Philippa was learning fast and on the run. In retrospect, she would refer to these early travels abroad as her college education. But she was enjoying her stays, practicing long hours and learning new pieces. The press had a field day following her, inventing stories about her strange eating habits. One Cuban reporter wrote that she ate the outside skins of pineapples, and when Philippa demanded a retraction, he only embellished on the story.

During one of her prolonged trips to the Caribbean, Philippa received a telegram from her mother. Possibly hiding behind Jean Tennyson's skirts, Jody had asked her daughter to return to the States for a concert engagement.

(Tennyson had been, for some years, a financial backer and producer of some of Philippa's concerts.)

"STAYING. AM COMMITTED," Philippa cabled back on September 4. "PRACTICING EIGHT NINE HOURS DAILY. EXCEEDING WORK SCHEDULE. LEARNING PIECES. TENNYSON CANT RULE MY LIFE. SHE RULED TOO [LONG] ALREADY. CARAMBA. THATS BEEN TROUBLE. EVERYBODY WANTED MAKE ME PUPPET."[2]

On November 6, after a three-month absence, a happy Philippa returned to New York. "Well," Josephine wrote to Carl Van Vechten, "it was a successful experience for her. Of course, she left a trail of clothes and jewelry around the Caribbean but she got back with more than I expected both in money and in possessions. And she is still a virgin. So that is something of a miracle."[3]

❖ ❖ ❖

Philippa liked the Dutch West Indies best. And it was in Willemstad, Curaçao, one sultry evening in 1952, that she had an encounter with the ancient occult ritual of the tarot cards, which would become ingrained in her daily routine.

Philippa had been wandering, lost in dreams, among the "turbaned East Indians, yellow-skinned Chinese, panther-like Negroes, arrogant Creoles, and heavy cheese-like Dutch,"[4] when she was dragged into a car and whisked off. No bystander had lifted a finger to help her.

There were four men in the car. She tried to scream, but one of them quickly put his hand over her mouth. After what seemed a long time, the car stopped in a narrow alley, and they got out. Furiously, she began to explain to her abductors who she was, flinging at them newspaper clippings. Only then did she realize they were American servicemen, out of uniform, on leave from the naval airbase in Trinidad. And they had mistaken her for one of the fancy Venezuelan "girls" who fly over to Curaçao periodically to ply their trade. With profuse apologies, they drove her back.

"I saw them carry you off," she heard a man say in English with a heavy Dutch accent. He was standing in the shade of an awning; she couldn't see his face.[5]

"Why didn't you help me?"

"I knew you would be brought back unharmed. I foresee things. There is a vibration about you that attracts strange and exciting events. Yet there is, too, a spiritual protection that saves you."

Herr van Kleed, as she called him, was from Indonesia, of a Dutch father and an Indonesian mother. Offering Philippa a cup of green Chinese tea to calm her nerves, he led her through a long, narrow, ill-lighted shop, piled high with Eastern bric-a-brac, and then through a bead curtain into a back room lit by a pair of flickering kerosene lamps. As he was brewing the tea, van Kleed told her about his lifelong preoccupation with the occult and, sensing an interest, offered to show her how to read the tarot. The twelve-by-six-centimeter cards were produced from an elegant orange box. They were thick and heavy, seventy-eight of them.

"The tarot cards are the Book of Life," he began. "Mat, the Zero, is the disorder that confuses men. Tarot I is Le Bateleur, the slight-of-hand trickster; Tarot II is La Papesse, the all-seeing woman. Tarot III is the Empress of Fixed Destiny. Then comes IV, the Emperor of Temperal Power, followed by Le Pape, the Spiritual Force, the Lover, the Chariot, the Hermit, the Wheel of Fortune, The Woman with the Lion, the Hung Man, Death, Temperance. After the Moon and the Sun comes Tarot XX, the Day of Judgement, then the Nude Virgin, framed by the four fixed signs of the zodiac. The Ox, Lion, Eagle, and Angel are Tarot XXII—they are called The World."

"What are the others?" Phillipa inquired.

"Ah, they are the 'minor arcanes,' arranged in four suits of swords, cups, deniers, and batons. Our modern card suits of spades, hearts, diamonds, and clubs descend from them. The fourteen cards in each suit begin with ace through ten, the Valet, Cavalier, Queen, and King then follow. The suits denote, respectively, struggle, spiritual riches, material riches, and worldly affairs."

Philippa shuffled the cards and cut. With a mixture of thrill and anxiety she watched him place all seventy-eight of them on a round baize-covered table in a great circle of twelve packs. Lighting a brown cigarette, van Kleed read her tarot-horoscope slowly, in a flat, unemotional voice. He told her that her life would be perpetual change . . . constant newness . . . continual travel . . . voyages upon voyages around the world. "Your ninth house of foreign trips is extraordinary: it holds all four horsemen."

She was now looking apprehensively at the tenth house. It held the Emperor, three batons, King of Swords, King of Batons, Star, King of Cups.

"Oh, it is good," he was assuring her. "Honors. Your star rises. You receive a decoration from an emperor. You are honored by royalty."

"What about love?"

"Are you sure you want to know?" he paused, using a time-honored gambit of soothsayers. "Your fifth house has cards of violence. Nine of Batons, Papesse Reversed, Hermite Reversed, all treachery, delay. The Hung Man Reversed is success without pleasure. Lover Reversed is duplicity."

"Will I marry? Should I? When?"

"After a long delay, you will triumph. The sun favors you.

"I shall give you these cards as a present; then you will not have to seek people's advice; they will guide you. Do not thank me. You will need them."

They sat for a while cocooned in silence. "Would you like to look in the crystal ball?" he asked. Van Kleed brought out a large glass ball and placed it on a bronze stand with gargoyle feet. He lit a burner that filled the room with acrid vapor. She felt giddy; her mouth was dry. The floor rose and sank like a ship in a heavy sea. At first her eyes saw only a hazy blur. Then the crystal ball, haloed in mist, began to stare back at her. A frightening procession of elephants, lions, and leopards materialized. They looked up toward a flaming sky, in which a silver airplane, one wing cracked open, was hurtling to the ground. There were screams of passengers being burned alive. Philippa fainted.

When she came to, she was sitting in the cool fresh air on a stone bench under a bent eucalyptus tree, outside the shop, her purse beside her. "I put some rum in your tea," a voice said. "I hope you don't mind. It will soon wear off."

❖ ❖ ❖

Philippa returned to Latin America during the summer of 1954. Arriving in Buenos Aires, she was met at the airport by a dapper and exuberant Sr. Navarro. "Ees fan-tas-ti-co to see vous! I 'ave arrangeda ehvrrything! I am greates' eempressario een ze woorld!" he shouted as he embraced her. And indeed her concert with the Buenos Aires Philharmonic, under the baton of Carlos Cillario, was all one could have wished for. The Teatro Opera was splendid. Philippa played Gershwin's Concerto in F and Saint-Saëns's G Minor Concerto and drew rave reviews.

The following morning, Navarro decided that Philippa should play a piano recital at the Opera House in ten days. He was going to plaster the city with colorful *afichas*—two thousand of them—to bring in the crowds. But it rained incessantly for the next three days, and the posters could not go up until the very morning of the concert. The few people who came were lost in the immensity of the house.

The night before, Navarro's assistant had appeared in Philippa's hotel with a stack of music by Argentine composers. No one had thought of telling her that a law required every public solo concert in the country to include an Argentine work. So, by now almost inured to the Latin way of doing things at the last minute and resigned to long hours at the keyboard, she went without sleep, practicing and memorizing a sonata by Alberto Ginastera.

Eventually, Navarro arranged for two recitals over Radio Splendid in September and for two broadcasts over the continent-wide hookup Radio el Mundo in mid-October. Meanwhile, Philippa was discovering the charms of Buenos Aires, enjoying its fine food, and attending soirées by Gilbert Chase, the outstanding American musicologist.

With trepidation and against the advice of her mother, who was concerned she might miss her November 5 concert date in Boston, Philippa decided to accept an engagement in Chile. Arriving in Santiago on October 20, after a two-day delay in Córdoba due to engine trouble, she discovered that her concerts had been poorly promoted. Nonetheless, she was taken up enthusiastically by the musical community and by the Chilean "fronda aristocratica."

Santiago enchanted her, the transparent jade waters of the lucid Mapocho river, overlooked by the aloof rocky shapes of Cerro San Cristóbal and the snow-swirling peak of Aconcagua. But there was another side: Surrounding the colonial opulence of its residential areas and the twentieth-century modernity of the business district were fetid slums. Never had she seen so many crippled, blind, legless, and armless people in the streets of a major city.

Returning after dark to her hotel from the Teatro Municipal one night, she lost her way. She found herself in an irregularly shaped courtyard where seven decrepit men congregated around a cracked stone fountain. She noted that one had a black beard, another had a face covered with scabs. A legless cripple sat on the ground next to a man with a belly swollen like a wineskin. She was quickly relieved of her scarf, gloves, and briefcase.

The music they found in the case puzzled them. One of the Indians lit a kerosene lamp. "Que es esto?" he muttered. "It's the music I use—I am a pianist."

When they finally understood her Castillian, the Auracanian said in a mocking voice that if she were a musician she had better play something. A small, four-octave hand organ with two pedals stood on the mendicant's cart. They set it on the ground and shoved a wine barrel toward it, for her to sit on. Philippa pushed a pedal and touched a key. The instrument had a flutelike tone. The rag-clad men gathered around her, passing a wine bottle. She began the Triana, by Isaac Albéniz but it went beyond the extent of the keyboard. She moved to a Bach fugue. It sounded sad, meditative. Phil played a number of passionate songs from the Pampas and from Patagonia, and when she came to some Chilean pieces the men began to sing. More Indians emerged from the shadows.

This might have lasted for hours, had not some sudden screams, followed by what sounded like a shot, caused her audience to flee in all directions. One of the Indians gave her back her scarf, gloves, and briefcase. He pushed her toward a door she had not noticed before and hit the bronze knocker. The door opened. A woman, candle in hand, let them in. After an incomprehensible exchange of words, the two led her down a long hall to another door which opened onto a busy street. He pointed to the right. Philippa walked a little way, and found a taxi that took her to her hotel.[6]

❖ ❖ ❖

Her next trip to Brazil, Argentina, and Uruguay (five months later) was under the auspices of the American National Theatre and Academy (ANTA).

Chartered by Congress, ANTA had brought foreign touring companies and artists to the United States and sent its own abroad. While it had some government funding, it was mainly a conduit for monies raised by popular subscription.

ANTA's financial support rarely amounted to more than a contribution toward travel expenses. Yet the sponsorship did two things: It provided the artist with an official stamp of approval, and often it allowed the local U.S. cultural attaché to secure a theater or hall for the right date. The company (or artist) still had to make all its own arrangements—which as a rule required an agent—before it was eligible for ANTA assistance. Thus Navarro, again, had to arrange the South American tour for Philippa.

Much of the agency's work of strengthening international cultural ties was laudable. One program, though, turned out to be problematic: the sponsorship of American companies and soloists touring Latin America and the Third World.

The academy's prestigious offerings sometimes preempted the most desirable theaters and concert halls and deprived the native artists of what they thought was their rightful forum; American artists, understandably, were resented. In solidarity, local critics would sometimes combine a grudging acknowledgment of the visiting performer's excellence with a tirade against the American embassy. Philippa was not immune from this.

She arrived in Brazil in April 1955 and stayed in South America until the end of June. It was winter in the Southern hemisphere. Philippa was taken with Rio de Janeiro. She stayed at the Copacabana Palace (where both her father and Marian Anderson had been denied rooms, as had Joe Louis and his entourage before that).

From her window, she could see the Corcovado, topped by its ninety-eight-foot-tall floodlit concrete statue of Christ, its arms outstretched. Elegant couples walked on the undulating mosaic sidewalks of the Avenida Atlantico, and far across the broad white-sand beach, near the ocean's edge, men built sandcastles by candlelight.

But she also viewed the sad side of a glamorous South American city. Phil climbed the steep paths to one of the many squatters' colonies that clung to the sides of the hills, overlooking the same splendid Guanabara Bay as did the villas of the rich. Called *favelas*, they were clusters of primitive shelters constructed of packing crates and flattened gasoline cans. The shacks had no running water, no electricity, no sewers. The terrain was so steep that one could stand in front of one shanty and step on the roof of the one below. Water was carried up from a hydrant a half-mile downhill by women balancing the burden on their heads. At noon, the sun beating down elicited odors from the ground saturated with urine and feces.

After playing two successful concerts in São Paulo to packed houses, she returned for a scheduled performance with the Rio Symphony. But the orchestra was on strike so she went south to the coffee capital of Santos. The weather was unseasonably hot; the humidity hung thick in the air; it was torture to play on keys so swollen by dampness. Adding to her misery, Phil was suffering from food posioning.

Her next stop was Uruguay. It was the first week in May and by contrast the air in Montevideo was cool and brisk. Phil played a program of standard European classics. The recital was poorly attended. Depressed, Philippa went that night to a party hosted by the U.S. cultural affairs attaché. A friend of the Schuyler's happened to be at the same party—Elton Fax. He too had been sponsored by the State Department, "going in and out of schools, sketching and speaking."[7]

There was something about Philippa that evening that deeply disturbed Fax. His personal diary for May 7, 1955, has the following entry: "How stiff and mechanical Philippa seemed, as though she had been wound up and set in motion to function obediently and efficiently. The cultural attaché's comment that her concert had drawn only 100 people or less possibly because her program was made up of the standard European classics rather than works of American composers, is interesting. . . . Philippa left quite early. She has a radio commitment for tomorrow night."[8]

As distinct from Phil's recital, her radio concert was devoted entirely to works of American composers. The talented but temperamental conductor Fabian Sevitsky heard it and invited Phil to return the first week of June to play Grieg's Piano Concerto in A Minor with his Sodre Symphony.

The following morning, Philippa took the flying boat across the Rio de la Plata to Buenos Aires. There, she gave two festivals of Gershwin's orchestral works at the Teatro Colon with Everett Lee, the underrated black American conductor. In dramatic contrast to Montevideo, four thousand people packed the hall for each concert; the seats had been grabbed up within hours after they went on sale.

Lee and Philippa traveled next to Córdoba to present the same program. Between rehearsals, Phil flew briefly back to Montevideo for her performance with Sevitsky. When the reviews finally caught up with her, she was delighted to read in the Montevidean *El Dia:* "Seldom have we heard an American pianist with so much artistic temperament—brilliant, passionate, yet pure, austere. Schuyler has at her command all the resources of the piano. . . . One of those rare occasions where one hears music recreated."[9]

While Chile and Brazil had telescoped the grinding poverty of South America, Argentina brought her face to face with political violence for the first time. There were rumblings of it in Córdoba where her performance was scheduled to start at 9:30 p.m. An hour later it still had not begun. Angry talk was heard in the audience, and someone was waving the Argentine flag. A man rushed backstage shouting to Mr. Lee: "You've got to play the national anthem, or there will be fighting!" With a great deal of help from the orchestra, Lee went out and stumbled his way through. At 11 p.m. their concert began. It ended at 1 a.m. By that time, tempers had cooled, and the artists signed autographs.[10]

Next evening, Philippa and Maria Luisa Desio, a young musician who accompanied her on this tour, left for the Argentine city of Rosario. On arrival, they were told the concert had been postponed for twenty-four hours owing to a nationwide strike. The morning after, Maria woke Philippa at dawn: "I've a feeling we should get out of here quickly," she said. They took the 7 a.m. train to Buenos Aires; two hours later the troops marched into Rosario to quell an uprising.[11]

It took the train eight hours to crawl two hundred miles. When they reached Buenos Aires, revolution was raging. They heard machine-gun and mortar fire, saw bombs dropping. Pitifully young soldiers in ill-fitting greenish gray uniforms rushed through the streets. The avenues were clogged with fleeing women and children.

Maria's husband, risking his life, had driven to the railroad station to find them. As he led them to safety, the lingering smell of fires and explosions permeated the air.

❖ ❖ ❖

While traveling throughout Central and South America, Philippa had encountered for the first time in her life an alternative to America's color obsession: The racial climate there was fundamentally different from that in the United States. To begin with, Latin America had an established three-color caste system: white, mulatto, and black. Latins almost unanimously considered the American one-drop-of-black-blood doctrine ludicrous.

There was solid historical rationale for this position. Much of the continent had been won by the Portuguese and the Spaniards, who had succeeded in colonizing the vast land through hybridization rather than through genocide and slavery alone. Miscibility, rather than mobility and might, was the process by which they compensated for their deficiency in numbers.

Philippa was also struck by another fact: Because their liturgy was social rather than dogmatic—a softened lyric Christianity with many phallic and animistic reminiscenses of pagan cults—Iberian colonizers recognized the sanctity of *any* issue-producing union of men and women, as well as the paternity of a child.

Thus, throughout Latin America, there arose early in the colonial era a recognized pattern of color differentiation which established mulattoes and mestizos (among others) as a distinct middle group between the white minority and the black or Indian majority. White planters acknowledged their mixed-blood children, providing for their education and often leaving an inheritance. Over the generations the percentage of mixed breeds increased. In some countries a mulatto elite emerged. In fact, a decided sexual preference for the mulatto developed in some areas of Latin America. The old Brazilian saying: "Branca para casar, mulata para fornicar, negra para trabalhar" (White for marrying, mulatto for fornicating, Negro for working) shamelessly expresses a popular proclivity.

Each country classified its mixed-breeds differently: The same person might be considered "white" in the Dominican Republic or Puerto Rico and "colored" in Jamaica, Martinique, or Curaçao. But what was firmly entrenched was this triracial system. Mulattoes had their own class; they were afforded a certain amount of social mobility; "passing," as understood in America, was not a transgression, and in some countries made little sense.

Throughout her travels in Latin America, Philippa was accorded the privileges of a beautiful mulatto. She was feted and entertained by her artistic peers and accepted by society. She encountered no barriers. In Ciudad Trujillo, Dominican Republic, she noted that their symphony orchestra was completely racially integrated: It had members of every shade and tint. This was unheard of in the United States at that time (and remains true today).

Despite all this, Philippa was not blind to the complexities and seeming contradictions inherent in the racial problems of Latin America. While there were many blacks and mulattoes in influential positions, one did not have to look far to find color prejudice and discrimination.

Philippa was also aware of the difference between her experiences as a mulatto in Latin America and her father's, as a Negro: "I didn't visit any of George's friends in Rio, or elsewhere," she wrote her mother. "I didn't have time. . . . Anyhow, staying at the Copacabana, I had to be very careful of who came to see me—and I thought it was a good thing that . . . some of the officials of the American Embassy came to see me for that was good for my CATEGORIA."[12]

For Philippa, personally, the discovery of a new *categoria,* one that did not view her as tragic, inferior, or strange, was a first step toward finding an identity.

III

Around-the-World Suite

17

United States and Europe

Philippa's first New York Town Hall recital,[1] originally planned for April 8, 1953, and marking her American debut as a full-fledged adult performer, had to be postponed. A viral infection—caught possibly in Central America—made it necessary to hospitalize her.

Fully recovered by May 12, she was given a fair trial by her fellow New Yorkers, who had been listening to her perform since the age of four.

"She has the markings of a remarkable artist," *New York Post* critic Harriet Johnson wrote the following morning.[2] "She does a work like the Schubert *G flat Impromptu* with a beauty and simplicity that are quite touching and a piece like Ravel's *Jeux d'eau* with a cascading of crystalline tone that is enchanting," said the *New York Times* critic, who continued: "and she can play a fugue as if she had worked it out by herself. That is a lot for a 21-year-old."[3] *Musical Courier* was taken not only by her musicality—"exceptionally gifted, technically skillful and emotionally intelligent"—but with a touch of benign sexism, added, "That she makes a lovely stage picture is a great asset."[4] And the *New York Journal-American*'s George Sokolsky wrote in an article entitled "An Interlude with Genius" that the young Negro pianist "performed with rare maturity and artistry. She undoubtedly will find her place in the musical world and it will be in time, high on the ladder of accomplishments."[5]

The *New York Herald Tribune*, the paper that had introduced Phil at two and a half as a child prodigy to the world, was less enthusiastic: "Her playing suggested a marked talent in transition," wrote Francis Perkins. "It told of innate musicality and of a technique that was confident, generally mature, often brilliant but sometimes uneven; a basic interpretive understanding which missed certain details."[6]

The *Times* also pointed out what its critic considered her major limitation, her incapacity to play with passionate intensity. But the *Musical America* reviewer felt differently: "She showed a mature grasp of the inner meanings of the compositions she played. There was fire and sweep in her playing of the

Chopin . . . , and Schubert's *Impromptu* had a beauty of tone and soaring lyricism that were deeply moving."[7]

By all accounts it was a highly successful debut. Though it was sponsored by a black organization, Olive Abbott's Musical Arts Guild, the audience was broadly mixed.

Philippa was delighted with her reviews although she continued to fear that neither they nor the accolades she had garnered south of the border would be sufficient to open other doors for her in her native land. Only a solid success in Europe, she felt, would catapult her into fame. Mother and daughter agreed that more musical guidance was needed before performing in the Old World.

Jody, who had already begun to arrange a European tour, decided to reengage a former teacher, Herman Wasserman, now living in Los Angeles. Although Phil was still studying with the well-known Belgian artist Gaston Diether, Jody thought that Wasserman would be the better choice, despite recent reviews attesting to the success of Diether's pedagogical techniques. She had taken to heart the *Times* remark about her daughter's lack of passion. But instead of facing the possibility that this might be the result of her stultifying control over her child, Jody clung to the idea that a different teacher would set all that straight. Wasserman had prepared Philippa for her successful debut at Lewisohn Stadium, when she was fifteen, and Jody felt he, not Diether, could help Phil interpretatively. Many years earlier, Jody had fired Wasserman because she had found his request of profit sharing in her daughter's future success unacceptable. But she had always respected him as a teacher.

Wasserman consented to take Phil for two to three months; Jean Tennyson would pay for lessons and contribute toward daily expenses.

In Los Angeles, Philippa stayed with a black family (living in an all-white neighborhood) and worked with her teacher almost daily; she practiced twelve to fourteen hours a day.

Although she had virtually no contact with the black community, she did take her teacher to his first "colored nightclub." The club was honoring Andy Rozaf, and in attendance was a long roster of "Who's Who in Colored America." Wasserman, as Phil later reported to Jody, "was all agog—all eyes and ears. I'm sure he expected it to be a dive, a real joint, but it was swank—real gone!!"[8]

By September, Philippa had seen a great deal of Los Angeles, meeting Arthur Rubinstein and members of the Hollywood set: "I played a surprise concert on Sat. at the home of Mr. and Mrs. Joseph Cotten (He is the movie star)," she wrote to Olive Abbott. "Surprise means surprise to me! . . . However, it was very successful! Cole Porter was there, Adrian the dress designer and his wife, Vidor the producer and his wife!"[9]

Philippa also found time to finish a piece for orchestra, chorus, and piano, entitled *Hymn to Proserpine*, based on the poem of the same name by Swinburne, and another, smaller work for chamber orchestra and soprano based on T. S. Eliot's "Love Song of J. Alfred Prufrock."

❖ ❖ ❖

In the fall of 1953, when she was twenty-two, Philippa visited Europe for the first time. Traversing the northern tier of the continent, she performed twenty concerts in seven weeks—in Sweden, Denmark, Norway, Finland, and Holland. She then appeared in London for the first time, playing with the BBC Orchestra and at the Royal British Academy Gallery. The intimate Salle de la Rotonde of the Palais des Beaux-Arts in Brussels closed the tour, and by the middle of that month Philippa was back in the States.

Although George's connections were helpful with the Brussels concert, it was Jody who had secured the other engagements. Her method was simple and effective. She would identify a concert agent and bombard him with material about Philippa until he agreed to promote an "introductory recital."[10] To arrange this concert tour was no mean feat. Excellent pianists lacking world-class recognition are, as the Dutch say, "thirteen to the dozen," and impresarios rarely made money on them. They preferred not to waste their time.

Her agent in Amsterdam, Ernest Krauss, a former manager of Pavlova, would become an inspiration to Philippa. Tall with a kindly face and a distinguished white beard, he was in his late sixties when she first met him. Krauss showed her that an impresario could also be a "poet, and a gentleman . . . a true, generous, faithful friend."[11]

The pace of her tour was furious, the images kaleidoscopic. She was enchanted by Holland. In Zaandam, a small town where the elders still wore wooden shoes, Philippa played in an old Dutch Reform church. The floor was covered in white sand, and charcoal stoves warmed parishioners' seats. After each piece, the entire audience rose and bowed to her, in silence.

From Stockholm, she wrote her mother: "Women have full rights here, and have had for much longer than in the U.S., yet the men still act with culture and courtesy. The Swedes look better than the Americans. People know how to eat here. . . . You fall over no scrofulous beggars in the streets (which are so clean you could almost eat off them), and no dope fiends lurk with knives in dark corners. You can walk anywhere completely unmolested, and everybody reads voraciously . . . *good* literature."[12]

Yet she would also write that "the general aura of Scandinavia, with the exception of Copenhagen, is aloof, distant and rejecting," adding jauntily that "in the cold and dark . . . one must either drink, commit suicide, go mad, or be hearty."[13] Nonetheless, she genuinely liked the Scandinavian people.

While in Oslo, Philippa met a young woman by the name of Gerd Gamborg, who was the assistant to the impresario handling the Norwegian and Finnish legs of her tour. Ms. Gamborg, a petite, energetic woman, remembered, thirty years later, her first impressions of Philippa. "I was simply overwhelmed by this elegant, exotic, slender young woman coming toward me. She was a very beautiful young lady, who wore no make-up except some light lipstick; a warm, open, impulsive person, who would laugh at herself as well as others. She had an excitement about her that others felt instantly. When people met her, they felt strongly attracted to her."[14] Gerd (who physically reminded Phil of her mother) became one of the few women in Philippa's brief life with whom there was a bond. And over the years, as she got to know her American friend better, Gerd, too, began to understand Philippa more:

> She was a very searching soul—always looking for something . . . especially through religion. She could never settle down. . . . Philippa was obsessed with making a success with her life. . . . An artist in a way has to conquer the world to be happy. And this was drummed into her head by her mother. . . . I think her mother had made an ideal of her, that she should reach fame. She was very anxious to get her absolutely to the top and perhaps Philippa thought she couldn't live up to that . . . because, you see, Philippa was very human. She struggled so much between the black and the white parts of her. They were always making war inside of her. She wasn't one person, one character, she was a thousand. . . . I think she was working very hard with herself as a human being. . . . she believed in so many things . . . and that must have been confusing. She was a kind person, with a beautiful smile and warm eyes.[15]

London also appealed to Phil. But of all the cities she saw that fall and winter, it was Brussels she liked best. "It is the most captivating city in Northern Europe," she wrote.[16]

Most of Philippa's European reviews were stunning. She was hailed as a remarkable pianist, a superb mistress of the keyboard. Conductor Rudolph Dunbar, a West Indian expatriate living in London, wrote, "Philippa Schuyler certainly will not have to fight a revolution against the maxims of success because her star is in the ascendant. It is only a question of time and that uncanny thing called 'luck' before she will be placed among the giants of the keyboard."[17] Even the highly critical *Times* of London referred to her "keen sensibility of tonal colouring . . . firm technique . . . not an ugly sound."[18] Almost all the reviewers were impressed by her solid technique and interpretive skills, and not a little overwhelmed that one so young could reach such heights. But they were careful to mention that she had not yet matured—she had what one called "the light pulse of youth." With few exceptions, they agreed that with time she might become one of the great pianists of the century.

Northern Europeans and Scandinavians considered her a "rare breed," "exotic," a "beauty"; they were intrigued by the woman with the "Dutch name and the jade [*sic*] complexion."[19] The Brussels reviewer and poet André Gascht (whom Philippa met on this trip and with whom she was quite taken) wrote, "She has passed through here like someone from a fairy tale: young, smiling, beautiful, an ethereal queen."[20] It was not just fine reviews that were important to Phil. Equally significant was her total acceptance as an artist. It was true that some of the papers mentioned her racial background, but mostly in passing or as a historical footnote.

Constantly on the run, Phil's communications to Jody were brief, and her mother resented it. Under the pretext of making it easier for her daughter, Jody began sending long lists of questions with the instruction: "Please check yes or no!" Pages of questions would follow, from "Did you get the blue nylons?" "Are you sending some photos?" "Have you had your clothes cleaned?" and "Have you been to the hairdresser?" to "Do you have plenty to eat?" Phil answered flippantly, "God no." And to "Where do you practice?" just one terse word: "Why?"[21]

Spending money was a source of irritation between daughter and mother. Phil lived and traveled on a virtual shoestring: "I, being economical, careful, and disciplined have $150.00 left of the $287.00 with which I left the U.S. [four weeks ago]. I shall spend nothing in Belgium, Holland and London. Thus, you need not send me money—save it. Put it in the *bank*. Don't buy me *anything*. In every letter you list 10 things for me to buy. . . . People here aren't like those insane . . . Latins. Nobody has much money and they put on no show at all. It is what kind of person you are and how good an artist that counts."[22]

But Philippa also knew how to yank her mother's chain. After twisting her ankle on a cobblestone street in Denmark, she wrote home:

> I got up to go to the Ladies Room and discovered I could not walk on my right foot at all. I sat down and practiced another hour and it had gotten so painful that I could not concentrate. Finally I got up to telephone and I could not walk. I had to slowly and painfully CRAWL on hands and knees—to the other side of the building. I knew I couldn't crawl out on the stage in Denmark so I thought it better to have it bandaged so I at least could walk some. . . . I know you don't believe in doctors . . . but it hurts more and more. I am very hungry but just to get to the opposite side of the room is an effort. . . . I know you think dreadfully of my letting this happen—but I couldn't help it.[23]

❖ ❖ ❖

On February 19, 1954, a new Steinway grand was moved into Philippa's bedroom at 270 Convent Avenue. This was the first piano she had bought with her own money, and it was telling that she had been able to afford it because of her European and Caribbean tours, not her U.S. ones.

Two months later, on April 29, Phil gave her second New York Town Hall recital—again sponsored by Jean Tennyson—almost a year after her debut.[24] The reviews were mixed, certainly less glowing than her recent European notices. A young Harold Schonberg of the *New York Times* called her "an altogether talented young lady," but then he added: " . . . her talent inclines more toward the music than the instrument. That is, her musical ideas and inclinations were healthy, she had spirit and vitality, but she did not have the pianistic discipline to put those ideas into controlled effect. . . . [But] she has what many more skilled pianists lack—an aura of musical excitement, a belief in the music and an ability to communicate that belief to the listener. It is unfortunate that all this is not harnessed to a mechanical proficiency of equal strength."[25] *Musical America* also criticized her technical prowess but praised her musical sense: "She showed a notably vigorous style and conception of musical works that was temperamental and dramatic. On the other hand, her technical finish of performance did not, on this occasion, appear to have kept up with her growth in a musical feeling and digital dexterity. Cultivation of a smooth legato and better equalization of weight between the hands seemed to be among the young pianist's present needs. Some of her best work was done when she restrained her impetuous manner."[26]

Francis Perkins of the *Herald Tribune*, who had reviewed Philippa's first Town Hall recital, echoed her previous sentiments: "Transition was also suggested in yesterday's playing, which revealed talent and musical temperament. But it did not fully disclose the variety and distinction of styles represented in the program."[27] Unlike the two previous reviewers, Perkins wrote that her "technical powers were noteworthy."[28] Finally, *Musical Courier* gave her a warm review, citing her playing of Bach and Scarlatti as "superb" and criticizing only what they termed a "slight carelessness" in the Beethoven and Brahms, "which may have been the temporary result of overplaying." The critic concluded the review: "Few prodigies evolve into great artists. Here is the exception that will prove the rule."[29]

Yet it was Deems Taylor who put the frequently quixotic world of music reviewers into perspective. "The notices are fine," he wrote her. "If the critics give you an occasional rap on the knuckles, that's fine. It means that they are considering you as a mature artist rather than a child prodigy who should be indulged. As usual, I'm proud of you."[30]

But Philippa got not one adverse word following her November 7 recital, presented by a black fraternity that winter at Jordon Hall in Boston—a city that previously had not been kind to her.

"Philippa Schuyler . . . is one of the most enjoyable pianists this reviewer has heard in a long time," wrote Klaus George Roy in the *Christian Science Monitor*.

> She is an accomplished artist who seems able to do everything equally well. Not many players would have chosen a program so big and demanding, so unconventionally put together: 18th- and 19th-century music in the first half, nothing but 20th-century in the second. The result was consistent fascination.
>
> It takes an admirable command of style to play Bach's towering *Chromatic Fantasy and Fugue* as impressively as Mendelssohn's masterly *Variations Serieuses*, to feel and communicate the power and originality of Charles Griffes' great *Sonata* . . . as keenly as the iridescence of Ravel's *Jeux d'Eau*.
>
> To all pieces Miss Schuyler brought a superbly singing tone, a virtually flawless keyboard technique and pedal control, and a natural musicianship which gave life to every phrase. . . . One may expect a promising future for an artist so richly endowed and well prepared. Her poise, her apparent modesty and restraint, her mature taste and refinement give one hope for an outstanding career.[31]

Thirty-five years later, Klaus Roy still remembered her Boston concert with crystal clarity. "She was simply first-rate," he told me. "But of course she had two strikes against her: being a woman in classical music and being black. The problem of being a woman in the concert field is pretty much solved today, but being black is not. You still have to be twice as good. . . . And even now, it is still unusual to see a black playing a violin or piano on stage. I don't know why, but there seems to be a built-in resistance. . . . Little has changed—except in opera—in almost four decades."[32]

18

The Lion of Judah

At age twenty-four, during the autumn of 1955, Phillippa went to the African continent for the first time. ANTA had agreed to sponsor her concerts in Cairo and Addis Ababa, where she had been invited to give a command performance for the Lion of Judah,[1] Haile Selassie, Emperor of Ethiopia. Around this commitment Jody and Phil organized a European tour.

Philippa continued to be generally unsuccessful in getting bookings with white organizations in the United States, and scattered throughout a notebook where Philippa kept track of her practice schedule, tarot readings, and horoscopes, were many expressions of bitterness.

❖ ❖ ❖

In a decidedly captious mood, Philippa left for Europe in early October, stopping in Iceland, where she gave three concerts. One of them was in Akureyri, Iceland's second-largest city, on the opposite side of the island from Reykjavík. The eighteen-hour journey to Akureyri, in a small car belonging to the American embassy, was memorable. The terrain looked like a low-lying rubble-strewn moonscape. Occasionally a beach would appear, a monochrome of jet black volcanic sand abutting icy gray waters. A few scattered bleak houses stood like "propped up corpses," and they passed dugouts where peasants lived "like vampires in . . . a crypt."[2] As darkness fell, the weather turned harsh, and in the long arctic night lightning slashed across the purple sky, the wind "howled a requiem."[3]

The concert took place in a drafty, dimly lit hall. But the house was full, the people cordial and hearty, and the applause warm.

Philippa continued on to the Continent, recording solo recitals for Hamburg, Munich, and Frankfurt radios (before live audiences). Then to Brussels, where she met Paul Fabo and André Gascht again. The two men accompanied her to Amsterdam for a solo recital at the Concertgebouw (October 30). Her reviews—in Iceland, Germany, and Holland—were once again outstanding. It was a successful beginning.

Egypt, however, was a disappointment. The Heliopolis airport was cheerless; her room at the Semiramis Hotel gray and depressing. It took almost two hours for an unappetizing breakfast to arrive. Her one concert scheduled for the next day disappeared and she feared that Africans were ruled by the same disorderly star as the South Americans.

In the evening, on a whim, she decided to visit the Great Pyramid of Giza and the Sphinx. Her laconic account to Jody reads: "I saw your pyramids and the Sphinx—it cost me 6 dollars. . . . I have a rock from King Cheops' tomb for Birdie [their cat] to scratch her teeth on. . . . Well they did look very eternal, and big and all that. I went out at night and it all looked just like those Biblical movies Hollywood makes."[4]

Her quasi-autobiography, *Adventures in Black and White*, published five years later, in 1960, gives a totally different account of her trip to the pyramids—forcing one to question how much of her book might be fictional: It was a long trek through the dark city and at one spot the taxi driver stopped to pick up a Dragoman, much to Philippa's surprise, but she refrained from saying anything. At the destination, "the Dragoman grabbed my purse, threw me out of the car, and the two drove off leaving me stunned, by the roadside. I got up. It was terribly cold and I was very frightened. I reassured myself, recalling all the millions of people who had passed that way since 3700 B.C. Above, the stars blazed coldly, remotely, just as they must have done then."[5] Pacing to keep warm, she wondered if she could last the night. After an hour, Philippa saw the headlights of an approaching car. The driver turned out to be an Englishman, a doctor, and he offered her a ride back to the hotel.

Phil's next stop was Addis Ababa, the capital of Ethiopia. But on arrival, Philippa was stunned to hear that the American embassy had allegedly cabled both ANTA and Jody to cancel her appearances. In complete panic she ran to the embassy, where she was assured that her performance for the emperor would take place as scheduled. Relieved, she spent the next few days practicing and enjoying the hospitality of Ethiopia's elite.

Only after returning to America did she learn of the drama that had been played out behind the scene. Philippa found a report written by a man in the foreign service (name unknown) who had been transferred to Addis Ababa at about that time.[6] It was addressed to his superior at the State Department with a copy, for some reason, to George Schuyler, and it described in detail how the long arm of American racial prejudice had stretched a third of the way around the world to place obstacles in Philippa's way.

In brief, the writer of the report stated that soon after his arrival in Addis he was informed by the American ambassador, Dr. Simonson, that Miss Schuyler was making a goodwill tour of the Middle East under the sponsorship of ANTA, and that the Department of State had suggested she should play for His Imperial Majesty. The ambassador asked him to talk the details over with a "Mr. A," in charge of the U.S. Information Service.

"It was from the first moment clear" the report continued,

> that Mr. A had [every] intention of sabotaging Miss Schuyler's coming and concerts. He started the conversation with the . . . sentence . . . "Do you really think . . . you have the right to bring a first-class pianist with first-class music to this country and those people would understand and enjoy it? I already wrote Washington that I am against Miss Schuyler coming over. I would not even have Miss Marian Anderson here!" Five days later Mr. A asked him, point blank, to cancel Miss Schuyler's visit.
>
> Philippa arrived from Cairo in the afternoon of November 5th. . . . [There was a] party at the house of Mr. Ingalls [a prominent American businessman] which we had arranged. Present were about 150 people (about 90 Ethiopians). Dr. Simonson was [there]. There were six Ambassadors and about half of the cabinet. I was told it was the first time so many Ethiopians had stayed for such a long time at a private house of an American. This party was a big success [not only] for the United States [but also] for Miss Schuyler, since only a few people [knew her]. But all knew that she is a famous American colored pianist.

Philippa played to a small group at the Armenian Club on November 8; and despite Mr. A's endless machinations, the embassy had arranged a concert for her in the new YMCA auditorium on the eleventh; tickets were selling well.

The command performance for the emperor remained a cliffhanger. Despite herculean efforts by the emperor's private secretary, definitive arrangements remained elusive. Finally confirmed, it conflicted with her public concert. A compromise was struck: Philippa cut her appearance at the YMCA short and was rushed to the palace.

"Miss Schuyler gave an excellent performance," the report said. "She played for more than an hour because His Majesty asked for encores. . . . [Afterwards, champagne was served]. . . . His Majesty presented Miss Schuyler with a gold and silver medal and [gave] her $100.00 US."

By all accounts she looked beautiful—wearing a blue chiffon and black Chantilly lace dress with five petticoats and sweeping skirts, and her diamond lavalliere.

❖ ❖ ❖

From Addis, Philippa flew to Paris, stopping in Beirut and Istanbul to give recitals. Having just come from the brilliantly clear, dry openness of Africa and Southwest Asia, she was far from impressed with the shadowy immensity of the Queen of Cities.

Her debut in Paris was more than three weeks away. It had been arranged by Navarro; she was scheduled to play Gershwin's *Rhapsody in Blue* and his Concerto in F with Albert Wolff and the Pasdeloup Symphony. Phil checked into an inexpensive hotel on the Left Bank. Knowing virtually no one, she felt lost. But uppermost in her mind was finding a good piano to practice on; and not until she had made contact with her father's friends Gaston and Therèse Monnerville was she able to begin regular practice sessions. Monsieur Monnerville, a black lawyer from French Guiana, was then president of the Senate. His wife, a white Parisian, was a painter, composer, and pianist. (Later, Philippa would take one of her compositions around the world.) They lived in a splendid suite at the Palais Luxembourg, in the wing where Napoleon had made love to Josephine and planned the fall of Europe.

Just as Philippa began to like Paris, practice daily, and meet other musicians, she received a distressing letter from Jody, suggesting that Navarro should come over from South America to arrange a concert tour of Spain after her Paris debut. Jody proposed that Philippa pay for his passage. She was furious. What could he do at this late date? Why did *she* have to pay his fare out of her dwindling ANTA money? She could manage very well without him: "I'm not running a Salvation Army!"[7]

If this were not enough, Jody had also decided that Everett Lee, with whom Phil had performed so successfully in South America, should come to Europe—at Phil's expense—where Navarro could book both of them in some "Negro" concerts and Gershwin festivals. Philippa protested in vain. Navarro arrived the beginning of December and Lee early in January. Ebullient as ever, Navarro set to work immediately: they would take Spain by storm, he promised.

Two weeks before her Paris debut, the prestigious music-publishing house Salabert et Cie began hounding her for a rental fee of 50,000 francs ($140 U.S.) for the Gershwin pieces—even though she was renting no music from them, Gershwin himself had given Albert Wolff rights to perform his *Rhapsody in Blue*, the music she had brought with her was U.S. government property, and she was traveling under ANTA sponsorship.

It was a messy affair: Salabert questioned Philippa's and ANTA's rights, and sent nasty letters to the U.S. embassy calling her a liar. Phil had visions of "getting dragged off to one of these awful French jails, where you molder for months gnawing that dreadful long French bread and drinking water with bugs in it."[8] Needless to say, she communicated this vision to her mother. The fracas was not resolved until Albert Wolff's son Pierre paid Salabert 40,000 francs, which he vowed to recover through his lawyers.

Her Paris debut finally came off without a hitch. Four days later she gave a solo recital at Salle Gaveau, under the patronage of the Monnervilles. The Parisians greatly admired her style; one reviewer called her "l'ange noir de la musique."[9] A week later, Philippa and Navarro departed for Spain, for a "wonderfool toor."

In Madrid, she first played privately for Segovia, a longtime friend of Navarro's. The legendary guitarist was complimentary and Philippa was elated. But everything else in Spain turned out to be a fiasco. Some of her recitals were canceled; others had never been arranged. Phil ran out of money, and by the end of December, she had exactly nine dollars to her name. She cabled her mother to send one hundred dollars with Everett Lee.

Lee arrived several days before their concert with the Madrid Symphony Orchestra. Phil was relieved, but the night before their recital, at a rehearsal, he collapsed with a high fever. A doctor pumped him full of penicillin, and Phil fed him lemon juice and honey, brewer's yeast, and vitamin pills. Although weak, he was able to conduct the next day; the reviews were adequate.

Philippa's financial situation did not improve. Her orchestral fees were modest, solo recitals now sparse. In addition, Navarro, with typical bravado, had added his own expenses to her bills. Phil's patience, like her stamina, was almost gone.

"I was starving," she recalled. "My stomach seemed like a bottomless abyss. I lost ten pounds. . . . My turquoise tulle gown . . . hung loosely on me now."[10] Philippa felt weak at her next concert, and even though the orchestra was exceptionally good and she got a standing ovation after her performance of the Grieg piano concerto, she was too hungry to appreciate it.

The next morning, at dawn, she simply left. One dollar remained in her purse after she paid the taxi to the airport. Philippa was on her way to Scandinavia. Gerd had previously arranged several concerts which Philippa had been planning to cancel if Spain turned out well. Now she was glad to have the airline ticket, purchased months before in Beirut.

After concerts in Sweden, Norway, and Finland, she flew home on February 9, 1956. Jody was alarmed over how much weight her daughter had lost. But Philippa was happier than she had been almost all year. Despite the fiasco in Spain, the rigors of travel, threatening notes from publishers, disappointments, and obstacles put in her way by the USIS, she had been doing what she had been trained for: performing.

Cogdell home in Granbury, Texas, ca. 1905, where Josephine was born and raised. *Courtesy Schomburg Center for Research in Black Culture, NYPL.*

Josephine Cogdell astride her favorite horse, Texas, ca. 1918. *Courtesy Schomburg Center for Research in Black Culture, NYPL.*

George Schuyler, first lieutenant, World War I, ca. 1917. *Courtesy Schomburg Center for Research in Black Culture, NYPL.*

George Schuyler's home, 311 Walnut Avenue, Syracuse, New York, 1906–7. *Courtesy Schomburg Center for Research in Black Culture, NYPL.*

Josephine and Philippa, age two, on their roof. *Courtesy Schomburg Center for Research in Black Culture, NYPL.*

ephine and Philippa, age four, with her orite dolls, one black the other white. *urtesy Schomburg Center for Research in ick Culture, NYPL.*

Philippa, held by J. A. Rogers, at her sixth birthday party. George and Josephine are in the front row, at either end. *M. Smith Studio. Courtesy Schomburg Center for Research in Black Culture, NYPL.*

Philippa, around age six. *Courtesy Schomburg Center for Research in Black Culture, NYPL.*

Philippa with Mayor Fiorello La Guardia, ca. 1938. *M. Smith Studio. Courtesy Schomburg Center for Research in Black Culture, NYPL.*

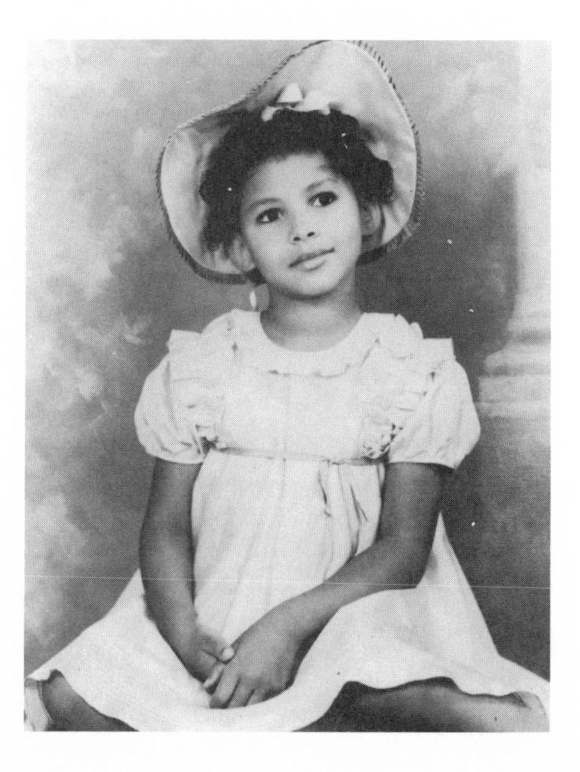

hilippa, age seven. *James Latimore Allen. 'ourtesy Schomburg Center for Research in lack Culture, NYPL.*

Philippa Duke Schuyler Day, June 19, 1940, New York World's Fair. *M. Styron & Co. Courtesy Schomburg Center for Research in Black Culture, NYPL.*

Philippa, age eleven. *Courtesy Laura Ja Musser.*

Philippa dressed as Spanish dancer. *Carl Van Vechten. Estate of Carl Van Vechten. Joseph Solomon, Executor. Courtesy Schomburg Center for Research in Black Culture, NYPL.*

'hilippa, age fourteen. *Carl Van Vechten. Estate of Carl Van Vechten. Joseph Solomon, Executor. Courtesy Schomburg Center for Research in Black Culture, NYPL.*

Josephine making a portrait of Philippa, age thirteen. Her recent oil of George hangs in the background. *Courtesy Schomburg Center for Research in Black Culture, NYPL.*

Philippa, age fourteen, walking home after her Lewisohn Stadium debut where she appeared as both composer and pianist. *Cecil Layne. Courtesy Schomburg Center for Research in Black Culture, NYPL.*

Philippa with Herman Wasserman (*left*) and Otto Cesana. *Cecil Layne. Courtesy Schomburg Center for Research in Black Culture, NYPL.*

The Schuyler family playing dominoes, ca. 1944–45. *Courtesy Schomburg Center for Research in Black Culture, NYPL.*

George, Philippa, and Josephine relaxing at home with one of their numerous cats, ca. 1946. *Courtesy Schomburg Center for Research in Black Culture, NYPL.*

(*Above, left*) Philippa, age eighteen. *Carl Van Vechten. Carl Van Vechten Estate. Joseph Solomon, Executor. Courtesy Schomburg Center for Research in Black Culture, NYPL.* (*Above, right*) Philippa in Haiti, 1950. *Cecil Layne. Courtesy Schomburg Center for Research in Black Culture, NYPL.* (*Left*) Philippa, age sixteen. *Carl Van Vechten. Estate of Carl Van Vechten. Joseph Solomon, Executor. Courtesy Laura Jane Musser.*

Philippa, age twenty-three, after concert in Argentina. *Courtesy Schomburg Center for Research in Black Culture, NYPL.*

Philippa, age twenty-one. *Pach Brothers. Courtesy Schomburg Center for Research in Black Culture, NYPL.*

George, Christmas 1946. *Courtesy Laura Jane Musser.*

Josephine, August 1947. *Courtesy Laura Jane Musser.*

Philippa with Ben Enwonwu. *Courtesy Schomburg Center for Research in Black Culture, NYPL.*

Dennis Gray Stoll. *Courtesy Schomburg Center for Research in Black Culture, NYPL.*

Philippa with André Gascht. *Courtesy Schomburg Center for Research in Black Culture, NYPL.*

Albert Maurice. *Courtesy Schomburg Center for Research in Black Culture, NYPL.*

Maurice Raymond. *Courtesy Schomburg Center for Research in Black Culture, NYPL.*

Tonino Ciccolella. *Courtesy Schomburg Center for Research in Black Culture, NYPL.*

Philippa with Georges Apedo-Amah. *Courtesy Schomburg Center for Research in Black Culture, NYPL.*

Ernie Pereira. *Courtesy Schomburg Center for Research in Black Culture, NYPL.*

Philippa with President Tsiranana of Madagascar. *Cecil Layne. Courtesy Schomburg Center for Research in Black Culture, NYPL.*

Folk festival in Philippa's honor, Togo, ca. 1961. *Stéphane Têvi Nuvi. Courtesy Schomburg Center for Research in Black Culture, NYPL.*

Philippa in Tananarive, Madagascar, with two Malagasy girls, 1960. *Courtesy Schomburg Center for Research in Black Culture, NYPL.*

Philippa with Bishop Kiwanuka of Masaka, Uganda. *Courtesy Schomburg Center for Research in Black Culture, NYPL.*

Philippa with George Rudiki III, king of Toro (Uganda) and Mrs. Kalibala. *Cecil Layne. Courtesy Schomburg Center for Research in Black Culture, NYPL.*

Philippa with King Prempeh II of Ashanti and unidentified teacher, at Kumasi College of Technology, Ghana. *Courtesy Schomburg Center for Research in Black Culture, NYPL.*

George Schuyler with Richard Nixon. Julius J. Adams at center. *Courtesy Schomburg Center for Research in Black Culture, NYPL.*

Philippa with President Kwame Nkrumah, Ghana, ca. 1961. *James Campbell. Courtesy Schomburg Center for Research in Black Culture, NYPL.*

Philippa with Albert Schweitzer, 1959. *Courtesy Schomburg Center for Research in Black Culture, NYPL.*

Philippa plays for Queen Elisabeth of Belgium, 1959. (*Left to right*) Countess Carton-Wiart, Baroness Kerchove, Philippa, the queen, Paul Fabo, and Mrs. Gilbert Chase. From *Adventures in Black and White. Courtesy Robert Speller.*

(*Above, left*) Philippa on her first tour of Vietnam, 1966. *Courtesy Schomburg Center for Research in Black Culture, NYPL.* (*Above, right*) Philippa in cheongsam, Taiwan, 1966. *Courtesy Schomburg Center for Research in Black Culture, NYPL.* (*Right*) Philippa, several weeks before her death, in her Vietnamese disguise. *Courtesy Schomburg Center for Research in Black Culture, NYPL.*

19

Winter's Night

Philippa's twenty-sixth year was an emotional nadir. She spent eight depressing months at home in 1956 and eleven the following year—her longest single stretches in the States since she had begun traveling the world. She read a lot, occasionally went to the movies, mostly alone, and compulsively practiced the piano. (No matter how depressed Philippa became, she always seemed to have a tremendous capacity for work.[1]) Her journal, which she continued to keep, contained expressions of deep unhappiness and her yearning for a meaningful romantic relationship.

Her concert schedule was comparatively light—scattered concerts in America, a third Town Hall recital, again courtesy of Jean Tennyson. A month's trip to England, Belgium, and Holland in November 1956 provided a welcome interlude.

Whether as a catharsis or in hopes of writing a successful novel, Philippa began working on a book. It would ultimately blossom into her quasi-autobiography, *Adventures in Black and White*, published in 1960. But now it was something else: a partially autobiographical and largely fictional roman à clef. She called it alternately *Appassionata* and *Scherzo of the Hearts*. It became an outlet for her dreams of glory, her sexual fantasies, her frustrations with her mother's relentless domination, with other people who had been running her life for too long, her anger at scurrilous and grabbing impresarios, and her unsatisfactory relationships with men. It was also a vehicle for describing the exotic locales she had visited.

When she finally reworked the book into an "autobiography," many of the fantasy-filled stories were dropped or brought more in line with facts. Some were too good to be excised—and hence the autobiography retains a liberal sprinkling of apocryphal material.

The heroine, Rolande Arras, a beautiful black-haired, amber-eyed young concert pianist of Chilean origin, stays in Paris at an old club, the Societé des Amateurs de Fauré, in a house on rue Balzac once owned by the notorious marquise de Malfoutent, seventeenth-century sorceress, murderess, and devotee of the black mass.

At the club, Rolande meets a variety of characters—poets, writers, musicians, actors, artists, spies, Algerian and Moroccan revolutionaries—and has a passionate love affair with a handsome and cruel British author, Rudolph Holmes, born in Argentina, whose cattle ranches there had been confiscated by Perón. "Powerful vibrations of icy disdain, intellect and pride emanated from him, and attracted me like a magnet, lending spark, interest, motivation to my mind," Rolande says of Rudolph, possibly revealing Philippa's own predilections.

Rolande travels the world. She is raped and robbed, but also decorated by an emperor. A broken-down impresario lures her to a concert tour in Spain with a drunken conductor and a tempestuous diva. The three trick and fleece her, leaving her starving in Madrid. Rescued by a gallant Spanish nobleman, Rolande returns to Paris where her love affair with Rudolph comes to a sad and violent end when he abandons her. Alone and pregnant, Rolande has an abortion in Scandinavia (in a subsequent draft this is changed to a miscarriage). Soon thereafter she triumphantly tours the United States, yet she continues to long for Rudolph, who has returned to Argentina.

Now wealthy from her concerts, Rolande resolves to follow him. In Buenos Aires, she finds Rudolph married to beautiful Maria Dolores Santo (patterned after her Argentine friend Maria Luisa Desio). Maria's father has bought back Rudolph's estates and given them to the couple as a wedding gift.

It is a gothic tale peopled with characters who are clearly recognizable, others purely imaginary. It is also garnished with bit players, such as Potiphar Gubbins, gin-drenched British civil servant; Lucretia Wrung, the pregnant bride; and Oedipus, the champagne-drinking loris monkey, all of ten inches tall, with enormous eyes leaving virtually no room for other facial features.

❖ ❖ ❖

Parts of *Appassionata* may offer insights into Philippa's psyche not otherwise readily available. The fancied tour of America offered by a premier New York concert agency allows Philippa to indulge in dreams of financial success, which her native country had denied her so consistently. "By June," Rolande writes, "I had played 120 concerts in the U.S. and after paying off the managers, agents and impresarios that clustered around me like flies round honey, I would still have $41,000 left. I was offered a two-month tour of South America, and I accepted."[2]

In Buenos Aires, after playing to thunderous applause and standing ovations, she meets Maria Dolores, who has married Rudolph.

> "Oh, you must be very happy!" Rolande says with all the strength she can muster. But Maria begs the question: "You are the one who should be happy! I will have children, and be just like every other woman, but you! You have the marvelous power to move the World, to live in the midst of great art, to travel to foreign lands, to be important in your own right, to stand for something and to express something."

> There was a Cathedral nearby, and I entered it after we parted, my satin train dragging on the cold stone; I knelt, to pray and think.
>
> And it came to me . . . that no one knows the thing that is best for him; that life has much richness in store, and that acceptance of Destiny is the greatest wisdom of all, and the most eternal.[3]

Philippa's novel was also a vehicle for expressing her anger at Josephine. In *Appassionata*, Rolande's mother (who like Jody is her daughter's business manager) is domineering, manipulative, and undermining. The resemblance is so strong that on several occasions Phil absentmindedly typed "Jody" instead of "mother" in the manuscripts.

"I am very angry with you," begins one letter from Rolande's mother to her.

> I don't see why you are remaining in Europe so long. It is ridiculous. Why don't you cancel your tour with Henmer and come home? Dr. Fakirsky is here and says that your technique and interpretation have to be entirely reorganized. He says you have no sense of style whatsoever, and must stop performing for two years, and take three lessons a week from him. He says you are totally unsuited for Bach, Beethoven, Chopin, Ravel, and all composers except Schumann. You must build your entire repertoire around Schumann from now on, he says, and play nothing but Schumann [A later draft used Bartók].
>
> Also, Mr. Ferrett, your manager, is very displeased with you. He says concerts in Europe mean nothing, unless they are with the big orchestras.
>
> Your concerts in Africa are going to fall through, I think. They are having an epidemic in Liberia, they don't want to pay a fee in the Gold Coast, and that manager hasn't written from Belgium about your tour in the Congo. Also, you won't get to play in Jamaica. . . . I wrote to a manager in Canada for you, who hasn't replied. Everything is going about as badly as possible for you, I think.
>
> Who is this Rudolph you keep writing about? Is he rich? If he's just another unsuccessful European intellectual, drop him at once. I insist. He probably just wants you to sponsor him into the U.S. . . . They had an earthquake in Beirut. I think your friends were killed. . . . Your friend, Mrs. Mackerall, just died, the weather is terrible, and I do not feel at all well. Love, Mother.
>
> P.S. Please send post-cards to *each* person on the enclosed list of 580 people. Remember, they helped support your last Town Hall recital, which otherwise would have been a fiasco."[4]

But perhaps most revealing are the glimpses she offers into her own sexuality and her difficulties with men. And indeed at one point in *Appassionata*, in the arms of an ardent young Levantine diplomat, she asks herself: "Why must I always be a lonely stranger, seeking fruitlessly for love and tenderness, someone to caress and embrace me with heart and meaning. . . . And when I seem to find him, I feel torn by conscience, cold and unwilling?"[5]

But sex is often more cruel than romantic in *Appassionata*, with Rolande invariably the victim. Yet her dalliances have shades of the "unabridged" *Arabian Nights*, read at perhaps too early an age.

Appassionata clearly reflects Philippa's inner turmoil; and by the time she had reached her mid-twenties, her music almost inevitably reflected it too. Her astute friends began to sense this even before critics did.

"Technically she was superb, but she lacked something which really touches," Gerd Gamborg recalled wistfully. "She was more the Beethoven or Brahms type. Her Bach was splendid and correct, but didn't get to the heart of it."[6]

"I didn't get any of the tenderness," Elton Fax commented after hearing her in Montevideo.[7]

Yohanan Ramati, a man she would meet in 1957, felt Philippa more suited for the bombastic works. "She was at her best not in quiet contemplative music," Mr. Ramati, himself a musician, said many years later.

> This was not the kind of music she could play best. What she played brilliantly was music which expressed turbulent feelings. And that she played as well as I've ever heard it done by anyone. Mussorgsky's *Pictures at an Exhibition*, for example. . . . There was also music of Chopin that she played very very well and there was music of Chopin which she did not—where what she put into it, was not what pianists normally did. I'm not talking about interpretation, but it was less quiet than it should have been. . . . But she had an intuitive grasp of music, a great capacity for conveying feelings in music, especially the more stormy kind. . . . When she sat at the piano and started playing, people listened, and she did convey the mood of the music . . . a feeling of movement . . . she had a good touch, an excellent technique.

After a long pause, Ramati ventured: "If she'd had a little more peace of mind, she could have been a great pianist."[8]

Both Gerd Gamborg and Yohanan Ramati commented on how, invariably, Philippa best performed works by the Romantics. Technically these pieces are more difficult than the Classical ones. But bravura, glistening scales and impeccable pedal technique can camouflage deeper weaknesses. A Mozart or a Bach often requires more maturity or life experience to interpret, as there is little to hide behind.

Philippa's temperament demanded pieces with a strong wattage which she could plug into. She played not so much with great feeling as with great intensity. "Sometimes I thought her playing was a need more than an art," summed up an Italian friend.[9]

Contradiction seemed to envelope her. Critics commented on how fragile Philippa looked when she walked center stage, and yet how forceful her playing. Some touted her style as superb and tonally brilliant, others as intensely neurotic. One critic, John White, called her the most intellectual pianist in America. A few of the writers, reflecting the vagaries of musical criticism, even began to question her status as a mature artist, which at twenty-six she was entitled to seek. "It will be interesting to hear her in, say, ten years time," said a London journal.[10]

While the causes of her critical setbacks are buried deep, some of them can be traced to her musical training as a child prodigy and young adult. Josephine's constant switching of teachers and techniques had after all never been in Philippa's best interest. "Her training did not prepare her for the scrutiny and criticism handed out to public figures," Sister Carroll, one of the nuns from Philippa's childhood who had continued to follow her career, would write to me. "She needed further discipline and maturity."[11]

Another factor was Philippa's excessive hours of practice. Achieving technical perfection protected her (she thought) from revealing too much of herself.

But the strongest causes of her pianistic setbacks were her own emotional state and her continued labyrinthine relationship to her mother. Philippa had virtually stopped growing; at times she began existing as the puppet she had both feared and hated, the automaton Elton Fax had observed in Montevideo.

"The effects of one's family life are vastly underrated in the music field," the venerable Klaus George Roy said to me. "No matter how great one's gifts, an artist cannot pull them all together if there are personal problems."[12]

It was her Dutch friend Lineke Snijders van Eyk who expressed the overarching problems in Philippa's life most cogently. Lineke had first met Phil in 1955, when she had just begun working for Ernest Krauss. Lineke was five years younger than Philippa but they became instant friends. In fact, she would be the only "girl friend" Phil had that was even remotely close to her age. Shortly after they met, she wrote to Lineke thanking her for their friendship. Coming from a large family, Lineke found it strange, at first, not knowing how isolated Phil had been in her childhood and adolescence.

In the years to come, Philippa, Lineke, and Theo, her husband, spent a great deal of time together. They went to movies, on picnics, to museums. They exchanged world views and talked about politics. It was a warm and comfortable friendship.

But Lineke also worked with Philippa's mother, and was keenly aware of her relationship to Jody, on both an emotional and professional level. "Certainly in the first years, Phil was very grateful to her mother," Lineke said to me. "Her mother was a tremendous promoter and it all worked for her, and Philippa was very careful to send home any item of good news there was on her. . . . I think Philippa, of course, was brilliant and talented, but she had been pushed into this career by her mother," she continued, understanding yet not understanding her friend. "I think maybe if she could have stayed at home and not traveled so much she'd have been a much happier person. . . . I didn't feel affectionate toward her mother. She was sort of living through her, it seemed. I always had this strange idea of her mother like a spider, constantly manipulating Philippa. . . ."[13]

Curiously it was Theo who provided an insight into another facet of Phil. I asked him what he thought her greatest ambition in life was—fully expecting the answer to be "success." Instead he surprised me by saying, "Sex—and not of the nicest kind." Adding somewhat whimsically, "That's why she enjoyed going to places like Italy, where they pinch girls' bottoms in the streets."[14]

But only a few were privy to glimpses of Philippa's inner landscape. She was adept at separating her private and public personae. When she performed and met her audiences at receptions, Philippa was invariably gregarious, witty, and charming. The public Philippa was extroverted. Most people never suspected that the inner Philippa, surrounded by a protective carapace, was isolated and so full of quiet fury.

Her media image, carefully nurtured by Josephine and George, only underscored this dichotomy. Publications such as the "white" *Musical Courier*, featuring an alluring picture of her on the cover, portrayed Philippa as a world-class, globetrotting pianist in constant demand. The magazine heralded her as a "sensational pianist . . . whose scintillating charm . . . finely drawn face, and . . . superlative musical gifts have been factors in her success as a 20th-century jongleur."[15]

To the outsider, hers was an enviable life—traveling from one exotic location to another, performing for an emperor, encountering handsome men, and pleasing sophisticated audiences. And indeed, to those who followed, in the press, Philippa's wanderings, her life did seem like a fairy tale.

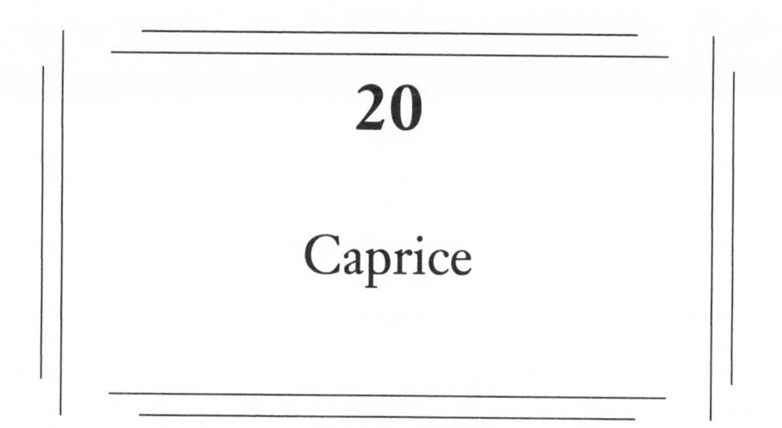

20

Caprice

By the fall of 1957 events took a decidedly upward turn. A glorious tour of Africa and Europe had been arranged for December through March. It promised to be a stellar event for Philippa, and *Musical America* was planning to feature her on the cover of its December issue.

To top it off, Phil received a letter that fall from Rudolph Dunbar in London about a mutual acquaintance, Ben Enwonwu. Ten years older than she, Ben was a Nigerian artist living in England. He had studied at Oxford and at the University of London's Slade College of Art.[1] Recently he had been commissioned by the Lagos government to do a sculpture of Queen Elizabeth. Phil had met him some eight years earlier, when he had come to America to show his art, and found him a "brilliant though erratic genius."[2]

Photographs from this period show him as not much taller than Philippa, powerful looking, pitch black in color, with wide-set eyes and a proud face.

The purport of Dunbar's letter was to suggest a union between Ben and Phil that was more than simply "artistic." Phil was both surprised and flattered. They began a brisk correspondence, often quite intimate. Ben and Philippa shared their feelings of loneliness, agreeing that being artists meant living out-of-sync with time, simultaneously being a witness and a prophet.

In his somewhat repetitive, even garbled style, Ben obliquely addressed their future together:

> Marriage in itself is a difficult thing—especially for a creative person such as yourself. I have found it extremely difficult to find a match among my own race which I prefer, for many reasons that I shall tell you when I see you. When marriage is good—but when you have indulged the sanctuary of artistic achievement—it is necessary to cling to your own race, however diverse [it] may be culturally speaking. . . . The height that one has reached in one's profession can be an impediment though, to one's happiness. . . . Sometimes, I wish I were only a lawyer or a doctor, or even a journalist, for life could be simpler. I'd have settled down long ago.[3]

Dunbar encouraged Phil to consider Ben as a spouse, "for the luster of his achievement is going to shine upon you."[4]

Philippa was attracted, at least intellectually, to the idea of marrying Ben. There was something romantic about his proposal, even daring. Himself an artist, Ben would understand Philippa's life and her needs. She was twenty-six and terribly lonely. Living in England was not her first choice—France or Belgium would have suited her better—but Ben had many connections in British society and marrying him would certainly help her career.

Although England had not been on Phil's original concert tour, it was soon agreed that she should spend some time there. Ben would introduce her to important people in London, have a reception for her at the house of the famed sculptor Sir Jacob Epstein, and, if possible, arrange a concert. Ben knew her time in England would be short but he hoped to catch up with Philippa later in Lagos, his home city, on her planned tour through Nigeria in early 1958.

Carl Van Vechten had advised Jody years earlier, perhaps when Prince Orizu had asked for her daughter's hand, that Phil should marry a European rather than an African (or American). But Ben seemed different.

Phil spent much of November getting visas, being inoculated against various diseases, and collecting paraphernalia. Loaded down with her vitamins, brewer's yeast, bone-marrow pills, writing materials, passport, tickets, traveler's checks, music, and hopes for a new life, she left New York the last week in November on the *Queen Mary*, bound for Southampton.

❖ ❖ ❖

At first, the Atlantic was beautiful, still and glassy; but as the sleek 81,000-ton *Queen* sliced eastward at better than thirty knots, the ocean turned violent, as if remembering that the time was November and the place the North Atlantic. But Philippa was barely aware of it; she spent most of her time in one of the ship's deserted art deco salons, practicing on a Steinway.

On the second day out, looking up from the keyboard, she noticed a man watching her. Encouraged by her smile, he came closer. She had just finished playing a piece by Villa-Lobos; the tall stranger thought that "she had rendered its turbulent feeling as well as anyone he had ever heard. . . . 'You are a concert pianist, aren't you?'" he said, without meaning it to be a question.[5]

Yes, she had played since she was a child. She was at the beginning of a four-month tour of Europe and Africa. As they talked, Phil watched his sensual face. There was a bit of sardonic sharpness to it, as though he saw through everything, himself included. And he in turn thought how very beautiful she was, how alive, how thoroughly charming.[6]

His name was Yohanan Ramati. Born in Warsaw, educated at Oxford, he was ten years Philippa's senior. In 1949, Ramati had moved with his family to Israel, where he became involved in business and politics. He had been editor of the *Israel Economist*, director general of a ceramics enterprise in the Negev, and an influential municipal counselor in Jerusalem. Ramati had also written a book on the life of miners in England, and another on the problems of minorities. Like many of his compatriots he had fought in the October 1956 war against Israel's Arab neighbors. Though a world traveler, he was returning from his first trip to the United States.

For the rest of the voyage, whenever Philippa was not practicing, she and Ramati were talking. She quickly discovered that Yohanan was as conversant with the writings of the Existentialists as he was with the intricacies of a machine gun. He had an extensive knowledge of music and was in fact a serious composer himself—in the vein of the neoromantics. It was not at all surprising that Philippa became quite taken with Ramati, and that the fascination was mutual.

But Ramati sensed that beneath Philippa's charm lay a troubled core: a deep need to prove something. It was when she showed him pictures of her parents and told how she had grown up in America between two worlds, truly not belonging to either, that he understood she was trying to prove to herself and to the world the great contribution she, a child of a mixed marriage, was capable of making and felt destined to make.[7]

Before long, Philippa confided in Ramati that she was seriously considering marriage. "An Englishman?" he asked. "No, he is a Nigerian. His name is Ben Enwonwu, you may have heard of him."

They were having dinner by candlelight at a small table, and for a while there was an awkward silence. When Ramati finally spoke, his voice was urgent.

> "You can't marry an African," he said. "He would be constantly unfaithful without even the grace [of being] hypocritical about it. He will have learned a Western skill, but that will not have altered his attitude toward women. How many tribal wives do you think he has had and discarded? How many children he has not told you about? Also, he will think of you as a servant, not an equal. . . . His relatives will not like you, for they will [object to] his marrying outside the tribe. . . . They would object to everything about you—your customs, intellect, profession, nationality, the fact that you have traveled around the world alone. . . . You can't mix the jet age with the canoe age."
>
> "But my friends think that it will be marvelous for me to be married to another artist."
>
> "A visual [artist] . . . greatly differs from a [performing] one. Only Europe has developed music from a minor art to a major one, so only in Europe will you find your music really loved, and you valued because you play it—in Europe and to a limited extent among those of European descent elsewhere. While Africans will attend your concerts once or twice out of curiosity they won't really understand them or you. . . . He probably made his proposal because he needs a wife suitable to introduce to British aristocracy, and thus impress the only people who can afford to buy his paintings. Is he a Moslem?"
>
> "I don't know. What difference does it make?"
>
> "A lot."
>
> "You're prejudiced."
>
> "No, I refer to the Moslem attitude toward women, stemming from the Koranic [belief] that a woman is only worth half of a man. . . . Also, the Islamic world is now in a stage of exceedingly intense nationalism, generated by a long period of

> subjection to others, and reinforced by historic traditions of military prowess. These feelings are less acute in Persia or in Turkey because these countries have been independent for many centuries. Elsewhere in the Moslem world, the newness of independence, or coming independence, serves to bring forth all the long suppressed dreams of grandeur. . . . By-products of these dreams of grandeur and ultra-nationalism will be hankering back to the glorious bellicose past—and certainly will include hatred of the West, contempt for the female, and an essential unwillingness to understand classical music, the three things you stand for."[8]

Their last day at sea, the Atlantic became turbulent and the ship could not dock. After a day's delay, all the passengers were lightered to shore on a trawler, whose open deck was lashed with a biting spray. Customs was bedlam. They were told that the weather had been terrible in London as well, where the worst railroad accident in sixty years had just occurred.

Feeling uneasy, Philippa made her way to Victoria Station. Suddenly, out of the shadows, Ben appeared, wearing a black overcoat that covered his powerful frame, his beard bristly, his face blazing.

Turning to stinging satire, of which she was a past mistress, Philippa describes in less than flattering terms her encounter with Ben in her *Adventures in Black and White:*[9]

> "Where have you been?" he yelled. "Where? I have met six trains. I was frantic. For two days I have waited! . . . WHAT? Not those bags? They are impossible! Your bags are too big for my car! Too big. Take a taxi to Kensington. I'll trot along after a while. . . . Why ask when I can come to [visit you]? How do I know? Everyone who knows me knows that if I *eventually* arrive, it is a great thing! Feeling terrible tonight, you can't imagine. I have ulcers, heart trouble, liver trouble, kidney trouble."

Perplexed, Philippa made her way to the Musical Club in Holland Park where she had stayed before; Ben arrived many hours later.

> "I seduce you now," he said.
>
> "No, Ben. Have some tea?"
>
> "No, gin!"
>
> "Should you drink if you have heart trouble?"
>
> "Heart trouble? I don't have heart trouble. I come to show you my latest press notice. What do you mean you want to ask me some questions? Why should you ask me questions? Of course I have children. Three! Oh, they are wonderful, oh, they are splendid—all different colors. Each one's mother was different, you see."

❖ ❖ ❖

The following night, December 5, Enwonwu gave a reception in Philippa's honor. It was a great success. Ben was urbane and resplendent in white tie and perfectly fitting tails. Save for his wild uncombed hair, he was a model of sartorial suaveness. Philippa wore a silver lamé gown with a long train.

Dunbar wrote a very anxious Josephine a letter several days after the event:

> We had a smashing party for Philippa. . . . the address where the party was given is a very exclusive and aristocratic section of London. We had a large gathering of

distinguished people in the arts, music and in the academic world. Philippa looked as beautiful as the mystery of sleep, and take it from me, she ravished the senses of the guests and stirred up pleasurable emotions by her playing. It was the more noteworthy because she played on a horrible grand piano. Everybody was delighted beyond measure.

She is now the toast of everyone—invited everyday for lunch and supper. One of the Queen's Private Secretaries . . . has invited her to lunch next Sunday, but alas, she would not be here. I am so sorry about this because such an invitation has immense possibilities. . . . When Philippa returns to London to give another recital, her stock is going to soar to great heights.[10]

Even Lee Mortimer's gossip column in the London *Daily Mirror*, "New York Confidential," carried an item about the affair.

Finally, Philippa herself wrote about it. Her story is somewhat different.[11] Although the party was, as reported, a smashing success, the chaos preceding it was worthy of a vintage Marx Brothers routine. When Philippa and Ben arrived, the day of the party, at the converted mews where the event was to be held, they found it thick with dust. The piano, an aged Broadwood, had rusted strings.

As Philippa practiced, Ben retired to the kitchen with the honorable hostess to make sandwiches, emerging every so often to blow dust from one bookcase to another and to shout at Philippa: "Cawn't you play a little softer? I cawn't stand that racket!"

When it became too dark in the room to see the yellowed keys, Phil turned on a lamp. It flickered for a minute, and then the light snapped off with an explosive sound, blowing the fuse. Ben ran in, shouting: "She did it! Philippa did it! I didn't do it!" He grabbed hold of the lamp, jerked it off the wall. The candle he was holding fell on the carpet, igniting first it and then the curtains. All three rushed around with bowls of water, throwing them on the windows and walls with more enthusiasm than accuracy. Soon the blaze was extinguished, the room cleaner than it had been in months, and the trio, sodden and limp, were able to accuse each other of causing the fire. At this point, Philippa slammed the door and walked back to the Musical Club.

The "courtship" came to a precipitous close on the evening of December 20, two and a half weeks after Phil had arrived. Ben appeared at her place at nine-thirty in the evening. After the usual opening gambit, "I seduce you now," which she declined, he said:

"You must marry me tomorrow."

"No, you don't need me."

"I need someone to do my housework very badly."

"You could mention love—"

"Nigerian man never says 'I love you' to his wife."

"No! I'm flying to Istanbul tomorrow. No, I won't cancel it. The boys are expecting me. And I won't let them down."

"Boys? What boys?"

"Twelve hundred boys at Robert College—"

"What! No! Twelve hundred boys. I won't permit it! It's unthinkable. I kill you."

That evening, Philippa wrote her mother a brief but pointed letter:

I will *NOT* marry Ben!
1. He has two illegitimate children.
2. He eats terribly—and never heard of vitamins—and his health is just what you might imagine.
3. Dunbar now agrees with my decision totally.
4. Ben hates London weather. . . . Cold weather makes him ill. Beginning in 1959, Ben says he will settle permanently in Nigeria.
5. He is not intellectually quick on the uptake like me. I would get horribly bored, restless.
6. Dunbar says he's had several scrapes with women.

NEED YOU KNOW MORE?[12]

Although Phil's letters to her mother were for the most part accurate, parts of her autobiography were so exaggerated that it is only fair to assume the London episode was partially persiflage. The reality, however, was the same: their relationship had come to an end, there and then.

21

Suite Africaine

From London, Philippa moved on to Istanbul and Beirut to give a series of concerts and meet old friends. Then, at the personal invitation of Haile Selassie, she proceeded to Addis Ababa.

On Christmas Day, a brief reception was held in her honor in the Throne Room of Selassie's new Jubilee Palace, a lavish hall with golden doors and murals of the queen of Sheba. Despite the opulence of the occasion, the atmosphere was subdued. Neither the emperor nor Empress Menen gave any outward sign of grief, but life had changed for them. Their favorite son, Makonnen Haile Selassie the duke of Harrar, who would have worn the crown when the emperor died, had been killed May 13, at the age of thirty-four; his car had crashed on a mountain highway near the capital.

During her week's stay in Addis Ababa, Philippa played two concerts and this time there was no USIS official to lodge roadblocks in her way. Unlike her first trip, she also found time to roam the city—haphazardly scattered at eight thousand feet over several hills—her heart often pounding from the unaccustomed lack of oxygen.

She visited the venerable orthodox church of Debro Damo. She wandered through streets crowded with men in once-white jodhpurs, with shamahs (thin homespun blankets) draped over both shoulders, most of them carrying a seemingly indispensable long staff in the tradition of shepherds. Women wore wispy white batiste dresses, embroidered at the edges in pastel colors, trailing in the mud at their bare feet; their greased jet black hair, piled high, shone through gossamer veils of magenta and luminous green.[1]

She thought the Amhara to be as beautiful as any people she had seen. Their color was deep golden ocher; they were tall and spare as a result of the innumerable Coptic fast days, rigorously observed. "We are God's Chosen People," a young Abyssinian told her. "When God made men, He moulded them from clay. He put the first batch in the fire, but left them in too long and they came out burned and black. He threw them away down South. The second batch He took out too soon and they were pasty-white. He threw these away to the North. The third batch came out just right, and He put them in Ethiopia."[2]

But this was no Garden of Eden. Life was hard, for women even more than for men, and in the narrow streets of the market she saw huge bundles of eucalyptus branches brought back from distant mountain villages by women bent double like human pack animals—their silver crosses swinging between their withered, tattooed breasts—old women, no longer much use for anything but heartrending toil.[3]

❖ ❖ ❖

The Ethiopian Airlines DC-6 which took Philippa from Addis to Khartoum on New Year's Day 1958 was no ordinary plane. Two huge red-and-yellow lions adorned its nose, and the interior was tooled in powder blue leather. But Khartoum was not the expected—golden sands, turquoise sky, and deep Nile waters. Instead, the city was a steaming sprawl of yellow dust and ramshackle white houses.

Despite the oppressive heat, and except for an extended noontime siesta, Khartoum was feverish with outdoor life. Philippa saw hundreds of coffee-houses, which until deep into the night were studded with black faces—laughing, quarreling, talking.[4]

But all these were men. As distinct from Christian Addis, there were no women, except for a few in the *suq* wearing heavy black veils. Being alone among hundreds of men at first fascinated but soon appalled her.

A 1,000-mile flight south from Khartoum took Philippa to equatorial Uganda, where she spent nine busy days. Then, on January 13, she winged 2,500 miles west over savannah and primeval jungle to West Africa. In the ensuing four weeks, she would visit, maintaining a furious pace, Nigeria, Benin (then still known as Dahomey), Togo, Ghana, Ivory Coast, Liberia, Senegal, and Morocco. It was her first trip to sub-Saharan Africa, and the sights and sounds were mesmerizing.

Phil landed at Uganda's Entebbe Airport and stayed in Masaka at the home of a Mr. Basudde, a well-to-do brown-skinned planter, and his very beautiful wife. Her concert, played to a jam-packed house—normally a Hindu cinema—was on an English Challen, a vestige of colonial grandeur. Members of the Buganda, Toro, Ankole, and Hima tribes in full regalia were there; Hindu girls in bright saris, Hindu men in dhotas, veiled Indian Moslems. The bishop of Masaka and several seminary brothers were there, as well as some British men and women.

The next day she had a recital in Kampala, at Makarere University. Her concert drew the largest crowd the college's Musical Society had ever seen, and many races were there—African, European, Asian. Even King Mutesa II of the Buganda came and later invited her to visit him at his Lubiri Palace.

Another houseguest at the Basudde's was a young African ethnomusicologist, Joseph Kyagambiddwa. At her request, Joseph drove Philippa to the St. Thomas Aquinas Seminary in Katigondo, where Africans studied for the priesthood. Sitting motionless in the last pew of the church, Philippa listened to the austere strains of Gregorian chant sung in organum style. To hear the familiar music in such a dramatic setting was enthralling. Since childhood, Phil had been visited by a recurring dream—that she was living in medieval Europe—and now, gazing around, she felt transported back to the Middle Ages, "when people still had a pure, simple, unchallenging faith. . . ."[5]

Afterward, a large group of Buganda women gathered outside the monastery to look at Philippa. They were wrapped in bright, exotic long gowns in red, scarlet, purple, and orange—as brilliant as the plumage of macaws. Several of them rushed forward to touch her white dress. "Do you know, it is very important you coming," Joseph said in his peculiar English. "Now all the Buganda are saying, 'Is wonderful Miss Philippa. Can a young girl do all that? . . . God loves her. Perhaps then we should not despise our girls? Perhaps a girl can learn as can a man, and then should we not teach our girls?'"[6]

But possibly the highlight of her stay in Uganda was the visit to the famed Queen Elizabeth National Park on the western fringe of the country, known for its exotic species.

"We passed villages of round thatched mud tukals, and slow, tranquil herdsman driving flocks of black and tan goats. . . . We drove past giant red ten-foot ant hills. . . . Mostly we saw trees, magnificent, wild, tangled green masses of foliage. 'How wonderful to be where the earth was as God made it,'" she thought.[7]

The entire animal kingdom seemed to be on parade. They saw olive green talapoints staring at them with deep reddish green eyes; white gazelles with gentle brown eyes and ebony-tipped tails, white-bearded gnus, kudus, and elands. At one point a warthog, living gargoyle with scimitar tusks, scooted across the road. Later, a heavy-maned lion approached. "He had a tired and pompous dignity of a Latin-American president, who had just escaped with his country's treasury," Phil wrote. "He neared the car and stared at us through the closed panes. His orange eyes looked deceptively amiable. We dared not move and in a few moments he trotted away."[8]

Philippa and her hosts remained overnight in a comfortable camp on high ground, and early in the morning she was captivated, like every traveler before or since, by the African sunrise—"when the sun seeps over the horizon, turning the morning from lavender to pink to the flame blue of African daylight."[9]

In the middle of January, Philippa flew into Lagos, Nigeria's capital. The oppressive, disorienting heat and humidity hit her like a hammer as she emerged from the cool airplane. Just walking the few yards from the terminal to meet her hostess's car at the curb seemed like an athletic achievement.

"How do you stand the heat here?" she asked.

"Well, we spend one-third of the time in bed, one-third of the time losing our tempers at the servants, and one-third of the time counting the days to our next leave!" said the English lady.[10]

But despite the cruel heat and humidity, which "sat on the brain like a red devil,"[11] Philippa was intrigued by the sights of West Africa. Yorubas, dressed in all shades of blue, milled incessantly in the maze of narrow streets, on both sides of innumerable bridges strangled by trucks, taxis, and private cars of all vintages. Frederick Franck, an American physician traveling Africa at the same time, painted this word-picture of Lagos: "Yorubas swathed in cerulean skirts, cobalt blouses, ultramarine scarves, indigo head-ties in all combinations and patterns, and always in contrast to the deep mahogany of their radiant faces. One has the feeling that these great females dominate the city and tie it to the earth triumphantly. The men, on the other hand, riding their bicycles in their blowing curtain robes or chattering in amorphous groups, seem incidental extras without substantial weight."[12]

In the streets of Accra, too, the matronly "mammies," loosely draped in colorful wraps which emphasized their impressive curves and bumps, were much in evidence. Surrounded by their broods, the mammies of Ghana monopolized all enterprise, occasionally amassing substantial capital in their ramshackle market shops.

"Mammy-wagons"—small open buses mounted on the chassis of prehistoric cars—with big signs on them which read "GHANA BEST IN THE WORLD," or "WE LOVE GREAT GHANA," were ubiquitous. Some signs were more earthy than others: "ARE YOU READY?" "TAKE ME AWAY, OH GOD." "QUO VADIS?" "NOT TODAY," or "NEVER TRUST A WOMAN."[13]

❖ ❖ ❖

Throughout West Africa, Philippa was received warmly and treated with great respect. In Lagos, she had stayed with Chief Bologun, the minister of research and information, and his wife in their new home. It was designed to be air-conditioned but the equipment hadn't arrived yet from America. Not having the open porches and thick walls that made traditional dwellings a bit cooler, the Bologuns' house was blazing hot.

It was a world full of contradictions. The Bologuns' six-year-old son was betrothed to the seven-year-old daughter of another traditional chief, and the boy's father had already commenced to pay his son's dowry. But for now the engaged couple were chasing each other around the living room, playing leap-frog over chairs and couches, the young bride kicking and pummeling Master Bologun without the slightest reverence.

Philippa was also escorted to the elegant stucco villa of Adah Obi, an African princess who had come to the Schuyler home some years earlier wrapped like a cocoon in Nigerian robes and staunchly defending the traditional position of African women. Adah was now discovered dancing to the loud sensual beat of a rock-and-roll record, hair straightened, her trim black body only scantily covered by a skin-tight white-and-gold sheath, her shapely legs in sheer nylons and high spiked heels.

In steaming Lagos, Philippa played three concerts, each to capacity crowds. It was a welcome relief to travel to the cool and dry mile-high city of Jos, where she stayed in a room overlooking a private aviary. Philippa gave two concerts in

Jos, both on an ancient Pleyel; at the end of her second, two African chiefs climbed the stage to present her with a pair of magnificent ebony bookends.

On January 22, a black Renault pulled up in front of Chief Bologun's house in Lagos. In it were a chauffeur and a secretary to Georges Apedo-Amah, the finance minister of Togo. They had driven in from Lomé, the capital, to escort Philippa to her next destination.

Phil remembered Mr. Apedo-Amah as a member of the French delegation to the United Nations who had befriended her father. She was barely out of her teens then but recalled him as the best-dressed African she had ever met. "He was never other than dapper, well-groomed and stylish."[14] Seventeen years her senior, "he was nutmeg-brown in color, of medium height with precise elegant features and grey white hair."[15] An ultraconservative intellectual, he had been weaving in and out of her life for years—seeing her occasionally and corresponding now and then. Over time, Georges had formed a deep affection for her.

Philippa, the chauffeur, and the secretary set out on the coastal road to Lomé. They crossed Dahomey first, a long wedge, shaped like an ice-cream cone, its base, along the Gulf of Guinea, only seventy-eight miles wide. A francophone nation, it would gain its uneasy independence in 1960, assuming the name Benin.

In its capital city of Cotonou, they had lunch with the minister of social affairs. "What a marvelous Olympian heaven for the stomach after the horrible British cooking in Nigeria!" she wrote home.[16]

They reached Lomé at dusk. Georges was waiting for her in his apartment above the finance offices. Phil performed that evening, and later he gave a party in her honor. What struck her most were the women, wearing sumptuous Paris styles. Champagne flowed.

Philippa played two recitals in Togo, for mixed audiences, and attended more parties. Her stay, however, was cut short when a message from Ghana arrived via courier requesting her to go to Accra as soon as possible.

Again, Apedo-Amah placed a car and chauffeur at her disposal and Phil left at dawn along with the Ghanaian messenger, Shakespeare Ka Bagesui.

When they approached Ghana, Shakespeare pointed to the fields and said, "Look! Look! These are Ghana fields! Those are Ghana cows!" This kind of chauvinism was typical of that country in 1958. When the Gold Coast established independence under Kwame Nkrumah in March 1957, it was the first colony of the sub-Sahara to be liberated. Many developing African countries looked toward Ghana as a model—to be emulated or avoided. Ghana was making every effort to become the star of Africa south of the Sahara, and pride—especially the kind Shakespeare showed—was sweeping the country. Yet Phil was surprised to discover that Shakespeare was an avowed animist. He had attended a missionary school but proudly opted for the religion of his ancestors.

Philippa was in Ghana for three and a half days giving recitals in Accra and in Kumasi, the old capital of the Ashanti. Amy Garvey,[17] the widow of the formidable Marcus Garvey, to whom George had taken an ideological dislike thirty years earlier, had settled in Kumasi; and it was through her efforts that the king of the Ashanti not only came to hear her, but also presented Philippa with a bolt of handwoven Kente cloth as a token of his respect and admiration. This gold-threaded Kente was very rare. Only three existed in the world: one made for King Prempeh, one for his chief wife, the queen, and one for Philippa.

Philippa also had the opportunity to reacquaint herself with the prime minister. She had met Nkrumah—another one of the many African intellectuals who stopped in at Convent Avenue to talk with her father—when she was a child. She thought him one of the most charming men she had ever met, "realistic yet idealistic, with humility tempered by realization of his mission as a leader."[18]

Nkrumah saw her as a cherished product of the African-American diaspora. He came to her second recital in Accra. Before she began he went backstage to wish her good luck, and at intermission and the end, returned to congratulate her.

But while the West African tour had been a triumph for Philippa, it was also a soul-searching journey. As an American of black heritage, she was both drawn to and repelled by one of the old slave export terminals near Lagos, where human cargo once awaited the infernal "middle passage." In one of the decayed wooden buildings, among a jumble of iron objects piled in a box, she noticed a thin chain with pencil-sized links attached to tiny handcuffs.

Chains for children, the caretaker had said.[19]

She went to a museum to see the Benin and Ife bronzes and ivory carvings. Cast by the lost-wax technique as early as the eleventh century, the bronzes easily established Yoruba and Benin artists as the West African equals of the Greeks and Romans. The skill was never lost and a small contemporary Benin bronze graced the Schuyler living room on Convent Avenue. Sent by an African friend, it was a symbol of the Negro's cultural heritage in a sea of derision.

But Phil was totally unprepared to learn that some of these exquisite objects were used not so long ago in rituals of human sacrifice fully as bloodthirsty as those of Mayan kings. It was difficult to imagine these bronzes caked with congealed human blood. And yet this is how a British punitive expedition found them as late as 1897, when it was sent to Benin City to stop the pervasive barbaric practice.

❖ ❖ ❖

After Ghana, Philippa was scheduled to play in modern Abidjan, the largest and busiest port in West Africa. But no planes, trains, or boats could get her there on time, and Prime Minister Nkrumah graciously offered one of his cars and two chauffeurs to make the twenty-four-hour drive along the coast.

By the time Philippa found her hostess and unpacked, she had gone thirty-two hours without sleep except for occasional catnaps in the car. And now there was no time to rest, for she had to give her recital and go to a reception in her honor. She crawled home at two in the morning and was up at dawn to fly to Liberia for another concert that same evening. On the flight to Monrovia, she thought to herself: "How can I stand this? It's hell playing classical music in soggy, damp tropical heat." The almost daily performance schedule had been too fast, and there was no end in sight.

Despite what she had heard about Liberia (and the fact that her father still was persona non grata there), Philippa was treated well. Dr. Kermit King, president of the University of Monrovia, met her at the airport and drove her over the long, uneven road into the city. She had two concerts in Monrovia, both at the university and both benefits for the promotion of the Student Welfare Program. President and Mrs. Tubman entertained her, as did Vice President Tolbert and other dignitaries.

But Philippa had difficulties coming to terms with Liberia. In part, she liked what she saw: There was no communist threat, no overpopulation. Urban woman seemed to have more dignity of position than elsewhere in Africa. But she was disturbed by the country's appalling lack of amenities, basic material goods, and health care.[20] Female circumcision was still widely practiced, and one of her hosts reported that cannibalism existed in the isolated interior forests. Tribal animosity was rampant. Phil remembered her father describing that while on safari in 1930, he had boys from half a dozen tribes with him, but they would not eat with or speak to each other. Her host confirmed that little had changed in that respect, despite the government's strenuous efforts.

In many ways Philippa shared her father's conservative and rather unpopular view about Liberia, that whatever the exploitation and indignities of colonial rule, colonialism had brought material assets to many African territories that were lacking. The descendents of African-American settlers, who had founded the country as a home for freed slaves, had remained as separate from their indigenous countrymen as had whites elsewhere, but without bringing to Liberia the technology, capital, and expertise of the Europeans. Liberia had no "mother country" to plan and finance roads, dams, airfields, agricultural research stations, schools, hospitals, or communications. Nor were there "the centripetal effects of a rising anticolonial nationalism releasing energies which elsewhere, with independence, [would be] put to excellent use. In some ways, Liberia suffered many of the negative, divisive effects of colonial rule, although not by Europeans, without receiving a share of its benefits."[21]

Phil's next stop was Dakar, the Paris of Africa, where she had three concerts in as many days. Elegant and sophisticated, the city was as French as it was African. It was also the cradle of négritude, a movement that stressed the black African experience. Co-founded by Léopold Senghor, a distinguished poet and statesman who after liberation would become Senegal's first president, négritude upheld the importance of one's African heritage. Senghor exhorted his compatriots "to assimilate, not be assimilated."

On her last day in Dakar she gave a private recital at the residence of the French haut-commissaire. Although the following dinner was sumptuous, the lobster looked gray. She ate it, against her better judgment, and hurried to catch the 10 p.m. flight to Casablanca. The plane ride was excruciating: Phil had gotten food poisoning.

She was picked up by her host and his wife, and driven to Rabat. The ride seemed endless, her nausea constant. At her hotel she was hardly able to lift her head from the pillow, and after twelve hours finally ate several tablespoonsful of charcoal, a trick learned from her mother. The retching stopped, but it had left her extremely weak.

Philippa had a series of concerts in Morocco over the next several days. Despite illness, she performed her engagements, although her playing was mediocre at best.

❖ ❖ ❖

The kaleidoscopic and exhausting tour had come to an end. Africa had opened up its complexity and beauty to Philippa. She had seen its primordial forests, and its parched savannahs; she had heard music unlike anything she had imagined; she had visited cities as modern and beautiful as Paris, and villages unchanged over millennia.

She had met men destined to lead their vast continent into its uncertain future—the Nkrumahs and the Senghors; she had encountered chiefs and kings, jealously guarding their hereditary privileges, and ordinary tribesmen, firmly bound to the soil and their ancestral ways. And she had heard the strong African voice, ominously predicting that the colonial powers' failure to prepare the African for the day when liberation could no longer be delayed would spell disaster.

How would Africa fare after liberation? And what would happen to the women of these countries, she wondered, some so strange and secretive in their long, dark shapeless robes; others dressed in Western clothes and trying to make the adjustment from rural to urban life; still others, like the mammies of Ghana, who she was convinced could rule the world.

Africa had also revealed its most tragic side to her: the omnipresence of crippling and disabling disease. Every fifth or sixth person in the streets was blind, led by a begging child. Cripples with amputated limbs crawled on the ground or hopped on one leg. Arms twisted like olive trees and legs swollen by elephantiasis protruded from rags; sufferers from yaws and leprosy looked out from partially obliterated faces.

Modern medicine could control these scourges. Why, she asked, was progress so slow in coming? Was there insufficient incentive to tackle the task as long as the colonialists deemed people in tropical lands expendable? or as long as greedy African chiefs were willing to conscript labor to work on white men's ambitious and often hazardous projects?

But she knew that these were not the only impediments. There were also cultural barriers. Ancestors' wrath and witch doctors' powers often kept the Bantu away from hospitals; and the belief that it was Allah who sent the suffering made the onlooker in Moslem lands first fatalistic and ultimately callous.

Philippa was both fascinated and confused. In the back of her mind she had vaguely nursed a hope that encountering Bantu Africa might help her sense of identity. But this had not happened. The cultural differences were too great.

Despite her black heritage, Philippa was clearly the product of a white Western world.

But Africa had challenged and inspired her. Genuinely moved, she decided to compose a piece dedicated to "all the men and women of color" she had met on her journey. She would call it *Suite Africaine.*

22

The Lamb Who Smoked a Pipe

Despite her busy schedule, Philippa went out of her way to listen to the indigenous sounds of Africa. From the "arabized melodies of the north, to the dry, archaic, embellished music of Ethiopia's Coptic Amharas, to the pentatonic polyphony of Buganda's complex themes . . . to the 'high life' of West Africa,"[1] Phil found their music startlingly diverse. "African music," she wrote, "has the widest possible variety, matching [it's own] kaleidoscopic range of races, colors, cultures and creeds."[2]

In northern Sudan, Phil heard music in a harem where a bride was being prepared for her wedding with a groom she had never seen. "One girl beat a funereal accompaniment on a round drum, another wailed like a banshee, while the 15-year-old bride, whose hands and feet were painted black and whose head was covered with an impenetrable black wool veil, shuffled back and forth lethargically, without lifting her feet."[3] Men's music, on the other hand, was lively and bouncy.

In contrast, she saw the Christian Amhara women join with their men in music and dancing.[4] Both sexes played the eight-string *begena-lyre* (whose origins are attributed to King David); a six-string *krar* banjo; and a one-string fiddle played with a horsehair bow. Flutes were often used as embellishment. She quickly became aware of how repetitious, dry, completely nonsensual and without personal emotion their music was. Yet it would be these iterative, hypnotic sounds of biblical Arabia that soon influenced Phil to experiment with minimalism in her own compositions.

Though stylized, the Bugandan music had a wider, more sensuous range, like the Ivory Coast war dances or the "hep and happy melodies" of the French Cameroons. It was the music of the Buganda that intrigued her most, and the ethnomusicologist Joseph Kyagambiddwa—whom Phil had met in Masaka and discovered to be a graduate of her alma mater, Manhattanville College of the Sacred Heart—became her guide.

She listened with Joseph to songs with such titles as "The Lamb Who

Smoked a Pipe," or "Eating the Delicious White Ant." Joseph interpreted the plots of others: "Mmomboze," commemorating the Great Trek of the Ganda tribe from Ethiopia centuries before, and their sufferings on the way. Or "Ssanya," which painted the ferocious days of the 1880s when a monster named Mwanga under the influences of his vices and his Arab councillors had reigned over a weeping people; about Mwanga's royal page boys, who had converted to Christianity and, refusing to renounce Christ, were bound together into human mats and burned to death in a huge bonfire. Phil listened with fascination to both the music and the words.

The Bugandas' scale is pentatonic, like the Amharas', but equally tempered—each tone 1 1/5 tones away from the next—making it impossible to exactly represent their melodies in Western notation. But Phil devised her own, and broke down their rhythmic patterns as well. In the years to come, she would appropriate many of these melodies into her own compositions, giving her later piano works, in particular, an eerie and unearthly quality.

But the musical highpoint of this African trip was hearing a performance of King Mutesa II's private royal orchestra—given specifically in her honor, a rare privilege for an outsider. "It is the richest music I have ever heard," she wrote Jody.[5]

The eighty-odd musicians were all raised from infancy to do nothing but play the king's music. They displayed their virtuosity on drums hundreds of years old, named after Bugandan heroes slain in battle; or on the *Mujaguzo* (Jubilation), the largest drum in all Uganda. When not in use, the king kept all the instruments in an elaborate house of intricately woven elephant grasses and reeds, to protect their spirit. "They contain our tribe's soul," the king explained to her.

Drums were only a small part of the orchestra, which traditionally contained stringed, wind, horn, and percussive divisions, similar but in no way an imitation of the Western orchestra.

"Their xylophones are raised to an importance unknown in the West," Philippa wrote,

> and are played as ensemble and solo instruments, with incredible brilliance, fantastic skill. There are two types. . . . both have large white keys about 7 inches across. [One] is played by six men, the [other] by three. . . . Other fascinating instruments are the eight-string *ennanga* harp, derived from ancient Egypt; the eight-string *endongo* bowl-lyre, 5,000 years old and originating from Babylon; the one-string *endingidi* fiddle of Ganda origin, and played with a bow (it resembles a miniature version of the mediaeval European *tromba marina*) and the one string *sekitulege* [always played outdoors with] a hole [dug] in the earth as resonator, and a repertoire of just one song, which, for centuries, has only been played on this instrument.[6]

Like most African music, the king's music was visual as well as aural—a performance as well as a concert—and Philippa wrote an animatedly vivid description of one:

> A couple came to the front of the stage—the man in a black Western suit with a leopard skin tied around his waist and posterior, the woman in a long crimson and scarlet Buganda costume with a black sweater encircling her middle. They began to dance hotly, furiously, frenzied pelvises switching torridly, behinds quivering . . . muscles twitching with orgiastic violence, as the drums' beat pounded and boomed. . . . The insistent repetition had an erotic fury. . . . Its throbbing seared into my brain and nerves. . . . Unaware of the . . . stares around me, I too began to move and wiggle, till the music abruptly stopped. . . . Now, a piece started about Kigula, the clown. Xylophones and drums began a short, staccato theme, and the man in the leopard skin leapt amazingly, lighting deftly on one foot in front of the *embuutu* drum. He beat it sharply, not losing a second, jumped again like a soaring antelope, landed before the next drum, beat it, flew up like an eagle once more. . . . Floating upwards with incredible virtuosity, the dancer leapt to the ceiling . . . like a winged deer, touched it with his stick, and landed just in time before the largest drum.[7]

What struck Philippa more and more as she traveled Africa was that its music was strangely without personalized sentiment. In attempting to explain this, it occurred to her that the native Africans live so completely subordinated to the will and traditions of their society—that is, their tribe—that singularization or personalization was frowned on, in art as in conduct:

> It is wrong to think of Africans as "free" or "wild," rather they are much more restricted and disciplined than Europeans or Americans. The tribal African does not "get out of line," have new ideas, or express *himself*. He is always subordinate to society. The African musician or artist is not a rebel against society, or a Bohemian on its fringe, as is so often the case in the West. Rather, he is an integral part of society, and an expressor of its collective traditional values. In my own concerts for the Bantu Baganda people of Uganda . . . I found 18th-century music far more appreciated than 19th-century composers, such as Chopin or Liszt whose romanticism seemed to them "not quite in good taste."[8]

But as if underscoring the dangers of generalization, on a subsequent trip to the rain-drenched blue-green Mountains of the Moon in Toro (western Uganda), Philippa would come upon something quite different: African music that expressed personal sentiment and profound individual feeling:

> Their unique seven-tone scale, in major and harmonic minor, is unparalleled in East Africa, and was evolved hundreds of years before any European influence. This traveller has never heard the seven-tone harmonic minor in any other authentic African traditional music.
>
> The saddest of Toro's dolorous chants are the "women's music," which could only be played and sung by females of the apparently unhappy polygamous households, and could never be performed by a male. Accompanying themselves with lute-like string instruments, and *entimbo* drums, a group of noblewomen, gracefully swaying in their long, green robes like reeds in a breeze, gave a concert of this touching music for me. . . . Worlds apart is this sad, melancholic music of the women of Toro.[9]

23

No Women's Country

From a relatively early age until her death, Philippa was an avowed feminist.

When she first read Pearl Buck's *Of Men and Women*, the book both fascinated and upset her. It mirrored her own concerns about reconciling marriage and career: From Los Angeles, where she had been preparing for her first European tour in 1953, she wrote Jody:

> There were some changes for the better since [the book] was published in 1941, however, the tragic dilemma she lays bare has not, by any means, been solved. It is that a woman now receives fully the education to prepare her for any walk of life she may choose—yet after she has received it, in most cases, she is not allowed to put it into action. Social pressures insist that a woman must marry or be a failure as a female. If she has the strength of will to ignore this and set out for a career in accordance with the education she has received, she is continually battered with invectives, jealousy from women and resentment from men. There can be no peace or serenity for her.[1]

Jody's response to this "manifesto" is not preserved, but though herself a liberated woman, she may not have had much useful advice to dispense, for Josephine had long since abandoned a career as an artist and writer to devote herself to motherhood, to the rearing of an extraordinary child.

❖ ❖ ❖

Philippa's reaction to the world's pervasive unequal treatment of women was almost visceral. She had seen appalling examples of their neglect, abuse, and exploitation in the poverty-stricken regions of South and Central America, and in the dogma-ridden Moslem and tribal societies of Asia and Africa. She witnessed the disproportionate burden falling upon women, young and old, in the Third World; she had seen the fear and sad resignation in their eyes.

It is therefore not surprising that she would want to write a book about the disadvantaged sex. She considered but discarded the idea of setting her story in Africa or South America. The average reader would not be able to identify with the plight of the "second sex" in the Third World; and she herself, an outsider, did not have the necessary intimate knowledge or insight to write perceptively. Why not set it in the West and step back a bit in time, she thought, for despite outward appearances, had there been a fundamental change in women's relative status in the past 150 years? Philippa named her new heroine Sophie Daw, and set her novel in Britain at the beginning of the nineteenth century; and like Dickens she focused on the lowest strata of English society.[2]

Sophie Daw, child of misfortune, is born into the world on the Isle of Wight in 1792, the daughter of Dickey Daw, who has a certain celebrity as a local smuggler, and Jane Calloway, the victim of his frequent drunkenness and long absences.

Dickey is captured on one of his forays and sentenced to prison; Jane and Sophie are sent to the workhouse, where conditions are wretched. The night before Sophie is to be sent to a neighboring farm to work—she is now thirteen—Sophie's mother takes her aside and advises her. She cautions Sophie never to sell herself cheaply and to be stingy with her sexual favors, avoiding poor men at all costs (she wants a life of "jewels an' fine frocks" in London for her daughter). She admonishes against ever falling in love, and instructs her never to believe anything a man says: "They all be rogues."

With these words tucked away in her heart, Sophie leaves for the farm of Squire Rowdle and his drunken, brutal sons. Two months later, she flees to Portsmouth with a man who offers her two pounds to deflower her.

In Portsmouth, Sophie becomes a chambermaid in a hotel and passes nine months fending off the unwanted advances of the locals. Ultimately, she meets two handsome naval officers, falls under the spell of their good looks and effusive charm, and impulsively runs off to London with them. There she is wined and dined and bedded for two weeks before they slip her a Mickey and disappear, abandoning her destitute to the appetites of the innkeeper, who forces her to work off the unpaid bill in his bed or choose debtors' prison.

Sophie sees her chance, after three weeks of compliance, to make her escape. Coyly flirting with the innkeeper's smarmy "watchdog," she gets him to drop his pants and his guard, and she renders him unconscious with the blow of a candlestick. Frantically wandering the streets, she pawns one of her last possessions and spends the night at a flea-ridden inn. Sophie, only fourteen, is lost and friendless in the largest city in the world.

The following morning, finding that she has been robbed of everything, Sophie enters "the most terrible period of her life." In desperation she turns to prostitution—"a life of revolting, ill-paid vice"—which supports her, but so appalls her that she punishes herself by prolonging it.

Then fate intervenes. During a blizzard, Sophie slips in front of a carriage and is kicked in the head by a horse. Unconscious, she is taken to the home of a Mr. Winkleton, a kindly man who treats her like a daughter and even finds her employment in a fashionable millinery shop, whose owner, Mme. Snager, treats Sophie well in hopes of luring Winkleton into marriage.

But after only three months, Winkleton is called away on business to Barbados and perishes in the Atlantic at the hands of French privateers. With her benefactor dead, Sophie's treatment by Snager degenerates into abuse and exploitation. She is consigned to a dark attic and reduced to a scullery maid with only table scraps to eat. She spends a winter with drudgery and despair as her only companions.

Then, in the burgeoning spring, while washing the steps in front of the shop, she meets a "dashingly handsome" Irish water carrier named Michael, who stops to comfort the weeping girl. Soon they fall in love, and one May morning they are discovered "doing the beast with two backs on the floor, in swooning, splendid passion." Mme. Snager thereupon throws them, half-naked, into the street.

Sophie and Michael move into his tiny warren in Stepney where they live in romantic blindness to their poverty, seeing sumptuousness in their ragtag sticks of furniture, imagining lace in the many cobwebs, and hearing stringed symphonies in the rustling of rats' feet. Michael is never discouraged or bitter about life, but is full of gratitude to God and ever ready to enthrall Sophie with Irish songs and tales.

But one day, Michael leaves for work and never returns. After a day of weeping and prayers, Sophie wanders out into a savage storm to search for him. She exhausts herself running from tavern to tavern along the waterfront and, in the end, stumbles, pitiful and crushed, back to their empty house, collapsing on the steps in a fever of fatigue and despair. When she awakens, she sees a shadowy figure standing over her as in a dream. But the tall man comforting her is not Michael. He is a friend but knows nothing of Michael's whereabouts.

Here the fragment ends, and the rest of the manuscript is lost.

Stylistically, "Sophie Daw" may be among Philippa Schuyler's best fiction.[3] The writing flows with a natural and graceful pacing, even in its fragmentary first draft; Philippa's strong suit is an eye for detail. In one of the most powerful passages in the novel, she describes Sophie's descent into whoredom:

> And it was cheap, revolting, ill-paid vice, that exhausted one, and made one's stomach turn with disgust. Sometimes, when I emerged at dawn into the melancholy dull light, or the thick yellow fog, I loathed myself and all of life, and longed to die, fling myself into the Thames. I hated the brutal, drunken men, and dreaded to get some foul disease, but I had not will nor knowledge of how to break away from this tread-mill in the evil mire of degradation.
>
> It terrified me to see the other girls and women who trod the dark, dirty streets, some young and brazen, others coughing with consumption and near death, some ancient flabby bloated hags with tumescent breasts and stomaches, some scrawny, ill-faced, covered with revolting sores.
>
> Sometimes, gentlemen came into the unsavory, dank district, with its fetid dampness and low carnal vileness. They would come in twos or threes, with whips, and, though at first they seemed superior to the coarse workingmen and drunken idlers, when one got alone with them, their refinement disappeared, and ghastly brutal perversion and cruelty took its place, and they enjoyed humiliating one, and sniffing at the filth in one's room, and forcing one to find ever newer devices for rousing their jaded passion. Yet I could not pull myself from the mire of this

> depraved, debauched existence, this hideous setting to malediction, suffering of body and soul, this grovelling in beastly sin.

To the biographer, "Sophie Daw" contributes insights beyond the feminists' concern. Despite Philippa's powers of observation, the characters that populate this novel are not always three-dimensional. In particular, Sophie never describes herself—she has no identity. Even when asked, she cannot give her own name—she spends her life pseudonymously. She is faceless, amorphous, without a sense of self-worth, a stranger to herself. She is not a force, she is always in reaction to forces—social, economic, and biological. Philippa herself must have felt only a modicum of control over her own destiny. Her focus on prostitution is also revealing—perhaps a reflection of her own dormant self-loathing.

❖ ❖ ❖

The more Philippa traveled, the more militant her feminism became. She discovered that violence against women, though often shielded from view, cut across geographic borders and socioeconomic barriers.[4] Michael Langley, in his book *No Women's Country* (which Phil read just prior to her first visit to the Sudan), had written: "The lives of Sudanese women are grim . . . they look dispirited drudges. Marriage in the Sudan was always a painful, wearisome ordeal for them. . . . Men's possessive hold and the dominant status of the male is maintained. . . . Sudanese women are [sexually] mutilated at an age when European girls are trotted off to their first communion . . . the most disagreeable country in the world in which to be born a woman."[5]

And indeed even casual contacts suggested to Philippa that Langley's assessment was true. One man told her that his sisters had gone to school for four years but were then removed so they could stay home "like good Sudanese girls." On her second trip to Khartoum she was taken to a wedding, where she sat apart from the men. "A crushed-looking Sudanese girl spoke to me. I asked her if she was married. She shook her head, mutely. I asked her if she was engaged. She laughed nervously, and said she 'didn't know.' 'What do you mean?' I said. 'In this country if a girl is engaged her father wouldn't tell her.' In other words, a . . . father doesn't even bother to *tell* her what plans he's made for her future."[6]

But the crime of gender that upset Philippa most deeply was the custom euphemistically referred to as "female circumcision." The operation, which removes all or part of the female genitalia, is a form of mutilation that robs millions of women of all sexual feeling. It also endangers their lives, both at the time of the operation and later during childbirth. The medical complications arising from the operation itself can be severe.[7]

Philippa never witnessed the ritual—few westerners ever had. Her Belgian friend Roger Nonkel, after considerable urging, described the secret ceremony to her. Nonkel had worked in the Congo for years, won the confidence of some of the Bwakwa, and had been invited to observe the operation:

It took place in a small glade in the midst of a deep forest. The young girls who were to be treated were kept apart, dressed as men, for they were considered as being asexuated till they were excised. In the glade were the witch doctor and his assistants. Then a young girl came up [about fifteen], accompanied by her parents or some intimate friend. . . . She wore a man's shirt and some breeches. Then she undressed, was seated, nude on the knees of a man who was, himself, seated on a primitive chair. Her back was to him, and her legs dangled over his. Members of her family kept her legs open—while the man on whose legs she sat took her round the breasts with one arm and pulled her slightly backwards. His other hand was kept free. The witch doctor now took a knife with a short [crescent-shaped] blade. With his other hand he threw some ngula-powder [made from scraping the wood of a tree called *ngula* and believed to help blood coagulate] on the spot he intended to work on. . . . The action was quick. The witch doctor took the vagina lip in one hand, and cut it off in a saw-like movement, then repeated the gesture with the second lip. . . . The assistant on whose knees the girl was sitting put his free hand over her mouth to stifle the cry. Afterwards, the wound was sprinkled again with ngula-powder, and another I could not identify; the girl was released and she walked away. A trail of blood was flowing down. She was supposed to do a few dance-steps to indicate her stoicism."[8]

24

Tour du Monde

During the summer and fall of 1958, Philippa had scattered engagements, including a successful Gershwin Festival in Hawaii. But most of her time was spent scheduling her first round-the-world tour. Having performed on four continents, she had never circled the globe in consecutive engagements. The journey—from the beginning of December 1958 through the end of March 1959—covered over forty thousand miles.

Philippa first went to the Dominican Republic and New Orleans. Next, she had a series of well-spaced concerts in San Francisco, Hawaii, Tokyo, and Seoul. At year's end she flew to Taipei and Hong Kong, continuing to Singapore the first week in January. On the last leg of her East Asian tour, Phil flew to Kuala Lumpur, capital of the newly independent Malaysia.[1]

January 14, a giant five-thousand-mile hop took her to Beirut and Jerusalem. Two weeks later, after a concert in Khartoum and a layover in (former) Ruanda-Urundi, she was on her way to the Congos (Belgian Elisabethville and Léopoldville, and French Brazzaville). From there Philippa proceeded to Lambaréné in Gabon, to see Dr. Albert Schweitzer. In the ensuing two weeks, she revisited West Africa (Jos and Ibadan in Nigeria; Accra and Kumasi in Ghana; Abidjan, on the Ivory Coast, and finally Dakar). On February 23, a very tired Philippa returned to Paris. The next four weeks saw her leisurely traveling and concertizing in Belgium, Holland, and England.

Musical Courier would describe her trip as "one of the longest world tours ever taken by a concert pianist. . . . She played in no less than 33 countries. . . . Her itinerary included universities, concert halls and royal palaces. . . . And she was doubtless the first American concert pianist to give recitals at the hospital of Dr. Albert Schweitzer in the French Congo."[2]

❖ ❖ ❖

Philippa was fascinated by the Orient. In Tokyo she stayed and performed at the modern Sacred Heart University, and one day an alumna came to visit: the charming and beautiful Michiko Shoda. A commoner, she was engaged to Crown Prince Akihito and had already received the traditional samurai sword from him with the ancient admonition to defend her chastity with it. Four months later, the young couple would be married.

In South Korea, only just recovering from the war with the North, Philippa was the guest of Ewha University in Seoul. Founded by Methodist missionaries in 1880, it was the largest all-female school of higher education in the world.[3]

On Christmas Day, Phil flew from Seoul to Taipei, where she lectured at one of the universities, gave a concert in the town hall, and played to students of the Military Music Academy. Taipei seemed gay and happy. "Gorgeous, sexy Chinese girls, like golden passion flowers were everywhere," she wrote, "their seductive cheongsams slit high to reveal sinuous, exciting brown legs."[4] Philippa bought several cheongsams for herself, which she would later wear in steaming Africa.

She felt an instant political affinity for this small island which had been a citadel against communism since 1949. The degree of freedom and equality women seemed to enjoy was surprising and she was particularly impressed that there were two women on the Supreme Court.

After spending New Year's Eve at Taipei's Grand Hotel night club in the company of several diplomats and their wives—Spanish, Turkish, Venezuelan, and Brazilian—Philippa flew to Hong Kong. Like all visitors to this part of the world, she was enthralled by the "Pearl of the Orient."

She was equally taken with Singapore. Her hosts there were New Zealanders John and Alison MacKenzie, and a friendship quickly formed. "You are one of destiny's children," John wrote Phil shortly after she left, "so always aim high, never settle for second best. May your art flourish and your heart have heavenly wings."[5]

In Kuala Lumpur she gave a command performance for Prince Abdul Rahman and his wife, on an excellent instrument in the airy Istana Negara Palace. The prince had become the first prime minister of independent Malaysia, after it emerged in 1957 from years of bitter guerrilla strife.

❖ ❖ ❖

The next leg of her journey typified the pressures and hardships she had to endure so often.

The Lebanon-bound plane was delayed for twenty-four hours in Kuala Lumpur. Scheduled for a radio concert with a studio audience on that day in Beirut, Philippa had wired to postpone it. The plane made a late-night stopover in Karachi to take on Pakistani emigrants loaded down with their worldly possessions. Although the weather at first was rough, it was a clear, cold dawn when they landed in Beirut. One of the pilots gave her a lift to the home of her hosts, the Etinoffs, who were standing on the balcony of their second-story flat, watching for her appearance.

No sooner did they light the small kerosene stove and begin asking questions about her trip, when the man from Radio Liban arrived to take her to the recording studio. The station had rearranged its programming to accommodate her travel delay. For them, it was an important event: Philippa Schuyler was the first American musician to be aired on their national radio.

On the way to the studio, Liban's American manager informed Phil, to her surprise, that she was going to Jerusalem in the afternoon where she had a sold-out house that night. He also informed her (to her annoyance) that the radio appearance in Beirut would be gratis. After thirty-six hours without sleep or a full meal, she had trouble keeping her eyes open while recording.

"Then I fly to Jerusalem [to the then Arab sector]," she complained to Jody,

> have to greet everyone. It's deadly cold, I have to insist on two stoves. . . . The stage is tilted, so to keep the piano from falling off they had to put it on high glass props, which meant my foot had to be straight vertical all the time I was playing, a great strain. It was an upright too—and it was so cold, I had to wear 3 prs. of stockings and my brown skirt and three slips under my evening gown—and I had not had a bath in three days—and after the concert—I had to sign about a hundred autographs—I signed them partly in Arabic which fetched everybody—then up at 7 am to catch the plane to Beirut. Saw the Etinoffs, rushed back to airport, plane late, got to Cairo, rushed in and out, got to Khartoum at 2 am.[6]

A week later, on January 28, 1959, after a concert in Khartoum and a layover in Usamburu (Ruanda-Urundi's capital), Phil was already performing on another stage—this time in the enormous and explosive Belgian Congo. In years to come, she would easily qualify as "an old Congo hand," but at this juncture Philippa was just beginning to gather impressions and information.

When she arrived in pleasantly cool Elisabethville (since renamed Lubumbashi), the Congo was in the initial stages of its uneasy journey toward independence. Three weeks earlier, brief but violent riots had broken out. Vans had been set afire, cars belonging to Europeans attacked and damaged, shops, schools and missions raided. The capital city of Léopoldville had been hit the hardest; everybody in the huge equatorial country was on edge, especially the Belgians.

E'ville, to employ the widely used abbreviation for the European-looking city, was the capital of the mineral-rich province of Katanga (now Shaba), and the focal point of Belgian Congo's incredible wealth. But her concert at the recently established University of Elisabethville was poorly attended—no doubt as the result of postriot jitters—and only by whites, despite the fact that the school was integrated.

She stayed at the home of Albert Maurice, the university's secretary general. Phil had made his acquaintance in Brussels five years earlier at the same time as she had met Paul Fabo and André Gascht, but had not seen Albert since.

Philippa was impressed with the tall, sharp-featured intellectual, in his white tropical suit and black-rimmed glasses. The forceful sense of manhood that emanated from him was not lost on her. For his part, Albert found his very intelligent guest in her favorite red cotton cheongsam, tightly fitting and slit high on both sides, much more attractive than he had remembered.

It happened to be Albert's forty-fourth birthday, and they celebrated it together in an intimate tête-à-tête. A fine dinner was served by candlelight, in which Stanley, the host's large cat, shared, nibbling alternatively from Albert's then Philippa's plate. After dinner, Maurice, who was also an accomplished artist, made several sketches of her, and they talked about her forthcoming book (*Adventures in Black and White*). Circumstantial evidence strongly suggests that they ended the evening by making love.[7]

❖ ❖ ❖

Philippa stayed in Elisabethville only forty-eight hours. The following five days took her first a thousand miles northwest to Léopoldville, the Belgian Congo's capital, and then across the majestic river to French Brazzaville.

She found the difference between the Belgian and the French Congos dramatic. "Léo" was a splendid city, modern and gleaming, with broad boulevards and tall white buildings. The streets were jammed with merchants in Moslem djellabas, selling exquisite ivory, mahogany, and ebony carvings. Despite some attempts, virtually no social contact existed between the 370,000 Congolese and 22,000 whites who lived there. Even at the supposedly integrated Luvanium University, black and white students had "only danced together once."[8]

Like Dakar had a year earlier, Brazzaville exuded an air of racial amity. The audience at her concert was mixed, and many a black she met had been educated at the Sorbonne or another major university in metropolitan France.

But of all the European countries Phil had visited, Belgium was her favorite, the one where she felt most at home. Belgium, she believed, had done a great deal for its Congo: it had brought first-class technical and organizational methods to that equatorial colony, creating a smooth-functioning state that claimed a relatively high standard of living and the lowest illiteracy rate in Africa. Yet, unlike France, it had completely failed to prepare its Congolese for independence—and the move toward independence seemed unstoppable. She feared for the future.

But her mind was now mostly on her next destination, Albert Schweitzer's famous "primeval hospital," near Lambaréné in Gabon. Although not originally on Phil's schedule, both mother and daughter thought it would be excellent publicity. Jody knew Marion Preminger (the ex-wife of the famed movie producer and director), who now devoted her time to raising funds for the hospital, and often stayed there. Phil had written Marion from Usumburu, but not hearing from her decided just to go—as it was common knowledge that all kinds of people descended on "le Grand Docteur."

On landing, a jeep transported passengers, luggage, and a load of bananas to the river's edge, where they all were piled into a canoe rowed by six lepers. Only a few miles separated the city from Adominanongo, the site of the hospital. The lepers sang under the ferocious sun, maintaining a constant rhythm on the strenuous paddle upstream. "Lambaréné is the end of the world," Philippa later wrote home.[9]

She first glimpsed the legendary Docteur in his "uniform"—white tropical sun helmet, khaki trousers, and white shirt open at the neck—as he walked nimbly down the hill to greet the visitors. Almost eighty-five years old, he looked much younger. "His face," wrote Philippa, "has the kindly, benign look of a practical saint."[10]

Learning that she was a concert pianist, he offered to see if someone could tune the new piano recently sent to him by a benefactor. Besides being a theologian, a doctor, and a humanitarian, Albert Schweitzer was a highly respected musicologist and keyboard artist.

Phil was led to her "chambers"—a room smaller than their bathroom in New York, covered largely with mosquito netting. The hospital had neither running water nor electricity, and she was provided with a pitcher of water for bathing and two kerosene lamps, to avoid tripping over snakes when it grew dark.

The gong was soon struck for lunch. (Life at the hospital was strictly regulated: bells signaled the time to wake, the time to eat, and the time to pray.) After lunch, and at Schweitzer's request, a somewhat terrified Philippa played a brief program: a John Field nocturne, a Schubert impromptu, and the Chopin Scherzo in B-flat Minor. The Docteur seemed pleased.

After dinner, and again the following evening, Phil performed other classical pieces, and when she suggested her own transcription of Gershwin's *Rhapsody in Blue*, Schweitzer said, "Yes, I would like to hear that . . . I have never heard it before. I like to keep up with what is going on in the world."[11]

On Philippa's final day, Marion took her to the hospital and the leper village. On the way, they walked through a rich grove of fruit trees that Dr. Schweitzer had planted with his own hands forty-six years before.

The village consisted of many small houses spaced at regular intervals. The adult lepers were clothed partly in tribal costumes, partly in Western clothes brought by Marion. Children were all around. Marion picked up several and embraced them. She had not the slightest fear of contagion. They all called her "ma mère Marion."

The pharmacy and operating room, both relatively primitive, were housed in buildings of hardwood frames and corrugated-iron roofs. Most of the "walls" were simple screens. She noticed that the hospital was filled to capacity.

When time came to leave, Dr. Schweitzer kissed her on both cheeks, advised her not to wear lipstick, and promised to write his friend, the Dowager Queen Elisabeth of Belgium, suggesting Her Majesty hear Miss Schuyler play.

Philippa was truly awed and humbled by Albert Schweitzer. Yet much to her surprise, whenever she subsequently mentioned his name in equatorial or western Africa heated arguments ensued. Some said that Schweitzer alone was receiving all the accolades while other missionaries working as diligently as he were going unrecognized. Others contended that his hospital was outmoded; that Schweitzer had overstayed his time; and worse, that he was a white colonialist. Why hadn't he started a school to train black nurses, for example?[12]

In time Phil concluded that the New African's quarrel with the Lambaréné hospital was in large part based on emotion, not reason. But it was an Oxford-educated black journalist talking to a white European colleague who put it in perspective: "Sure," he said, "we want housing, food, hospitals, and education, especially education. But above all we want human dignity. You never gave us that, did you? You took it from us. You didn't mind doing things for us. But not with us. You despised us really."[13]

❖ ❖ ❖

Philippa's plans now called for revisiting a number of West African cities; her next concerts were in Nigeria—Keffi on February 8, Jos the tenth, and Ibadan on the twelveth and thirteenth. Several connections had been booked in New York. But traveling to out-of-the-way places in Africa in the 1950s was difficult, at best, and this trip turned nightmarish.

The small plane to Douala in French Equatorial Africa was miserably uncomfortable. It stopped five times in 450 miles, loading and unloading passengers and freight. In Douala, she was supposed to catch a flight to Kano, Nigeria, with connection to Keffi. On arrival, she learned that the schedule had changed and the only weekly flight to Kano had left the previous day.

There was no way out—no plane, no train, no road through the thick jungle. Nor was there any reliable telephone communication between the French and the British territories. She would have to miss all her concerts in Nigeria. In tears, she sought out the airport commandant, a very handsome Algerian Frenchman who came to her rescue. He found a private citizen who was planning to fly his tiny plane to Fernando Po the next day, and at least agreed to "drop her off" in Tiko, British Southern Cameroon, even though it was unauthorized. Located seventy miles from the border, Tiko had "regularly scheduled flights" to several Nigerian cities.

After spending the night in Douala—at the amazingly first-class Hôtel des Cocotiers—she met her pilot and they flew into a cloudless sky. Passing over miles of jungle and swamp they landed in Tiko. Her "gallant" Frenchman hurriedly threw her baggage from the plane onto the landing strip, and zoomed away. The authorities rushed out, ready to arrest her. But even though she succeeded in pacifying them, she was stymied. The next plane to Kano or Jos was not for several days.

Philippa finally decided to fly via Port Harcourt; but arriving in that eastern Nigerian city, she discovered that the afternoon flight to Jos had been canceled.

Still hoping to get to Jos by the following day (she had canceled her concert in Keffi), Phil went in search of a car and driver. How far was it, she asked? One said "far," another "not so far." Estimates ranged from 300 to 900 miles. The map showed 380 miles.

Philippa was exhausted, hungry, dusty, dripping with sweat, and annoyed. But she persevered and found a man with eyes bright as beads who said that for only twenty pounds, he and his brother would drive her there. It was not far at all, he assured, perhaps a ten-hour trip. They would start at sundown, drive all night and arrive in Jos at dawn.

So Phil stopped at the Cedar Palace Hotel to get some food for the journey.

When she told the Lebanese manager of her plan, he insisted that she must not go; that the driver and his brother looked villainous; that they obviously did not know the way; that it was a most dangerous route and the trip was likely to take two and a half days and not ten hours.

After canceling the car and spending a "wretched night of solitary reflection,"[14] Philippa discovered another combination of flights that would get her to Jos in time for the concert.

Canceled flights, lack of sleep, an aching arm from lugging her belongings, and humidity-swollen piano keys were not her only trials and tribulations. Using George's and Philippa's network of friends and acquaintances and occasionally the resident U.S. foreign-service personnel, Jody had accomplished the monumental tasks of arranging the tour singlehanded. No place in Africa did she have a local professional manager to promote the events and to smooth the way.

This led to difficulties, misunderstandings, and not infrequently poor attendance at the concerts. Typically, Gwendolyn Shackleford, a friend who was to be Philippa's hostess in Accra, complained to Jody shortly before the scheduled visit that, although she had been led to believe that all arrangements had been made, "no one knew who was sponsoring the trip. . . . The Prime Minister knew nothing, not the head of Broadcasting, not the local USIS officer, or the head of the Arts Council." Mrs. Shackleford concluded her letter: "What a brief visit to Ghana [you are planning]. Why can't she rest a little between recitals and see people, places and things. Not only here, but in all countries she visits. Bless her. How can she ever get married, unless she takes a chance on someone she doesn't know more than a day or two. She is too precious for this."[15]

And from Abidjan, William Dunbar, the U.S. consul for the Ivory Coast who had had a hand in arranging the concert, in transmitting a check for $207, the net take, wrote: "Obviously $200 is a ridiculously low profit for an artist of Philippa's caliber." He forcefully advised Josephine to engage some local person with knowledge of the impresario business and of the area. "Personally I feel that [she could earn] . . . three times that amount with good publicity." He added, "We continue to be amazed at her vitality, cheerfulness and general adaptability. She faces and overcomes difficulties which no artist should even have to think about, and comes out smiling! Our hats are off to her."[16]

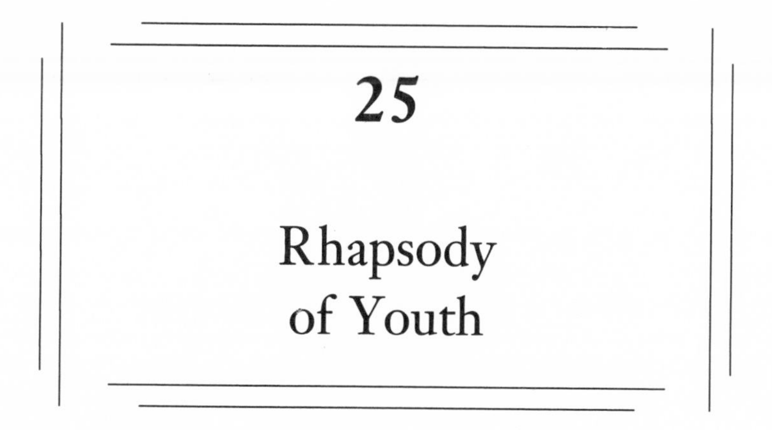

25

Rhapsody of Youth

Philippa arrived in Paris on February 23, 1959, and stayed with her recently acquired European godmother, the countess Goguet—unwinding, exchanging information on the obscure, throwing each other's tarot cards.

She was exhausted—from travel, performances, and changes in weather, food, and water. "I'm just sitting!" she wrote Jody. "How wonderful it is to be able to drink a glass of water out of the faucet! To sit on a toilet and be able to flush it and not have to squat over a hole in the ground and shovel dirt into it afterwards. I may not call a damn soul here in Paris. . . . You don't know how tired I am of being 'marvelous.' And of being alternately boiled, fried, and frozen."[1]

All throughout the African leg of her journey, they had argued over the date of her return. Josephine wanted her daughter to fly home immediately after the final concert: "I *cannot* leave March 15th," Phil insisted. "I've been working like a machine for 4 months. I am not a machine! Do I deserve no reward at all after the Labor of Hercules? Even machines have to be oiled. I've been through . . . hell. I'm tired of fighting. It's utterly unjust and unfair even to think of expecting me to rush back after crucifying myself. I want time to relax and see my friends in England and Brussels. I want to have some pleasures."[2]

Philippa decided to stay until March 21. Despite a lingering flu and a sore throat, she enjoyed shuffling between Brussels and Amsterdam; staying with Lineke and Theo van Eyk in Holland, and with the Gilbert Chases (her American friends from Argentina) in Belgium. She renewed her acquaintance with Paul Fabo; gave two performances with Everett Lee in Brussels;[3] played under the baton of Hugo de Groot in Amsterdam (John de Crane, her Dutch agent, attended the rehearsal. Unsolicited, he wrote Jody: "The orchestra was so impressed that they shouted 'bravo' after the rehearsal!"); and had long discussions on the state of American music with Gilbert.

Phil also spent intimate evenings with André Gascht. Time, she hoped, would bring them even closer.

Sometime between her Brussels and Amsterdam engagements, Philippa received a telegram asking her to give a command performance for Queen Elisabeth of Belgium on the twelveth of March.[4] Apparently Albert Schweitzer had been true to his word.

The invitation to play for the queen threw Philippa into a panic. Her clothes were limp from months of hard travel. The orange leather coat, in particular, was unpresentable. Hurriedly, she had a silk dress cleaned and borrowed an elegant fur coat from a friend of the Chases.

March 12 was a relatively clear but sunless pre-spring day in Brussels, the temperature in the fifties with a soft breeze from the west. Philippa was dropped off in the circular driveway of the Stuyvenberg Castle. The white, three-tiered building looked stark in the early afternoon light. Waiting in the luxurious drawing room decorated with paintings and drawings by some of the great twentieth-century artists, she could look out on a thickening forest to the right and a manicured lawn straight ahead.

Shortly, accompanied by her lady-in-waiting, the dowager appeared wearing a pale rose-gray silk suit, her hair in flawless waves, "her dignified regal grace [belying] her years."[5]

Though at first nervous, Philippa gave, according to reports, an excellent performance on the black Steinway concert grand. Her own ornate account would read:

> I began by playing the thirty-five minute *Pictures at an Exhibition*. Her Majesty honored me by praising my rendition. She asked me to continue. I played for three hours, going from the shimmering quasi-Oriental works of Griffes, to the romanticism of Chopin, from the glitter of Schubert, to the Celtic melancholy of John Field, from the snowy clarity of Beethoven's *Waldstein Sonata*, to the enchantment of Ravel's *Jeux d'Eau*, the Attic delicacy of his *Sonatine*, and the cruel, sardonic *Alborada del Gracioso*, from Aguirre, to Japanese *Cherry Blossoms* and Gershwin's rhapsodic blues, and two of my own compositions, *Manhattan Nocturne* and *Rumpelstiltskin*. Her Majesty urged me to play more and more, praising what I did with a warmth that made me feel inspired and most deeply grateful.[6]

A late tea with the queen and her guests followed; they all posed for pictures. When she departed, Queen Elisabeth kissed her and wished her good luck. "I truly felt it had been the crowning experience of my life, as though all else had merely been leading up to this wonderful occasion."[7]

The only gaffe in an otherwise perfect afternoon was reported to André by the very proper Paul Fabo, who had accompanied Phil to the castle. After tea, a fingerbowl was passed around. The queen handed it to Philippa, who instead of taking it just dipped her fingers into it while a somewhat surprised dowager held the bowl. Phil never realized her faux pas.[8]

❖ ❖ ❖

Nine days later, Philippa returned home and immediately began a twelve-city sweep of the Midwest and the Mid-Atlantic states, playing for black institutions—from St. Louis to Richmond to Washington (for the National

Council of Negro Women) and Philadelphia. On the heels of an extraordinary round-the-world trip, having played for dignitaries and royalty, the American tour could not fail to underscore the chasm between her colorblind acceptance abroad and the segregation in her native land.[9]

Once again, Philippa's depression returned—a now familiar refrain after each "homecoming" to America. But somehow her unhappiness was more introspective this time. On March 30, she wrote to herself: "Quiet time. I must have directness and self-confidence to triumph over my surroundings. God gave me a gift and he meant for me to use it. He did not mean for me to waste my time. . . . I must pray for help in difficulties I now face. I must have strength of character, courage and determination."[10]

It was a critical time for Philippa. She had a debut at Carnegie Hall coming up on May 17. It, too, was under the auspices of a black organization, the Church of the Masters. Its purpose was to celebrate the five-year-old integration decision of the Supreme Court, and the proceeds would be donated toward a youth center on West 122d Street in Harlem. Yet as a musical event, it would stand tall, attended by a broadly based audience and reviewed by major critics. Much, she felt, hung on its success.

Philippa had another tangible concern. She was not sure about the program. Would it serve to demonstrate her greater maturity and increased skill? Jody and Philippa had argued about the program in their letters. One of the arguments Jody had brought on herself. In a revelation of her insular mentality, she had "translated" one of the sections in Mussorgsky's *Pictures*, known as "Samuel Goldenberg and Schmuyle," to "A Rich and a Poor Man Argue." Phil objected, "No where in the world is that called that. It is not done."[11]

But by and large Philippa had left Jody in charge of planning the program. Her mother had turned to Carl Van Vechten for advice (though really more, it seems, to get his approbation than to follow his suggestions), sending him three versions. Van Vechten was not pleased with any of them:

> Program making is an art in itself. Some artists are born with the feeling for it. Still more acquire it painfully. Some of the greatest artists have no sense for it at all and some of the worst ones are very adept [at] it. It is a knowledge of what goes with what and why. Neither you nor Philippa seem to have any knowledge of it at all and your new programs are no better than usual. As I have often told you, the best way to arrange programs . . . is to study the programs of some one who makes good ones, or to go to the concerts of this artist constantly until something sinks in. I cannot tell you much but I can TELL A GOOD PROGRAM WHEN I SEE IT. . . . The trouble is that unconsciously a bad program makes a bad impression. . . . I am very patient about all this, and because you ask me I try to explain things to you, but Philippa has NEVER paid the slightest attention to anything I have ever said to her about these matters and I doubt if she even knows what I am talking about.[12]

Van Vechten was right. Neither Jody nor Phil had any sense of what good programs should contain. They often made them far too long, and without enough contrast. Failing completely to understand the temperament of the public, they sometimes programmed works by Philippa alone, or other unknown composers. Originally, Josephine had planned—unwisely—a two-and-a-half-hour concert, with no intermissions, for her daughter's Carnegie debut. Even the most seasoned concertgoer would be hard-pressed to sit through this, no matter how provocative the performance.

The final program was better than the three originally suggested by Jody to Van Vechten, but it was still too long (the intermission was limited to five minutes) and somewhat monochromatic:

Oriental Suite	John Kelly Collection
Pictures at an Exhibition	Mussorgsky
INTERMISSION	
The Roman Sketches	Griffes
Jeux d'Eau	Ravel
El Alborada del Gracioso	Ravel
Fantasy in F minor	Chopin
Scherzo in C-sharp minor	Chopin

Philippa had another cause for anxiety. She was feeling weak and tired much of the time, and seemed to be fighting off a constant headache. Prone to migraines as a child, she suffered more as she grew older. Days of valuable practice time were now slipping away.

Her debut only forty-eight hours away, Philippa gave a radio broadcast about her recent trip around the world. At home, she complained to Jody about feeling very dizzy, and in midsentence she collapsed. Her mother reluctantly called a neighboring physician, who diagnosed the condition as lingering effects of a virus picked up in Africa. Dr. Allen warned that to go ahead with the concert might permanently injure her health. Rushed to Sydenham Hospital in Harlem, she was given an immediate blood transfusion. Thus, for the second time Philippa had to cancel a New York appearance. Six years earlier—just before her Town Hall debut—she had also been rushed to a hospital.

The concert was rescheduled for June 7, 1959. The crowd was huge and international. The management reported it as the largest audience that season for a classical music event, with a box office gross of over seven thousand dollars. Friends and sponsors from the United Nations, Liberia, the Philippines, Korea, France, Haiti, Belgium, the Dominican Republic, Ghana, Lebanon, Britain, and Malaysia were there. Deems Taylor, Marion Preminger, Roy Wilkins, Carl Van Vechten, and Leonard Bernstein came. Old friends and new loudly applauded. Philippa wore a Hattie Carnegie gown of black Chantilly lace and blue chiffon. After the concert, she was presented with eight bouquets of flowers and a bust of Albert Schweitzer, delivered to her dressing room by Marion Preminger.

The reviews were strong and encouraging. The *New York Times* cited her "maturity and sureness . . . delicacy of touch and sensitive phrasing."[13] *Musical Courier* wrote that Philippa had "matured into an artist of major stature. She is a personable, graceful young woman possessed of fine pianistic technique, ample tone power, and a subtle skill in etching pictures in sound."[14] *Musical America* noted that Philippa "demonstrated her natural flair for the keyboard plus a maturing concept for refinements of tone and expressive nuances not always in evidence in her previous performances."[15] The *Times*, however, complained that in the Mussorgsky and the Chopin, Phil tended to rush and blur the more difficult passages. But both the *Times* and *Musical America* commented on Philippa's unflappability. A note failed to play in the *Alborada* and (wrote Salzman), "she carried on without faltering."[16] *Musical America* called her a "seasoned trouper."[17]

Van Vechten, although still disapproving of her program, wrote: "Fania and I agreed it was your VERY BEST concert. Parts of it were extraordinary."[18] But perhaps the review Philippa treasured most was a personal letter from Deems Taylor. "Your technique is impeccable. You have a real legato, one that doesn't need to be helped out by the damper pedal, and dazzling speed when it's called for. . . . You've grown up a lot, haven't you? You sound like a young Myra Hess; and that's my compliment for the week. Altogether, youngster, I'm deeply proud of you."[19]

26

What Is Africa to Me?

Nineteen fifty-nine was also the year when conflicts over race and politics surfaced in the Schuyler family. To have produced a highly gifted and beautiful interracial child was, to Jody and George, the fulfillment of their lives. Philippa, however, was troubled: On the one hand, she enjoyed her status as a role model for countless young African Americans, on the other, she remained unhappy and frustrated over the seemingly insurmountable racial barriers in her career, which created a deep personal ambivalence toward her own blackness. Yet she could not share her unhappiness with her father and could only obliquely hint at it to her mother.

Political conflicts ran just as deep. While all three Schuylers were conservative, significant differences existed between Philippa and George in their attitudes toward events in Africa.

The globe-circling tour of 1959 had started as a family affair, on a highly harmonious note. They had flown *en famille* from New York to the Dominican Republic. For not only had Philippa been invited to perform; George and Jody had also been asked by Rafael Trujillo to attend a banquet in their honor celebrating the "social achievement they had made by their mixed marriage."[1]

Family harmony, however, soon turned into discord. Seven weeks later, after George and Josephine returned home, Philippa, now in Khartoum, received a cable from her parents. They were deeply upset by an article that had appeared in the *Hong Kong Tiger Standard*—an interview of their daughter.[2] The contention concerned Philippa's allegedly anti-American sentiments. They were forwarding the article to her next destination.

A long chain of events had led to the *Tiger* story. When Philippa arrived in Hong Kong on New Year's Day, no one was at the airport to meet her. The cable sent to her agent, Harry Odell, had apparently never been delivered, nor had he received her latest clippings.

Lacking advance publicity, her performance on January 4 at Loke Yew Hall drew only a small crowd. The following morning the *South China Evening Post* referred to her as a "well-known German pianist." But the reviews themselves were outstanding. The *Hong Kong Tiger Standard* described the evening as "remarkable," Philippa as "dynamic, dazzling, brilliant."[3] Also reported was the fact that the concert began at 9 p.m. and did not end until well after midnight.

The following afternoon, she did a radio broadcast and then went to Mr. Odell's office. Waiting for her was a young man by the name of Ernie Pereira, writer for the *Tiger*. Half Portuguese, half Chinese, and exceptionally handsome, he had heard her play the night before. Impressed by both her technical and interpretive skills, and taken by her looks, he wanted to interview her. They chatted briefly and then decided to continue their conversation at a nearby teahouse where they talked for over two hours—mostly about reincarnation and romance.

Philippa was intrigued with Ernie, his erudition, his looks. Mr. Pereira certainly had some "future possibilities." They agreed to stay in touch, to correspond, and possibly to collaborate. As they parted, Ernie invoked an old Chinese saying: If there is divine affinity between people, they will meet though they be separated by a thousand miles, but if this is not so, they will never meet even if they live on the same street.

After she had left Hong Kong, Ernie wrote a profile of Philippa for his paper. He sent a copy to her parents, thinking they would enjoy it.

When the clipping caught up with her in Léopoldville, she felt relieved. The article's title "Nothing Could Stop Philippa," referred obliquely to the length of her concert, but more directly to Phil's perseverance. "About her career," Ernie wrote, "she says she was discouraged from the start, even by her mother [*sic!*] from pursuing it, as she was bound . . . to meet prejudices along the way. . . . 'You'll be knocking your head against a brick wall,' her mother had warned."[4] (Ernie had recently finished Ruth Slynczenska's autobiography, and when he asked Philippa if her mother had forced her to be a pianist, too, she had responded, "Oh no! Quite the contrary.")

The subject of integration in America came up because Philippa had told Ernie that the Africans "had a distorted view of how the Negroes lived in America, thinking that they had no freedom at all."[5] Speaking of the United States, she was quoted as having said that "considerable progress had been achieved in the [area of integration], but there were more subtleties . . . in any situation that exploded."[6]

Philippa immediately wrote to her father:

> Now that I have read the article, I don't see what is so bad about it. Surely it must be realized that I am not writing the article myself—that I am in the hands of whoever is doing it. . . . If I got upset by . . . everything that appeared in a paper—how would I feel when bad reviews came that completely misunderstood my playing? Josephine always assumes that whatever a newspaper says must be true—if it said I played dreadfully, therefore I did play dreadfully. . . . But it's not so. . . .
>
> As for America, the article doesn't say ANYWHERE that I said I hated America. . . . Usually when I have a conference with reporters, and they bring up

> the race problem, I just say I prefer not to discuss it. But you can't say that to a world-travelled professional. . . . Nor can you say the race problem does not exist to a person who has read all the important books on that subject and visited the country. . . . The article said there was a race problem, but progress had been made. . . . That's true—what else can one say? I'm not being sent by the U.S. State Department to represent them. When they sent me, I knocked myself out for them, and . . . got the shabbiest of treatment.[7]

While Philippa could perhaps brush off or explain away to her parents why her interview with Ernie had touched on the American Negro question, she could not really talk to them about her feelings on the subject. In fact, there was almost no one in whom she could confide. She could discuss race relations with her European friends, but they had little insight into the heart of the "American Dilemma."

But Ernie was different. He was a racially mixed person himself. In time Pereira became a good friend to Philippa, and to him alone would she confide her deepest fears, her bitterness over her treatment in her native land.

Only three weeks after the fracas over Ernie, she had another set-to with her family. This time, it was over an article her father had published in the *American Mercury*. Entitled "Khrushchev's African Foothold," it was mostly about Guinea and its swing to the left. But George had also written briefly about Ghana, describing Nkrumah, Philippa's friend and supporter, as "a demagogue with a Napoleon complex who dreams of a West African empire."[8]

Phil had worried a good deal about the damage her father's no-holds-barred writings and far-right politics might do to her career, especially in Africa. George's view was that Africa was a poorhouse except in natural resources and that it desperately needed European and American investment. She had on several occasions asked her father to temper his writings, for her sake.

Instead of writing her father about the *Mercury* article, she addressed an angry letter to Jody, knowing that her mother would report a watered-down version to George:

> I am miserable, positively miserable because of George's article in the *Mercury*. It is a stunning shock. It was utterly unnecessary to write those ferocious lines attacking Nkrumah. . . . Nkrumah has been kind to me. Why abuse my friends? Why attack the people who help me? Why not attack the people who hurt me? . . .
>
> Nkrumah is the best possible man for Ghana. None better could be found. People who have only just emerged from utter tribalistic slavish following of authoritarian chiefs, can't jump right into democracy. . . . How can an American understand what a progressive African faces in dealing with the backward, malevolent bush element in his own country? And was the U.S. "wonderful" 2 years after its formation? NO! It was god-awful . . . chaos.

Despairing of being able to change her father's mind, she added: "If for some reason George absolutely *cannot* refrain from criticism, he should sugar-coat it and mix it with praise. . . . As for Nkrumah, he is NOT a Communist. You have to have some socialism in a small country like Ghana—every first-class country nowadays has some socialism, which means social programs to keep women from being horribly exploited and the poor from being too scandalously poor—that's what makes them first-class."[9]

And Philippa concluded her letter to Jody with a bitter statement: "There's not much in the Negro race to be proud of—damned little. Why attack and pull down the little there is?"[10]

❖ ❖ ❖

Philippa had a love-hate relationship with Africa and, by extrapolation, with herself.

She had been brought up to appreciate the contributions of the Negro race, and like many Americans of color, had wanted to travel to Africa—to sense "the unremitting beat" of her "dark ancestral blood"; the blood of "strong bronzed men or regal black women . . . three centuries removed," as Countee Cullen sang in his poem "Heritage," which George had taught his daughter when she was a little girl.

And indeed Philippa wrote once about how she adored Africa: "Its gentle feminine women, its proud and patriarchal men, its merry bright and mischievous children."[11] But Philippa also despised the continent's "inefficiency and cruelty. . . . Just live around Africa for a while," she wrote her mother in an unguarded moment, "and you get rid of any kind of nonsense that all . . . races . . . have the same capacities."[12] Like her father, Philippa struggled much of her life with both her pride and her shame about her African-American roots.

Philippa's ambivalence toward her own racial makeup gave her a persistent sense of placelessness;[13] and her crippling relationship to her mother made it all but impossible for her to remain whole. "Wounds and hurts," she wrote to herself late one night alone in her Rome hotel, overlooking the dome of St. Peter's. "I am a beauty—but I'm half-colored so I'm not accepted anyplace. I'm always destined to be an outsider, never, never *part* of anything. . . . I hate my country and no one wants me in any other. . . . I am emotionally part of nothing and that will always be my destiny."[14]

Gerd had noticed this years earlier when she described Philippa as a woman of a thousand faces, struggling so much between the black and white sides of her. "They were always making war inside her."[15] Madame Cosi, an Italian friend, also remembered Philippa as dismembered—"forever searching, never satisfied in one place. Even if she had found what she was looking for, it would have been partial for she herself was never a complete whole."[16]

But in 1961, while traveling in Madagascar, a possible solution presented itself to Philippa. She had been eager to visit this large island in the Indian Ocean—peopled, in part, by immigrants from Malaysia many centuries earlier. According to family lore, George Schuyler's ancestors came from there.

Philippa had met Madagascar's president-elect Philibert Tsiranana in New York and now was his guest. Her travels on the island were like a fairy tale. Escorted by a large retinue, including a maid who attended to her personal needs, she gave concerts in Tamatave, Tananarive, Majunga, and Antsirabe.

She then accompanied Tsiranana on a state visit to the island's southern provinces. They traveled by private plane and rail, and a fleet of Land Rovers, reaching into remote villages. Everywhere along the route were large crowds, festivals, folk dancing. Tsiranana spoke with enthusiasm about the coming independence, but earnestly urged the nation to remain in the French Community. And everywhere Tsiranana made a point of introducing his guest.

Philippa had fallen in love with the country and its people. Sitting on the president's plane and looking at the map, she savored the names of the island rivers: Sifia, Betsiboka, Manamboa, Mangoro, Tsiribihina, Onilahy—echoes of the fantastic place names she had invented as a child.

But most important to her was the discovery of what she thought were her roots. And her roots were Asian rather than African. "This is not Africa. This is the Orient," she wrote her mother.

> It is so obvious that these people are of Malayan descent that you don't have to be told it. Everybody looks as though they have stepped right off the streets of Kuala Lumpur or the fields of Penang. Their language is a combination of Malayan, Sanskrit, and Polynesian words. Their music is like Philippino, Samoan and Hawaiian music—sweet, lazy, melodic, relaxed. No orgiastic frenzies. Totally unlike African music. Their faces have the delicate Malayan bone structure, and many have slightly slanted eyes. Their skin is of fine quality, like mine. Some are my color, some darker. Their hair is long. Some have straight hair, some slightly kinky. They have no barbaric customs. Their manners are like the Malayans—agreeable, sweet, happy-go-lucky, relaxed, non-aggressive. THEY CONSIDER THEMSELVES ORIENTAL, NOT AFRICANS. . . . This island isn't African any more than Cyprus is. . . . Actually, since my ancestors came from here, I HAVE NO RELATIONS TO AFRICA AT ALL. . . . So I am Malay-American-Indian-and European![17]

27

"Je meure pour un homme"

Life without a man troubled Philippa deeply, but she refused to admit it, often saying to newspaper reporters such things as "I find young men very jealous of my music. A concert pianist has to spend so much time away on tours—no one I have met so far could bear my career."[1]

Public posture to the contrary, Philippa, at age twenty-seven, was desperate to marry.[2] She was particularly interested in her friend the Belgian poet, André Gascht, and she assumed the interest was mutual. So, despite the fact that she had no concerts planned in Europe, Phil decided to go there for three weeks in the summer of 1959, in hopes that personal contact would enhance her marriage prospects. Convinced that she wanted to settle in Europe and marry a European, she wished to spend some unhurried time with her friends in Belgium, Holland, and France.

Jody was greatly opposed to the trip—inclined to view it as a wild-goose chase. When she finally gave in, it was not from a realization that her daughter, by now the family's major breadwinner, did not need her permission, but rather that Philippa would have been impossible to live with if she continued to object.

Phil flew to Luxembourg and arrived in Brussels, by rail, late July. She phoned André who hurriedly dressed, came to the station, greeted her warmly, and found her a hotel. He then took her to a café, where they talked into the morning. Thus began a lively social schedule. Albert Maurice—with whom she had not been in touch since their liaison in the Congo six months earlier—was also in town, as were Paul Fabo and the Gilbert Chases.

A spirited competition for her time and attention quickly developed between Albert and André, which she gleefully described to Josephine. After receiving a graphic account of the jousting match between her two "beaus," Jody wrote her daughter a concerned letter. While she would not directly call into question the seriousness of Albert's and André's intentions, she hinted at it. Jody thought that neither of them had the financial stability to contemplate marriage. She also picked at an old wound by questioning how free of race prejudice these two white Belgian gentlemen might be.

On that subject, Philippa could give her mother unequivocal assurance: her race did not stand in her way with either Albert or André. In fact, like many European intellectuals, Albert in particular shared with the countess Goguet a measure of negrophilia, she claimed.

Despite her fascination with Albert, Philippa was much more attracted to André and spent as much time with him as possible. He was her ideal: "His eyes are exquisite, kaleidoscopic . . . that shift from blue to green to gold. His hands are long, delicate, poetic and slender," she wrote her mother. "He is thoughtful, shy, always refined, elegant, aristocratic. Seems cold on the surface, but that is really shy sensitivity, not coldness. Underneath he is tenderly warm and so sweet."[3] Phil and André were deriving increasing enjoyment from each other's company, and they were getting more and more intimate, although they still had not made love. But it was from this growing closeness that Philippa drew hope and encouragement.

She had confided in him that it was impossible to establish herself in America because of racial prejudice, and told him of her desire to settle in Europe and marry a European.

While marriage may not have been exactly what André had uppermost in his mind, he continued to send out mixed signals. He had written a poem, commissioned for the annual blessing of the Soignée Forest in Luxembourg:

> The ardent, fiery young girl
> Who speaks hidden words to you.
> She will show you the pure waters
> Whose promise shines in you.
>
> She is the most living part of you—
> And your only reality.
> Offer yourself to her—
> And you will live in joy.[4]

It was quite natural for Philippa to assume that these verses, if not dedicated to her, had been inspired by their friendship.

There were other signs. In Catania—halfway through her Sicilian tour—she received a poem written, she believed, in her honor. In it André compared her to a beautiful red sacrificial flower from a painting by Gauguin.

Thirty years later, when I asked André about it, he thought the reference to Gauguin plausible—he was a great admirer of the painter. Yet adding with a smile: "Poets have a tendency to get carried away. I dreamed a lot then, wrote poetry about chimeric creatures, women who did not exist, about the ideal girl."[5]

Six weeks later Philippa was in Brussels again, and André took her to the World's Fair. He was wearing a gray tweed suit and a gray raincoat which made his hair seem more blond than ever. She was sporting her jaunty white beret.

They rode through the exhibition grounds in a little train, the giant "Atomium" structure towering over everything. They visited the Soviet and American pavilions, neither of which appealed to Philippa. "The American one had an attractive exterior," she wrote, "all shining glass, round airiness. Inside it was too empty. . . . The distasteful collection of modern paintings looked like they had been painted by a chimpanzee."[6]

In fact, the pavilion had been designed by Edward Durrell Stone. The entrance space was dominated by two large objects intended to be very American. One was a huge stump of petrified wood, the other an assemblage of automobile license plates from the forty-eight states. They were encircled by a large mural of America executed by Saul Steinberg.

The exhibitions that Philippa so roundly dismissed contained works by seventeen of the leading American contemporary painters and an equal number of sculptors. They included William Baziotes, Richard Diebenkorn, Grace Hartigan, Ellsworth Kelly, Robert Motherwell, and Ad Reinhardt. Among the sculptors were David Hare, Jacques Lipchitz, Theodore Roszak, David Smith, Alexander Calder, and Isamu Noguchi—hardly a herd of chimpanzees. Obviously, Philippa was not an aficionada of contemporary art.

The Soviet Pavilion fared even worse. "It was tasteless, overstocked. . . . Great machines stuck up . . . amidst trivia, like cacti in a desert, the implication being that no one else had thought of these quite ordinary machines. It was gloomy, humorless, pedantic, with an elephantine earnestness."[7] (The "quite ordinary machines" included a model of Sputnik 1, the first artificial satellite.)

She liked the Belgian Pavilion best for its magnificent display of tribal art from the Congo; and she faulted the otherwise "heavenly, stylish, chic"[8] French Pavilion for according so little space to the four million square miles of African territory that were in some way or another affiliated with metropolitan France.

André and Philippa were luncheon guests of the Sudanese cultural affairs representative, and thirty years later Gascht would recall with amusement that Phil had Coca-Cola with her gourmet fish course.

❖ ❖ ❖

André had seen Philippa in 1953, 1955, and 1956. But these occasions were brief and rather formal, and mostly in the company of others. "Then she was an international artist to me," he reflected years later, "and I was proud to associate with her."[9] Gascht went to all her concerts in Belgium and Holland and wrote three separate reviews extolling her as a great pianist. Now, she had not come for a specific concert and the two could afford to spend a great deal of time together: they went to museums, to theater; they talked about every conceivable subject; and they "flirted" (to use Gascht's word).

André was endlessly impressed with her knowledge of the world, of literature, her quick and unorthodox mind. He found her charming, but also at times quite baffling (*déroutant*). "One could not predict what she was going to say, how she was going to react. . . . One day, when we were discussing food and health she brought up out of the blue the question of intimate feminine hygiene." Seeing a look of puzzlement on my face, Gascht smiled nervously: "How shall I explain this? A woman must wash her sex organs, but, Philippa insisted it should not be done too often, because vaginal secretion is for the protection of the body."[10] Coming from a bidet-hopping culture, André had been quite taken aback.

He was thirty-seven years old at the time, but readily confessed that he "did not have the maturity of his age."[11] He had led a very protected life. His father, a military man, had been captured by the Germans in 1940, and spent three and a half years as a prisoner of war. An only child, André had always lived with his mother. He studied literature at the University of Brussels, and after graduation took a not too demanding job with the Ministry of Colonies. "My mother took care of everything, and I could dream, occupy myself with literature, write poetry, go to shows. . . . I had the reputation of a flirt, while in reality I was very timid."[12]

In fact, André did not have a sexual experience until his mid-thirties. He remembers a group trip to Portugal and Spain. Four of them—a friend, himself, and two women—were late in departing and had to make it on their own to Santiago de Compostela, to catch up with the rest. "It was a marvelous discovery for me."[13]

When his father died in 1956, his mother asked: "Are you going to get married, or are we going to continue to live together?" Since he had no marriage plans, he continued to live at home.

"My mother was an extremely possessive woman who kept me for herself. Whenever I found a girl she suspected might interest me, she placed herself immediately in the way, while outwardly maintaining the hypocritical I-wish-poor-André-would-find-a-good-woman posture."[14] At first Philippa tended to attribute the slow progress of their romance entirely to André's domineering mother. But perhaps she herself was too much for him.

Understandably, Phil also failed to recognize, or chose to dismiss, an entirely different deterrent. Paul Fabo's magazine *L'Afrique et le monde* regularly received courtesy copies of American Negro publications. Knowing that his friend André shared his interest in Philippa, Paul would pass on to him clippings about her.

The cumulative effect was both positive and negative. It boosted Phil's stature, but the more sensationalized articles—such as a 1958 *Ebony* exposé, "Why I Don't Marry," in which she shamelessly brags about her successes with men—while making André jealous, also frightened the shy, immature poet-bureaucrat.

It was Josephine, however, who may have struck the final blow to her daughter's relationship with André. In the spring of 1960, Philippa made a very brief trip to Brussels after her triumphant tour of Sicily and Italy. André seemed distant, almost angry with her. Phil was puzzled and hurt. Finally, the day before she left, André revealed that six months earlier he had received an anonymous letter with references to Philippa. He had concluded that it was from Mrs. Schuyler.

"How could you?" she raged at her mother,

> – and do it so obviously. Using the same kind of airmail envelope that you used to write to me care of Fabo. With the same typewriter? . . . André knows no one but us in America. Why did you not send the letter to someone we know in Europe and ask him or her to post it, so it would have a postmark from some other place, not New York? Why not use another machine? Another envelope? . . . Why write the letter at all? How embarrassing!
>
> While there was no evil in the letter—things are never done that way in Europe. And you should not have criticized Mr. Fabo and Albert. My God, they are his friends! Albert said the nicest things about me to André—it makes me ashamed to think that I did not say something nice about him in the book [*Adventures in Black and White*].
>
> And have you never heard about the fact that in the modern world—a courtship period precedes engagement. . . . Writing letters means nothing much nowadays. If I never have a chance to go out in a relaxed way—over a period of time with any one—of course I'll be still a spinster at 60.
>
> Now just understand this: I AM NOT A PUPPET OF WHICH YOU PULL THE STRINGS![15]

The anonymous letter has not been preserved, and Gascht, when asked, had no recollection of it. It's content is difficult to reconstruct. One can only speculate that Josephine accused André of breach of promise, and advised him not to listen to Albert Maurice, whom she may have characterized as an idle and worthless individual. Jody may have also suggested that André should stop equivocating about marrying her daughter.

Philippa's response is angry but it is in no way self-affirming: the fact that she could even suggest a better way of handling it—to mail the letter to a mutual friend in order to alter the postmark—only attests to the ferocious control Josephine still maintained over her daughter.

Jody went to great length to be conciliatory: "You said that you did not understand him, that he seemed to be neurotic. I think that what you mean is that he cannot make a decision when the time comes; that he evades and avoids taking a positive step. Isn't that his whole pattern? So why think he will ever change?"[16]

During the ensuing four months, Philippa saw André whenever she returned to Brussels. Things seemed on a better footing—at least in her mind. Then in the beginning of July they had another falling out.

By now, Philippa was desperate to marry André. In one final attempt, having just returned from Holland, she went to his apartment house early one morning. She did not ring the bell but waited for him to come down for the mail.

"I was surprised to see her," André remembered.

> We were already on bad terms. I suppose she had in mind that I should make a decision vis-à-vis marrying her. I don't think she put it exactly that way, but you

> know, I had no intention of getting married. I said a few harsh words and went upstairs. When I came down later, to go to the office, she was still there. She insisted on accompanying me, but I did not want her to. You see I had my work. I had not time to occupy myself with things of that nature. My mother taught me that there are things one must admit. So I walked ahead of her. I was walking fast. But she followed me. I remember she was wearing these high heels; she kept stumbling on the cobblestone. When we got to the Ministry we parted in rage. Perhaps it is more correct to say that I was mad. Philippa did not get mad. Hers was rather a sadness, a despair. That was the last time I ever saw her.[17]

Outside the hotel window, on Brussels's rue Royal, in December, thirty years later, a fine rain was misting the failing light. The tape recorder continued to run, but André had fallen silent. When I looked up from my notes, I saw that he was crying. "I was so cruel to her, so very cruel," he finally said. "I reproach myself that I was not more conscious at the time how devoted she was to me. . . . I was so cruel to her. . . ."[18]

❖ ❖ ❖

Emotionally devastating as the break-up with André was, the year 1960 became one of her most productive.

Philippa spent only nine weeks in New York. Determined to establish herself in Europe, she found a a new comfortable place to stay in Brussels: Mme. Beltjen's house at 55 rue Van Eyck, one block from the fashionable avenue Louise. The house was a narrow but elegant four-story building of gray sandstone with large arched windows. Probably the home of a well-to-do merchant at one time, it was now a rooming house. Phil stayed on the top floor; there was a piano on the second. The bathrooms were between floors, at the rear of the house, with their art nouveau stained-glass windows providing light for the stairwell. Georges Apedo-Amah, who in a recent shift of political fortune was now working for the European Economic Community, lived on the third floor, where Philippa went to read, iron, and listen to his excellent record collection.

In the meantime, an Italian agent had organized an extensive tour of Italy, including Sicily, for her. It began in Verona on March 7. Recitals in Bari, Brindisi, Lecce, Messina, Catania, Caltanissetta, Agrigento, and Palermo followed. She also played in Milan and debuted in Rome's Palazzo Antici Mattei at the invitation of the Italian-American Society.

Toward the end of April, Philippa flitted through Cyprus, Israel, Jordan, and Lebanon, alighting in Khartoum May 8. For the ensuing two months she trekked through Africa. Her first visit was to the island of Madagascar. From there she continued to the Sudan, Kenya, Tanganyika, Uganda, Nigeria, and Ruanda-Urundi, arriving in the Belgian Congo on June 7, exactly three weeks before the country would gain its independence.

Philippa's Italian tour was particularly rewarding, her reviews outstanding. Her big sound in pieces such as the Mussorgsky, and her lyric, singing lines, as in the Vivaldi-Bach-Murdoch Concerto in D Minor appealed to the Italianate sense of opera and drama. And they went crazy over her Gershwin, having long been aficionados of American jazz.

Quite unexpectedly, on March 12, after her sold-out performance in Brindisi, she met a young man by the name of Antonino Ciccolella. He was tall (six foot three), muscular, with curly brown hair and green eyes, and he was the son of a wine merchant and wealthy landowner.

Cultured and well read, "Tonino" went to all the classical music events in the area, and he attended Phil's concert not expecting much; he had never heard of her. But when she walked onstage with her blue-and-black lace dress, her long graceful limbs, and her hair pulled back to accent her high cheekbones, he was completely taken. At intermission he hung around the wings to get a better look; she was even more beautiful up close, he decided. After the recital Tonino again went backstage, this time to introduce himself and to tell her how much he had admired her playing.

Would she care to have dinner with him, or perhaps she preferred a walk on the beach? Tonino had fallen completely in love with her. The few days she had left in Brindisi they spent together.

Throughout her Sicilian tour, Tonino barraged her with flowers, love letters, and telegrams. He was obsessed with her. He wanted to take care of her, to "take [her] face in his hands and kiss [her] full on the lips, watching that expression in [her] lonely, lost eyes which at times seemed so terrified."[19]

Philippa had written earlier to her friend Lineke about her desire to marry a European and about her concerns. While protesting that "it's silly to presume a European man would not want to marry you," Lineke had added a precautionary note:

> There is one thing, though that makes it more difficult. . . . A man here in Europe—and I'm talking in general now—who marries, considers his bride as his possession, a woman who will always be there, who'll look after his house, prepare his meals, and is completely devoted to his well being. This is not a problem for the (European) bride since . . . the things he expects from her are exactly the things she would love most of all to do. But you are an artist, and that means you won't be home always, you won't be his private possession, since public fans, etc. would claim your attention. Do you get the idea? Artists, female artists, have much more trouble finding a husband than plain simple, not even attractive girls.[20]

But Tonino seemed different—completely devoted to the idea of helping her career, and desperate to marry. Josephine was all in favor of it, believing that this step would greatly enhance Philippa's professional success.

Especially after it became apparent that Philippa's relationship with André had come to a sad end, Josephine's manipulations went into high gear. She began corresponding with Tonino directly about a premarital agreement, brazenly insisting that certain real-estate holdings be transferred to her daughter. Although Philippa and Tonino rarely saw each other after the initial encounter, they became engaged. Their romance was largely on paper. But her mother's insistence that she marry Tonino drove Philippa to seek support from Ernie, who had now become a confidant. Ernie, who had recently met Philippa's parents in his extended stay in America, wrote back unequivocably:

> Your mother has no business trying to run your life and force you into a marriage with a man you dislike and cannot love. She should remember that when she wanted to marry your father and did, in fact, it was against the wishes of her family. . . . She must learn to accept and understand you for what you are, the person you have become, and not to see you as she wishes to see you. This is her fundamental mistake.
>
> If you need a business manager, then get one and put it on a strictly business basis. Your mother may be alright for the job, but . . . when reality hits her and she realizes how wrong she has been, she refuses to admit it to you, but rather seeks redress by taking a delight when others find fault in you. She is being pathological. . . .
>
> As a product of a mixed marriage, which has worked out, you have become a victim of circumstance and it is only on you that your mother can lacerate her frustrations, moments of unhappiness when she becomes nostalgic of her past and what else she might have been, forgetting that her union produced such a wonderful talented person like you.
>
> Your father gave me the impression of being self-centred and remote. But let him love you in his way, just as you should love him in yours. But he as well as Josephine must accept the fact that you have long grown up—they cannot chastise you because they wish to beef against the world and can't pick on someone else.[21]

In the end, Philippa and Tonino parted ways, both feeling terribly duped.

"You know," Philippa wrote her mother soon after the engagement was terminated, "sometimes I get to think about all the flowers Tonino sent me, to Africa, to New York and Europe—his eagerness to help my career, his happiness at the thought of marriage, and I feel guilty. I think that if I had only loved, I could have filled his void, and erased his loneliness."[22] It was as much a cry for herself as it was for Tonino.

28

Who Killed the Congo?

Constant travel was Philippa's anodyne. At first her globetrotting had been in pursuit of a musical career, but in the summer of 1960, just shy of her twenty-ninth birthday, she added another itinerant profession: that of a roving political journalist.

In part it was the result of being in the right place at the right time. Philippa had arrived in Elisabethville, capital of Belgian Congo's copper-rich Katanga province, on June 7, 1960, exactly three weeks before the colony would gain its uneasy independence. The atmosphere was ominous; one of Congo's most prominent political activists, Patrice Lumumba, allegedly said that come the liberation, the Congolese "could have all the white women they wanted,"[1] causing mass fears of rape. White men were shipping their wives and daughters out by car or plane, and people worried about the resurgence of primitive customs.

There were no American journalists in Katanga or in Congo's interior at that point. Two months later the place would be swarming with foreign correspondents. Robert Speller, publisher of Philippa's *Adventures in Black and White*, had recently started Transglobal, a fledgling news service in New York, and it was there that Philippa began filing her dispatches.

On June 25 she stood on the balcony of the National Palace in Léopoldville watching Lumumba proclaim Joseph Kasavubu, the idealogue of the liberation movement, president, in order to stop his tribe from seceding.

King Baudouin of Belgium arrived, looking earnest and young. Eighteen months earlier, he had made a momentous proclamation: "Belgium resolves to lead, without baneful delay, but without [haste], the Congolese population to independence in prosperity and peace."[2] But now he seemed dispirited as he listened to Lumumba: "We have known ironies, insults, and blows which we had to [endure] morning, noon, and night, because we were blacks. Who will forget that the blacks were addressed as 'tu,' certainly not as friends, but because the honorable 'vous' was reserved for the whites alone?"[3]

Five days later Philippa performed at the inauguration of Kasavubu and Lumumba. But it did not turn out the glorious occasion she and Josephine hoped for.[4]

She had waited at the hotel for three hours to be taken to the gardens of the National Palace where the festivities were to take place. Finally a journalist friend gave her a ride.

Though many good grand pianos were available in Léopoldville, the organizers of the program had forgotten to obtain one and had to drag out an old battered upright at the last minute. It now teetered precariously on boards balanced across the middle of an empty swimming pool. The only light came from the stars. Philippa asked someone to get a flashlight. He laughed and said, "I am going to do!" Then he ambled into one of the pavilions to continue drinking.

A Congolese sauntered to the microphone and announced, "Mees Fippuh, a journalist, will play!" ignoring the carefully written out biography and program. Her first impulse was to go up to the microphone and make the announcement herself, but she was afraid to slip on the wet boards and ruin her elaborate dress. So she proceeded to play, giving her best: Schubert, Chopin, Gershwin, and African music. There was little applause. The crowd, Philippa noticed, was too inebriated to care. Premier Lumumba and his followers seemed high on hashish. President Kasavubu's contingent was roaring drunk. At the end she asked the "master of ceremonies" to tell the audience who she was and what she had played.

"I don't know," he said.

"It's written on the piece of paper you are holding!"

"I can't read. . . ."

The celebration continued until early morning. After a dismal beauty contest and some laconic cha-cha-cha music, three guitarists sang songs proclaiming that though the continent was shaped like a question mark, Africa and the Congo had the answer to everything. To Philippa, nobody present appeared to have the answer to anything.

"Congo's independence will be tragic," she cabled to New York. "Was the Congo ready to govern itself—to become a democracy? . . . Probably not. It lacked two critical ingredients: an educated native elite and a significant middle class. . . . And besides, can democracy work in Africa with its tradition, millennia old, of non-individualistic civilization? . . . Is it that only three kinds of government can work in these nations: colonialism, monarchy, and dictatorship? When democracy is tried, it turns immediately to anarchy, dictatorship, or a new disguised form of colonialism."[5]

And indeed shortly after Independence Day, the Republic of Congo, as the new nation was called, began to disintegrate as ethnic and personal rivalries surfaced.

On July 1, everything seemed stagnant in the city. Independent Léopoldville had a hangover.

On July 2, there were riots in Orientale Province. On the fourth, mass rapes, attacks, and pillaging began in other parts of the country. In Léopoldville, Bayekas and Bakongos began to attack each other in a resurgence of intertribal conflicts. Philippa had tasted that tribal hatred a year earlier, when standing on a balcony in the native section of Luluabourg, she watched victorious Lulus parade the severed heads of their Baluba enemies on long bamboo poles.

By July 7, white refugees were streaming into Léopoldville, telling tales of brutal rape and humiliation.

All hell broke loose in Léo the next day. Two thousand whites fled across the Congo River to tranquil French Brazzaville, before this escape route was cut off. "One C-47 plane packed with refugees took off from Brazzaville for Belgium," Philippa reported to her agency. "One man, thirteen babies, forty-seven children and thirty-seven women were on board. There was no luggage; they had to leave everything behind."[6]

Philippa left for Ghana in the midst of the terror. In Accra, she found a sedate and orderly celebration of the country's independence from Britain. Two Americans had been invited by President Nkrumah to be featured artists at the festival, Paul Robeson and herself. But Robeson never came (possibly due to passport difficulties), and she carried on her part, performing American music.

Phil returned to Brussels the second week of July along with some ashen-faced, unwashed, haggard refugees who were pouring into its airport by the hundreds. Four to nine planes were arriving daily from Africa. Many refugees were penniless, some wounded. Many told shocking personal stories. They also told of good Congolese who had tried to protect them, but being outnumbered, had also been savagely assaulted.

An unmarried social worker from the coastal garrison town of Thysville told her of an attack by three Congolese soldiers. One could still see the bruises on her legs and the scar on her face. She had saved her underwear, worn at the time of the assault. The underpants, bra, and slip were ripped and thickly smeared with clotted blood.

The woman's brown hair and narrow gothic face made her look like someone from a fifteenth-century Flemish painting, not a refugee from equatorial Africa. In a calm, evenly modulated voice that made her revelation even more frightening, she reported that four-fifths of the women in Thysville had been raped.

The Congolese soldiers at Camp Hardy had mutinied, beaten the European officers, taken their guns away, and locked them up. The women were now helpless and the rapes began. One woman was raped twelve times in the presence of her children. Soldiers stripped one woman naked, plucked out her pubic hairs, and made her eat them. One Congolese wrapped sandpaper around his fingers and kept thrusting them brutally into her vagina till she fainted.

The refugee's Belgian compatriots were stunned. Medical personnel, nurses, and stewardesses tried to help. There was a national outpouring of sympathy. It was not exactly a good time to be a black African in Europe.

Philippa planned to return directly to the States toward the end of July, but her mother had cooked up a trip to London: a front-page interview with the London *Daily Mirror* about her Congo experience. Speller was supposed to have arranged it, but the trip turned into a fiasco. Apparently, nothing had been firmed up. In utter disgust, Philippa unleashed a tirade at her mother, beginning with a litany of sympathy-seeking complaints:

"Besides the moral and psychological strain of these events, do not forget the physical pain. Feeling awful from lack of sleep, head aching terribly, eyes burning [from a recurring eye infection]. . . . For several months I have had a pain in my left arm—a reaction from a long-standing overstrain of carrying bags. I don't know just when it started; it used to go away. Then, one day, I noticed that it did not go away anymore."[7] Crescendoing into a scream, she wrote:

> Do you realize what you are expecting of me? Are you aware of the pressures you put me under? Are you aware of the impossibilities you ask of me?
>
> To be a great pianist.
> To be a great composer.
> To be a great arranger.
> To be a great author.
> To be a great journalist.
> To always get marvellous reviews.
> To always pull off marvellous coups no one else could do.
> To get good photographs everywhere—even when it is always raining.
> To always make money, and always keep within my budget, and always pack so efficiently that I never lose anything.
> To always be a great beauty.
> This is beyond human capability.[8]

Jody replied promptly, in conciliatory tones:

> We have all acted here in good faith, and have tried to promote your name and talent with all our might, and have devoted a large amount of time to this. . . . Would it have been better just not to have tried an African tour, or a book? I don't think so. This has all made you much more important. However, I can see why you are tired, it was a great ordeal, a tremendous task, which you carried out with great triumph and ingenuity. Even though there were disasters, the outline of it remains pure and definite: you went, you played, you were there, Charlie. That in the end, is all that counts. YOU WERE PART OF HISTORY.[9]

But "in the end," Philippa may not have cared whether she had been "part of history." She was at the end of her rope.

Back in the United States she learned that her African and Belgian dispatches had attracted wider attention than she had thought. Lawrence Rutman, director of the United Press International Feature Syndicate, asked her to return at once to the Congo as a UPI special correspondent, and she accepted the challenge.

❖ ❖ ❖

Philippa arrived in Elisabethville the third week in August. A number of dramatic events had taken place during her five weeks' absence from the Congo. On July 12, their country in total disarray, Kasavubu and Lumumba had appealed to the United Nations for help. Two days later the U.N. secretary general Dag Hammarskjold was authorized to provide military and technical assistance.

Much of the initial U.N. force was hastily supplied by neighboring African countries. Escalating rapidly, it soon numbered four thousand. Its primary mission was to keep the Congolese from killing departing Belgians, to forestall civil war, and to prevent outsiders from interfering. But they also struggled mightily to maintain essential health services, man radio stations, operate airports and power plants, and keep communication channels open. Given the appalling dearth of qualified native personnel, the country would have hopelessly collapsed without this assistance.

A treaty of friendship and cooperation between Belgium and the new Congo had been worked out at a round-table conference in Brussels. It stated that the European administrative and technical personnel would remain during a transition period. But the treaty was never ratified and many Europeans had already fled.

Another crucial event had occurred on the fourteenth of July. The province of Katanga, under the leadership of its enormously popular provisional president, Moise Tshombe, proclaimed its independence from the rest of the Congo. He advocated a federation of autonomous Congolese states. Tshombe advised Hammarskjold that peace, order, and legality reigned in his province, and that he neither wished U.N. troops sent there nor had any intention of expelling Belgian troops or advisers.

Even though Katanga's setting itself up as a separate state enjoyed the sympathy of much of the Western world, which had a large stake in its phenomenal resources, Tshombe's secession caused an international furor. Lumumba was incensed and vowed to bring the province back into the fold by force. Since the U.N. would not support him in such an action, he turned to the Soviet Union, receiving first a hundred trucks and later a dozen Ilyushin air transports. Katanga had become a focal point of the cold war.

Dag Hammarskjold, strongly believing that Congo's turmoil presented a serious threat to world peace that could only be defused by Katanga's rejoining the Republic of Congo, arranged to meet Tshombe.

The meeting accomplished a tangible, though limited result. Tshombe permitted the entry of U.N. troops, provided they did not come from communist countries; Belgian combat troops would depart Katanga within three weeks; and for the time being both the Belgians and Tshombe would cooperate with the U.N. The question of Katanga's secession, however, remained unresolved.

When Phil reached Elisabethville less than a week after Hammarskjold's visit, the U.N. force was setting up headquarters there. Tshombe was living in the former governor's mansion, surrounded by European advisers and technicians with whose help Katanga's economy was still functioning. These circumstances easily created the impression of his being a puppet of the Belgians.

But he and the people in his orbit greatly appealed to Philippa. They passed the litmus test of being good Christians and active anticommunists. "To meet Tshombe is to like him," she wrote "and the common people of Katanga love him."[10]

❖ ❖ ❖

Katanga's charming capital, situated on a high plateau and enjoying a climate much more favorable than the rest of the equator-straddling country, was rapidly becoming an international dateline for a large gaggle of journalists. Most of them stayed at the Grand Hotel Leopold II, "the social, gastronomic, news, and black market center of Elisabethville," as one of them, Robert MacNeil, described the atmosphere. "United Nations press briefings were quicksands of bureaucratic nonspeak. Katanga government information was propaganda. . . . Finding action stories meant hiring small planes for the day to fly to the northern towns where fighting was reported."[11]

Philippa neither stayed at the Grand Hotel nor seemed to have difficulty finding trouble spots. She quickly learned from her previously established contacts that brutal tribal warfare was raging in the neighboring Kasai Province. In preparation for an assault on Katanga, Lumumba had been moving his troops into northern Kasai's Luluabourg region, using Soviet-provided transport. Poorly led and lacking logistical support, the soldiers had to live off the land. They immediately became entangled in a vicious tribal conflict, which the U.N. forces were unable—or perhaps unwilling—to suppress.

The families displaced from the Luluabourg area were streaming to the diamond-mining capital of Bakwanga, a hundred miles to the south. The refugees now numbered over sixty thousand, and starvation was rampant. The once beautiful town of Luluabourg was reportedly in shambles. Transportation there was unavailable, so Philippa decided to fly to Bakwanga, five hundred miles northwest of Elisabethville.

Four days later, on August 23, Lumumba ordered his soldiers to enter Bakwanga. "Carnage raged from there on," Philippa reported to her news service, before she escaped. "The dawn found hideous corpses in [Bakwanga's] streets, [some] slaughtered in ritual . . . murder."[12]

Before long, Lumumba's Bakwanga adventure backfired. It alienated Dag Hammarskjold and angered President Kasavubu, who on September 6 dismissed the prime minister. Lumumba protested violently both in the Congo and at the United Nations; but despite support from African and eastern-bloc countries, he failed in his bid for reinstatement.

Patrice Lumumba's fall from grace was complete when on December 1 he was arrested by the central government, together with nine of his close associates. They were held incommunicado and deprived of all civil rights in the military complex at Thysville, near Léopoldville. Despite strenuous efforts made by the U.N. on Lumumba's behalf, the conditions of imprisonment were extremely harsh.

His fate remained unknown until mid-February 1961, when Tshombe's government informed the world that Patrice Lumumba, having been transferred to Katanga for "safekeeping," was killed while attempting to escape. The story sent a shock wave around the world, and Kasavubu cynically accused Tshombe of murder.

Philippa was once again in the Congo, and tipped off by her missionary friends that the story was almost certainly apocryphal, Philippa spent much of the month of March trying to ferret out the true facts of the case. She reported her extraordinary findings in early April 1961 in the *Manchester (N.H.) Union Leader*. Her other American wire service, not believing her conclusions, declined to carry the story.

Philippa claimed that Katanga did not want to accept Lumumba in the first place when Kasavubu sent him there; that Lumumba had died on the plane en route to Elisabethville on January 18 from heart failure, beatings, unhealed wounds, and weeks of abuse; and that Lumumba's body was kept in deep freeze (to make accurate determination of his time of death impossible), while Tshombe and his advisers debated how best to conceal the truth.

Philippa's investigation was an exceptional journalistic feat. It was not until mid-November, seven months later, that a U.N. commission of inquiry, after hearing much conflicting evidence, published findings that essentially confirmed her story.[13]

International efforts to force Katanga to abandon its separatist stand continued unabated. A new and stronger resolution was passed in the Security Council on February 21, urging "immediate withdrawal . . . from the Congo of all . . . foreign military and paramilitary personnel and political advisors not under UN command, along with mercenaries."[14]

But instead of complying, Tshombe was now pursuing, even more vigorously, through negotiation with other Congo leaders, his idea of a federation of autonomous Congolese states. He seemed to be making headway when at the end of April he was arrested—on Kasavubu's orders—while attending a conference in Equateur Province.

Tshombe remained in jail for two months and was released only when he pledged to abandon his separatist stand. Once back in Elisabethville, he quickly repudiated his promise, claiming that his Parliament would not go along.

By midyear, a new U.N. commander, the Irishman Conor Cruise O'Brien, had arrived on the scene. He soon concluded that since peaceful and civilized methods were not succeeding, the U.N. should take Katanga over by force. Interpreting the February 21 resolution as a mandate, O'Brien's U.N. troops began at 5 a.m. on August 28, 1961, to round up the mostly European mercenary commanders of Katanga's security forces, having declared them "undesirables."

Nine days later, O'Brien asked for authority to react immediately and strongly to all incitements to disorder. This request worried Hammarskjold, and after spelling out once again in detail what he believed the authority of the U.N. force was, he flew, in a last effort to resolve the conflict peacefully, to Africa for a personal discussion with Moise Tshombe. His hope was that he might be able to convince him to meet with the central government under U.N. protection.

While Hammarskjold was flying over the Atlantic, O'Brien began an action in Elisabethville in disregard or misinterpretation of his instructions.

"At 4 a.m. on Wednesday, September 13, 1961," Philippa reported in her dispatch, "bitter fighting began in Elisabethville. Furious exchanges of machine-gun and rifle fire blazed down the main streets. Bullets whined, shattering store windows, wounding civilians. Eight hours of hand-to-hand combat ensued as the U.N. mowed down Katangan resistance. Tshombe and Munongo [his minister of the interior] fled, the latter to Northern Rhodesia."[15] By late afternoon, resistance had been broken. Martial law was proclaimed, and a cease-fire ensued.

With the aid of Sir Roy Wilensky (Rhodesia's prime minister) and after some tense delays, a meeting between Hammarskjold and Tshombe was arranged. They were to meet in Ndola, Northern Rhodesia, on September 17.

Albertina, the United Nations' white DC-6B, took off that morning from Léopoldville's airport at 4:15 local time into the sultry overcast with Hammarskjold's official party on board. Expecting to be away only overnight, Hammarskjold took but a briefcase. Including crew, there were fifteen persons on board. Having filed a fictitious flight plan to Luluabourg and observing radio silence, the craft flew due east to Lake Tanganyika, and then south, skirting the Congo border. After contacting Salisbury and obtaining clearance from Ndola, it overflew the Ndola field with its navigation lights switched on. While turning into the landing approach, however, the *Albertina*, with wheels and flaps lowered, brushed the treetops, cutting a long curved swath in the forest. After 250 yards its left wing touched the ground. The craft cartwheeled and disintegrated in a mass of flames. Hammarskjold was thrown clear of the wreckage and must have died very shortly after the crash. When found he was lying on his back near a small shrub which had escaped the fire, his face extraordinarily peaceful, a hand clutching a tuft of grass. His briefcase, found nearby, contained, in addition to a few official papers, a copy of Rainer Maria Rilke's *Duineser Elegien* and Martin Buber's *Ich und Du*, which he had been translating into Swedish.[16]

In early December, claiming that Katanga was planning an attack, the U.N. forces recommenced military activities, ending in a quasi-capitulation of Tshombe and his government.[17]

❖ ❖ ❖

Between 1957 and 1962, Philippa visited the Congo six times. Traveling by any available means, venturing into remote and strife-torn regions, staying at times with missionaries, she observed, photographed, reported for American newspapers, and gave recitals. Philippa met and interviewed the major political personages of the day and diligently gathered information from books and journals.

Not only did she become knowledgeable about that large and unhappy equatorial land, she also became deeply and emotionally involved with it, and in 1962 published a controversial book about the country's bloody struggle for independence. She called it *Who Killed the Congo?*

Governing and running the Congo, Philippa claims, had been, for decades, like chess in ancient India, a four-handed game. The players were the Belgian government, providing the administrative structure and the peacekeeping forces; the traditional ancient rulers, who claimed divine authority and to whom the local chiefs were beholden; the Catholic Church with its extensive network of missions; and the industrial complex, which included such giants as the Union Minière du Haut-Katanga and Unilever.

Each had its own agenda, of course, but they also shared a common objective: to maintain the status quo. Belgium wanted to keep its rich colony; industry wanted to perpetuate cheap labor; the native rulers jealously guarded their inherited prerogatives; and the Catholic missions, entrusted with educating the natives and providing the bulk of medical services, occupied enviable high ground from whence to spread the gospel.

The missionaries, the great majority of whom were Roman Catholic, had done an admirable job of raising basic literacy, but until the mid-1950s only a handful of blacks had the opportunity to be educated beyond the primary level. A scant 5 or 10 percent could hope to get into a middle school, or a technical or agricultural vocational institution; the remaining 90 percent languished.

The Belgian colonial government had opposed higher education for the Congolese. Only on the rarest of occasions would it allow a gifted African to study abroad. As a result, the Congo lacked, almost totally, an educated native elite. Thus at the time of the country's independence, the American press repeatedly stated that there were only seventeen college graduates among Congo's population.[18]

Even as late as 1936, a conference of Catholic bishops of the Congo expressed the view that it would be unwise "to elevate the African to an economic and social level too far above his moral development."[19] It was not until 1954 that two universities were created in the Congo.

This was in marked contrast to French and British policy. Many a French colonial could be seen at the Sorbonne or other metropolitan universities; or a British African, tribal "hashes" and all, at Oxford or the London School of Economics. It pained Phil that the attitude of the church was not in conflict with the restrictive policy of the Belgian government.

❖ ❖ ❖

Another serious handicap was Congo's ethnic fragmentation: deeply embedded enmity among the tribes, of which there were more in the Congo than in any other African land. This was aggravated by the country's having been divided for administrative purposes into provinces that often split tribes. It is difficult to say whether the Congolese tribesmen hated each other more than they despised the white man.

Rightly or wrongly, Philippa firmly believed that the ghost of Leopold II and the lingering memory of his crimes against the Congolese people was one of the roots of the tragic postliberation events in that country.

For twenty-three years, from 1885 to 1908, the Congo was King Leopold's exclusive fiefdom, and he exploited it for his personal gain, ruthlessly and hypocritically, through a series of concessionaire companies.[20] For the ensuing half century, the Congolese smoldered with resentment.

They took their revenge when independence came. They retaliated against the specters of the past, however, for the white-bearded royal criminal against human rights was long since dead, and so were his sadistic henchmen. In 1908, under the pressures of worldwide public opinion, the Belgian parliament had taken the Congo from the Belgian king. After that, the administration of the vast colony was more humane.

Tracing the taproots of the Congo debacle was only part of *Who Killed the Congo?* Much of the book is devoted to analyzing the Soviet plotting at the beginning of Congolese independence, and their subsequent subtle maneuvers; America's failure to support a separatist pro-Western Katanga; and the intervention of the United Nations. Surprising is another inclusion among her causes: the "expansionist pan-African conniving of Ghana's Nkrumah." This was in direct conflict with the position Philippa had taken so strongly against George three years earlier concerning his article in the *American Mercury*.

This accusation did not go unchallenged, and in April 1963, she received a letter from Kwame Nkrumah:

> I have been greatly distressed to read your latest book *Who Killed the Congo?* You must know that when people of African descent come to Africa from the Americas and West Indies, few of us can resist the genuine sentiment of joy and pride over the homecoming of our kith and kin, who have been torn away from the Mother Africa by the cruel accident of history to live a rigorous life in strange lands in conditions of subjugation, privation and indignities.
>
> When you came to Africa, we in Ghana were proud to have you with us as one who by intellect and talent was already a shining example of the best in our race, in spite of your comparatively tender age. . . . I am shocked to see in your book on the Congo that our confidence has been gravely misplaced. What on earth could have led you to believe that one of the greatest champions of our liberty and dignity that our continent has produced [Lumumba], was a Trotskyite, that Gizenga is a Communist, or that Ghana, which is well known for her policy of neutralism and non-alignment, is "leftists?" Have you ever found time to study the records of Ghana's international relations since our independence? What could have influenced you to take this most anti-African attitude in Africa's international relations? . . .
>
> Little did I suspect you to be such a forthright advocate of Colonialism in Africa and a friend of Tshombe who has done so much to betray the Congo and Africa to foreign finance capital and mercenary interests. . . . I have written to you in such

sincere and frank terms because I felt it to be my duty not to dissimulate my feelings in a matter that touches the very heart of Africa.[21]

The reviews of Philippa's book were mixed. Conservative applauded it, liberals disdained it.

Gwendolyn Carter, then professor of government at Smith College, gave Philippa's book a merciless panning. Writing in the *New York Herald Tribune*'s book review, she said: "This is a sensational, ill-substantiated and, indeed, highly inaccurate book. By piling one horror story on another, it panders to the emotions that have already been roused in this country over the Congo. It shows no awareness of the character of the issues involved in the Congo crisis or in the broader milieu in which it has taken place."[22]

Yet, with all its flaws, *Who Killed the Congo?* makes some significant, though partisan, contributions to the understanding of Belgian Congo's postliberation debacle. Her dire predictions, which her father also espoused, that democracy could not work in Africa and when tried turn quickly to dictatorship, have often come true.

Today, over thirty years later, Colonel Mobutu, a military man made president in 1965, is still the absolute ruler of Zaire. The country has been cited for serious violations of civil rights. Despite Zaire's rampant inflation (consumer prices are eighteen thousand times what they were only several years ago), a foreign debt that exceeds $10 billion, and a per capita GDP among the world's lowest, Mobutu is among the six richest men in the world. At last count he had amassed a personal fortune in excess of 3.5 (some say closer to 9) billion dollars—mostly stolen—causing a U.S. congressman to coin the term "cleptocracy."[23]

Nor would it be accurate to accuse Philippa of completely misjudging the New Africans. She understood that their superficial adoption of the European culture dramatically altered their traditional family life, resulting in both moral and spiritual decay. Political allegiance, once given to the chief as the representative of the ancestors, was being given to the new political leaders without the same conviction.[24] Trying to fit in with the European mold, the New Africans saw the old values as part of their regrettable savage past. But the new ones found no bedrock.

29

Jungle Saints

Philippa, nevertheless, believed that the future of Africa—in fact its salvation—lay in Christianity. Over the years she had become a deeply religious person, a turn that could not have been predicted.

The Cogdells had been far from a God-fearing family. They attended church only on "social" occasions—to christen a child or to celebrate a marriage. Josephine strayed even further from the house of God in her California days: she declared herself an agnostic, eschewing all organized religion or established orthodoxy. By the time she met George, however, her attitude had softened: she began calling herself a "free thinker."

But Josephine was not without "faith." She was devoted, if not addicted, to the psychic and supernatural. She firmly believed that some obscure power determined and controlled man's destiny; and that ways and means could be found to augur fate. She believed in signs and portents, studied numerology and dream interpretation, consulted the horoscope, and frequented seers all her life.

Philippa grew up to share her mother's fascination with the supernatural and even her belief in reincarnation. In fact she contended that her former lives explained her prodigious IQ, her ability to learn languages so quickly, her photographic memory, and even her musical gifts. Her faith was a mixture of Catholicism, mysticism, and fatalism. Neither her interest in the occult nor her belief in reincarnation struck Philippa as antithetical to Christianity. Christ, in her mind, had not simply been resurrected, he had been reincarnated.

George's childhood had been very different. The family believed in God; the Bible was read in their Syracuse home and grace said before the evening meal. Nonetheless, in years to come, George would claim that the family felt lukewarm toward organized religion, and that the Schuylers in fact "gagged at Churchianity."[1] Later he would slam the Negro church, calling it a "combination of vaudeville, theatre and gymnasium."[2]

Like his wife, George lost his Christian faith early, but unlike Josephine he never found an alternative spirituality. To the end he was a secular humanist—a self-defined "scientific man."

Given the background of her parents, it seemed unlikely that Philippa would become a deeply religious person, in fact a devout Catholic.

Yet the seeds for that had been sown early. As a child, she was attracted to Catholicism from her first days at Manhattanville College. The acceptance she felt in the halls of Sacred Heart was comforting, the closeness with nuns, such as Mothers Morgan, Stevens, and Saul, nurturing. The ritual of the weekly mass filled her with a sense of wonderment.

From the outset, Philippa took her religious studies seriously. After a year and a half at the Convent School, she began to insist that her mother buy her only long-sleeved dresses; she refused to wear anything else. At about the same time she demanded that the nude painting of Josephine hanging over her piano (a John Garth original) be removed. The parents were surprised; but, for once, her mother did not object.

As she matured, Philippa's belief in God deepened. He was there. She could feel him. She desperately needed to sense his love. A curious parallel between her secular and her religious life began to emerge. She constantly yearned for God's approval, just as she sought Josephine's. Here was an extension of her relationship to her mother as well as a powerful antidote to it. And the paternity of the church, the eternal fatherhood of God, filled a deep void in her life: the Catholic Church was the father that George Schuyler could not always be.

Philippa formally converted to Catholicism in the fall of 1958. Earlier that year, at the remote St. Thomas Aquinas Seminary in Uganda, where Africans studied for the priesthood, she had by sheer coincidence encountered Monsignor Cornelius Drew, pastor of the St. Charles Borromeo Church in Harlem. They were only casually acquainted in America, but now she confided in him her deeply felt desire to become a Catholic, and they spent many hours studying and discussing the catechism. Upon her return to the United States she received further instruction from Mother Saul, before being baptized by Monsignor Drew in November.

Philippa did not arrive at her decision lightly: "I was in my twenties when I saw the Light after patient research, a long period of study, of self-doubt and painful uncertainty," she once wrote. "The Light is so beautiful for one plunged in darkness! Perhaps only a few who are born Catholic are capable of appreciating the faith as do adult converts."[3]

Philippa worked hard at being a Catholic. She would sit, for hours at a time, in front of her piano, reading aloud passages from the Bible. It was her way of washing away the sins of a former life and clearing a path to the future one. She raised money for Catholic causes and performed at no fee for missions and churches—to the point that her mother, on more than one occasion, accused the church of exploiting her.

But her faith was unequivocal and unquestioning. When Ernie Pereira quoted, in one of his letters to Phil, Ivan from *The Brothers Karamazov*—"What if I have been believing all my life and when I die, I find nothing but burdocks growing in my grave?"—he elicited no response. But when he contended that "Christ had spoken too long to simpletons. Why had He never appeared before an intellectual?" Philippa's reply was quick.

"The intellect," she wrote, "must be the instrument of the soul, to express the soul's dilemmas and aspirations. Otherwise it is merely vain, empty, frivolous . . . a source of irritation to itself and others. The idea of the intellect as godless, faithless is passé. . . . Now most great intellectuals have religious faith. Many are and have been Catholic—such as G. K. Chesterton, Hilaire Belloc, or Graham Greene. I am a Catholic intellectual—and I intend to devote most of my writing from now on to spreading the glory of the Church."[4]

Religion was an avenue that led away from her seclusion toward a public life more fully involved with people. "When I became converted to Catholicism," she told a radio interviewer, "I got the feeling that I must do something with my life to relate it to people and humanity . . . sitting at the piano was no longer enough."[5]

While traveling Africa—performing, reporting to news media, gathering knowledge of the continent's music—her most important collateral activity was getting acquainted with the work of Christian missions. From 1958 to 1963, she visited over 150 of them, having decided to write a book about their work, almost as soon as she had finished *Who Killed the Congo?* She would ultimately call it *Jungle Saints.*

For the most part *Jungle Saints* is a personal account of the good works of the priests and nuns in Africa, of their hardships and heroism. Everywhere, Philippa found near-saintly, dedicated individuals who lived their lives in service to the African people, as well as to God. Why did they come? Why did they stay? What were their lives like?

But like *Who Killed the Congo?*, *Jungle Saints* is part travelogue, part journal of Philippa's adventures, part exposé of her own political views. Any book that attempted to cover so much ground in a scant two hundred pages would surely suffer for it. *Jungle Saints* does. It is uneven and cannot hold its parts together. However, considering that it tries to do the impossible, it succeeds surprisingly well in many places.

Her descriptions are visually arresting. One is treated to a succession of landscapes and faces—the desiccating heat of the sun, the moldering dampness of the underbrush, the rhythms and silences of the land.

But she also speaks of man's invasion of nature's sovereignty. Visiting Kolwezi in northwestern Katanga and observing in fascination an open-pit copper mine, she writes:

> Picture a great dusty terrain, hacked into an endless downward series of scarlet tiers . . . steps leading down into hell. Like giant insects . . . [or] microbes in a wound . . . tireless soulless machines move back and forth in this ghostly depression: scooping, carting, and conveying the malachite-bearing ore.
>
> A vivid kaleidoscope of color [splashes] over this giant [gash] in the earth. . . . Henbane green is there, and mysterious grey, smokey gold and glittering brown, burning yellow and muddy turquoise, bloody orange and pitiless black. I

> never knew the earth's body held such a dazzling palette of hues. . . . And when the setting sun touches the savage wound with a delicate, roseate glow, there is a moment of pure poetry that is incalculable and deathless.[6]

The book begins with a graphic account of the massacre at the Kongolo mission in central Katanga on New Year's Day 1962. It is told in the words of young Father Darmont, one of the only three priests to survive, whom she met by chance.

Father Darmont had remained in the large mission after independence in 1960, although many white and black missionaries had left. By the end of 1961 Katangan troops had also pulled out. Kongolo was caught in a no-man's-land between U.N. troops and warring political factions. On New Year's Eve day, Gizengist soldiers from nearby Stanleyville entered shouting "Lumumba is dead!" and "Exterminate religion!"

They gathered all the missionaries, ostensibly to verify that they were not "mercenaries," but then arrested all, sequestering them into separate cells for the night. On Monday, January 1, 1962, at 8 a.m., they were brought out to be interrogated. Red circles were drawn around selected names. Father Darmont was among those designated to be killed.

Then a strange event occurred. One of the soldiers, apparently bothered by his conscience, and wanting to save one priest, grabbed Father Darmont and led him back to his cell, proclaiming to the others that Darmont knew where the rifles and powder were hidden in the mission.

The massacre had a brutality almost unmatched since the early Christian era. Three times Father Darmont was brought out to be killed and three times he was saved by the same soldier, using the same pretext.

When Philippa met Father Darmont he was collecting books and supplies to reopen the mission.

But even if the life of a missionary was rarely that dangerous, there was no glory attached to it either. It was a life of frugality and service. The responsibilities were heavy and endless. It involved ceaseless travel over abominable roads. In the rainy season the Land Rovers might get stuck in the mud for hours or days. The priest might have to hack himself out of the jungle with a machete, or fell trees to rebuild a washed-out bridge. And there was always danger from leopards, serpents, insects, and infested water.

Jungle Saints is full of statistics about the schools and dispensaries that the missionaries had built, about the rise in literacy they effected, about their brave efforts to control disease. But Philippa is more in her element when she speaks of specific projects and accomplishments.

She writes admiringly of the large-scale works of Francis Mazzieri, who was bishop of Ndola in Zimbabwe for thirty-two years; or of black American Maryknoll Father Rogers, who directed the education of 160,000 children in Tanganyika. She also speaks of heroic efforts on a small scale: Sister Bianca, a young Franciscan nun, running a leprosarium at St. Theresa's Mission, treating forty lepers by herself; of the nuns at the Edelvale Shelter and Starehe Youth

Center in the most wretched sections of Nairobi, teaching useful skills to destitute girls and boys, so that they might have a chance to climb out of the maelstrom of oppressive poverty.

❖ ❖ ❖

While the Christian missions specialized in providing medical and educational services as an effective means of opening the way to spiritual ministry, Islam—the other monotheistic religion proselytizing in Africa—did not. Nonetheless, the followers of Allah had gained a stronghold in this part of the world.

This puzzled Philippa. She was correct in believing that the conqueror's sword had spread the crescent and halted the cross. But it was an Egyptian Copt—casually encountered by Philippa in Cairo, and no friend of Islam—who pointed out to her the real differences:

> Islam offers far less conflict than does Christianity with the African modes of thought. . . . Christianity is an obstacle to such key facets of African life as polygamy, the bride-price, the low position of women . . . [even] female circumcision.
>
> The primitive Animist is full of fears—fears inculcated by the mysterious malice and difficulty of nature around him, fear of the unknown, fears induced by terrifying pagan ceremonies. . . . Even a stone, a root, or a tree may be an evil spirit that must be propitiated. Islam soothes [his] fears. Christianity brings new [ones].[7]

Others have also pointed out that Africans are often confused by Christian spiritual standards, questioning whether Christian teachings add much to their moral precepts. After all, the injunction not to kill, not to steal, not to sleep with another man's wife were already an integral part of their credo. As for bearing false witness against one's neighbor or coveting his property, these were among the greatest taboos of tribal law.

Philippa had a great deal of respect for tribal societies—the solidarity of the clan members, which had enabled them to maintain a stream of human life, to defend themselves from a malevolent environment for centuries. She disdained missionaries who wanted nothing to do with "savage customs" and would not defile themselves by even learning about them, and she greatly admired two Belgian priests whose flock were the mine workers of the Union Minière in Kolwezi: Father Placide Temples, and Father Ceuteurick Paschalis, better known as Père Pascal.

Both believed that the missionary, in order to fulfill his vocation, must discover the "personality" of those addressed, to be a caring father rather than a civilizer.

Father Temples, a tall man with a flowing white beard, had studied the beliefs and customs of the Bantu. In 1945, he shook up the narrow and petty by publishing a controversial book entitled *Bantu Philosophy*. Bantu thought, he insisted, had a philosophical system, not limited to the mere physical. The Bantu had a profound aversion to all evil and believed that they had an inalienable right to justice. They wanted a "full intense life, fecundity both spiritual and physical; the union of beings on every plane." This he summed up as the essence of their *force vital.* The thought forms of the Bantu, he believed, were reminiscent of those in the Old Testament: the priest could readily assume the role of the biblical wise man.

But despite her sympathy for tribal customs, Philippa turned a deaf ear to those voices which claimed that Christian practices denied a man his rites of passage into adulthood, made him inaccessible to his ancestors, and disrupted the tribal sense of family.

IV

Flight-from-Self

30

Felipa Monterro y Schuyler

The cornerstone of Josephine's belief that her brilliant interracial child would be able to overcome the barriers of American segregation was proving to be a dismal failure. Although the civil rights activity in the mid-1950s had been dramatic—from the Supreme Court decisions on voting and desegregation in 1954 and the emergence of Martin Luther King, Jr., to the passage of the Civil Rights Act of 1957—for Philippa time had stood still. Her star as a "Negro virtuosa" kept rising and she continued to be a role model for young African-American women, but the doors to the white-dominated classical music arena remained firmly closed in America, despite her successes abroad. Years earlier, when impresario Sol Hurok had heard another brilliant black pianist, Hazel Harrison, he is reported to have said: "You are one of the greatest. You play like the masters. . . . But the [American] public is not ready to accept a Negro pianist. You are ahead of your time."[1]

If confirmation was ever needed that the racial climate had not improved, it was provided in June 1962 when Philippa took a trip to Houston to play at Texas Southern University. The concert had been arranged by Kathleen Houston, a forward-thinking professional woman, and an accomplished print- and photojournalist. She was a Cogdell, by marriage. It would be Philippa's second personal contact with members of her mother's clan, the first having been the nationwide tour in 1949.

Ms. Houston not only set up the concert but arranged for Philippa to stay with Josephine's sister-in-law Susie May (Buster's ex-wife)—all firsts for the "nigger from up north."

As in 1949, Philippa hoped some of the Cogdells would come hear her perform. But there were only ten whites in the audience, including three white music critics. One of the remaining Cogdells from Jody's generation had called Kathleen as soon as he heard about the concert and yelled, "I'm gonna leave town if that nigger comes here."[2] And he did. (Interestingly, one of the white critics, Hubert Roussel, writing for the *Houston Post*, described Philippa as "one of the most uncannily beautiful players of this instrument it has been this reviewer's pleasure to encounter in 40 years of attendance at concerts," and enamored of her Beethoven *Appassionata*, he added, "I have not known its equal in our music establishments from another player of this artist's generation."[3])

On her first night after the concert, Philippa and Kathleen went to an "elite restaurant in Houston. Everyone stared as we walked in," Kathleen reminisced. "The black waiter, however, recognized her immediately. He came up to Philippa right away with the menu and said, 'Please will you autograph it for my children?'"[4]

The following afternoon, Susie May, Kathleen, and Philippa visited a second cousin by marriage, Mrs. Foley, who lived in a rigidly segregated area. Susie May had been concerned about taking her niece there. Kathleen was unruffled. "Everyone thinks I'm nuts anyway so I do as I please," she had written Jody before Philippa's arrival.[5]

Mrs. Foley's house was elegantly furnished and quite large.

> She employed an elderly "nigra" woman as a maid who looked just like Hattie McDaniel in *Gone with the Wind.* When we arrived, the maid was deep in the kitchen. So Mrs. Foley went back to inform her that some company had arrived in need of cookies and drink. This maid got to the door between the living room and the dining room. She was carrying a big tray in her hands and it almost went to the floor when she looked at us. She was completely thunderstruck. "What's a nigger doing here being entertained by Mrs. Foley," she almost said. She set the tray down and literally ran back to the kitchen.[6]

Kathleen Houston was ninety-six when I interviewed her. Her mind was sharp and her memory undimmed. When I told her that Philippa had successfully "passed" in Europe and Africa she was rather puzzled. "Oh, but here, honey," she said, "Philippa was a nigger, she was a nigger all right."[7]

But Ms. Houston understood clearly that Philippa was being held captive by an anger masquerading as grief: "That plaintive cry Philippa repeated over and over keeps coming back to me: 'Why did my mother's family treat me so badly? Why did my mother's family treat me so badly?'" After a moment's hesitation Kathleen added, "To me she was a pitiful sight."[8]

Nor was Ms. Houston oblivious to Josephine's dilemma: "I believe at first Josephine was proud of her daughter, but I think later she got scared of her. In a way I felt sorry for Josephine because she was so ruled by Philippa."[9]

❖ ❖ ❖

The idea of shedding her African-American ballast, or even perhaps "passing" into the white world, began to cross Philippa's mind as early as 1959. A significant number of people in Europe had told her by then that she did not look Negro at all but rather southern European, Levantine, Indian, or Oriental. In sub-Saharan Africa she was generally perceived as a "European artist."

It was on her globe-circling trip, the end of January 1959, in remote Ruanda-

Urundi, a small country (now two) cradled in the mountains of east-central Africa just below the equator, that Philippa tried an experiment.

Perhaps because she would be completely unknown to anyone there, she wrote "white" under "race" on her entry form, "just to see if I could get away with it."[10]

She had made her mind up to stay in the best hotel in the administrative capital of Usumburu; it was for whites only. "When I got to the hotel," she wrote her mother, "they said there were no more rooms. I thought they were making it up—that it was really prejudice, and decided I'd make a fight. I told them about the important people I knew and brought out some clippings. Then they found a room!"[11] Rather, they had moved a bed into one of the conference rooms for her.

And it was also by putting "white" on her entry form, as well as taking the precaution of obtaining the visa in Rome rather than New York, that she traveled briefly to apartheid South Africa, where she quickly became the darling of a white society of music lovers.

❖ ❖ ❖

To escape one's self has been the fervent desire of many. Paul Klee perhaps symbolized it best when, having quit the Bauhaus in 1931, he produced a pen-and-colored-ink-drawing that he called "Flucht von Sich." It shows a line figure in full flight, its legs and arms flaying. Above it and slightly off-center is a congruent figure, somewhat larger in scale, decidedly thinner in line, and paler.

Inevitably, Philippa's "flight-from-self" required the creation of a new persona, and in the early 1960s, mother and daughter began this most difficult task.

Philippa applied for a passport under a new name: Felipa Monterro y Schuyler, suggesting an Iberian-American heritage. In Latin American and Spanish cultures the patronymic—in this case, Schuyler—is regularly dropped, making Philippa's official surname Monterro. She had selected the name after trying many alternatives and consulting her cards.

Miss Monterro would preserve all of Philippa's talents, but drop the ballast of her former self. Miss Monterro would be an independent writer without her father's extreme political views impeding her way; she would never be identified as "the daughter of the Negro journalist, George Schuyler." Miss Monterro would be a pianist and composer but without critics comparing her to an erstwhile child prodigy, wondering if she had indeed matured as an artist. Miss Monterro's life would not be burdened by the baggage of segregation.

But the crowning objective of the Monterro gambit was to break into white America as a classical pianist. Both Josephine and Philippa hoped that if Felipa Monterro could establish a solid reputation in Europe, she could reenter the American concert scene, as a white, and perform for audiences so far denied Philippa Schuyler.[12]

A corollary to the scheme involved Felipa becoming a citizen of a European country. At this point Portugal was her first choice, even though Roger Nonkel had suggested that Spain or Greece would be preferable.

The passport arrived in January 1963. The small identity photograph shows "Felipa" wearing a wig which frames her face in soft, smooth curls. She is smiling widely.

Philippa's "flight" would be a complex process. Much of 1962 appears as preparation for this escape. Even minutiae were carefully altered in order to obfuscate the past. Now when she completed forms which asked for her mother's maiden name, Philippa put Duke, not Cogdell. Her birthdate was changed from 1931 to 1933 (soon to be 1934, especially on legal documents as it was easier to forge a "4" from a "1"). She no longer mentioned her citation in *Who's Who in Colored America*, only in *Who's Who in America* and *Who's Who of American Women*, and she insisted that all her journalistic bylines omit any reference to George.

"Remember," she wrote Jody, "NOTHING in my Congo book must refer to any racial ancestry connected with me. NO photos either."[13] And when *Who Killed the Congo?* was finally published, and the book jacket described her as "Madagascan" (once an attractive alternative) she was equally furious.

A little later she wrote to Josephine again: "NOWHERE in my forthcoming book [*Jungle Saints*] do I want the word Negro or colored mentioned in connection with me, NOWHERE.[14]

Almost every letter home now commented on her negritude. When a cover article entitled "My Black and White World" appeared in *Sepia*, in June 1962, with the subtitle "World's Greatest Negro Pianist," Philippa was beside herself. (The article, although written by Philippa was heavily edited by Jody. Originally, Phil had not wanted to write anything for *Sepia* at all, but always in need of money, she accepted after a substantial fee was offered.)

And when her father included a profile of his daughter in his latest manuscript, *The Negro in America*, she exploded. With obvious pride, George had written a five-page essay on her, one of his longer entries. In a fury, typically, she wrote Jody not George:

> I was shocked by your news that George put me in that book despite my express instructions to the contrary. Apparently he wishes to handicap me. . . . Get me OUT of that book. Everyone here [Europe] thinks of me as a Latin, and that's the way I want it. Anyone who had any paternal sentiments would want a child to escape suffering. I look like any other of the Sicilians, Greeks, Spaniards or Portuguese here in Rome. I am not a Negro, and won't stand for being called one in a book that will circulate in countries where that taint has not been applied to me. It makes all future effort on my part to forge a worthwhile niche for myself in society where I will be accepted as a person not as a strange curiosity useless. I had 30 miserable years in the USA because of having the taint of being a "strange curiosity" applied to me, and I sure don't want to bring that taint along with me to a foreign country and thus have 30 more miserable years.[15]

Among the Schuyler papers at the Schomburg Center is George's voluminous manuscript of *The Negro in America*, which runs to almost fifteen hundred pages. Rather poignantly, the five-page biography on his daughter is pasted over with blank paper; each has not one but three layers over the original. (Philippa might have rested easier if she had known that although Avon had contracted to publish the book, it never did.)

Phil did not keep her plans for a new identity a total secret. No one in America knew, but some friends in Europe did, and their reaction depended on their understanding of the American racial climate. By the end of 1962 Tonino Ciccolella, Madame Beltjens, Roger Nonkel, and Lineke Snijders van Eyk had been apprised. Tonino said he liked the sound of the name; Lineke was not sure what to make of it. "After all the publicity and work to get yourself known as Philippa Schuyler? You'll have to explain. This does not seem sensible to me. Or perhaps your heart is involved?"[16] Madame Beltjens, whose address on rue Van Eyck would become Miss Monterro's official abode, was less baffled and wrote a sweet though somewhat superficial note: "I understand that the racial issue shames you very much. But rest assured, dear Philippa, in no way do you look like a black woman. Quite the contrary, you have the appearance of a Spaniard, even of an Andalousian; and many others think the same way as I. You have every right to add the name Monterro to your own, which should alter your life significantly. May it contribute immensely to your success, and to the desires of your unhappy heart."[17]

But it was Roger Nonkel who questioned the wisdom of adopting an alternate identity, although he understood her urge to be a European: "I think it would be advantageous for you to be a citizen of another country ONLY IF you really feel that, because of the colourbar in the USA you are barred from the success to which you are normally entitled. [But do] you *really think* you can come back later to the USA as a foreign artist with status of a non-coloured? Are you not too well-known, so that they will remember you later and quickly find out about your origin . . . ?"[18]

The entire experiment was a calculated risk, the future a matter of keeping the past at bay. It was also, Philippa thought, her only solution: "For remember one thing," she wrote Jody in a moment of utter nakedness, "it doesn't *hurt* when I'm rejected as Monterro. It hurts being rejected as the other."[19]

For years, Philippa had stumbled around in a landscape whose signposts all showed retreats rather than ways to selfhood. Her problem, like that of Nella Larsen's protagonist Helga Crane in *Quicksand* (1929), went deeper than (lack of) racial identity: at bottom was her inability to be a whole person, regardless of race, and this gambit would become not only a flight from self but an attempt to escape into wholeness.

❖ ❖ ❖

Josephine now opened up another front.

Using Madame Beltjens's Brussels address, she wrote to the John Birch Society about a "young woman of her acquaintance," one Felipa Monterro. Could they use her on a lecture circuit? An archconservative, she could talk on a variety of subjects, including "Red Terror in Angola and Central Africa," "Portuguese Africa and the United Nations," and "Christian Missions in Africa."

Felipa's biography, created for the *American Opinion* (the voice of the John Birch Society) was cut from whole cloth. Although Mademoiselle Monterro lived in Brussels, it began, her English was fluent. She had been tutored during her younger years, schooled in Europe, and had lived in Africa for nearly a decade.[20]

As an "introductory," Felipa's first article appeared in the *American Opinion*.[21] Entitled "Terror in Angola," it was a shameless defense of the Portuguese in Africa. By then negotiations with the John Birch Society to use Miss Monterro as a speaker had stalled. The delay was regarding Felipa's spoken English, and a sample voice-tape had been requested.

Fearing that her daughter might lose all opportunities for making money from a lecture series, Josephine submitted Philippa's name as an alternative—an obvious shoo-in, as George was a member in good standing of the John Birch Society. Jody was on the verge of mailing an acceptance (on behalf of Philippa Schuyler) when an express letter from her daughter arrived:

> DON'T contact the American Opinion Speakers! They have just written FELIPA MONTERRO, care of Mme. Beltjens, who sent it on to me, that they want *her* on a tour this summer. They're just as eager to have FELIPA MONTERRO, as they were to have Philippa Schuyler. AND IT'S MUCH BETTER THIS WAY.
>
> Now for the question of whether I will be recognized. I have run into half a dozen American Negroes here who never recognize me. One actually explained the race problem in America to me. I think that a lecture advertised as Felipa Monterro is not going to cause any Negroes to make the effort to come to it. . . . Also, the public . . . will be completely different. . . . As you know, my concerts have always been, in effect, segregated, almost always for no one but Negroes—whites in America hardly know me.
>
> Also, I could ask them to restrict the lectures to New England and upper New York state, and of course I would not speak in Manchester [NH], or Syracuse, or New York City. After all, I almost never have had concerts in New England. . . . Of course I will wear my brown hair, or even get a new hair piece like one Mrs. Nonkel has and tried on me, that looked just beautiful. . . . And IF, IF, IF someone should recognize me, I could always say that Miss Monterro developed laryngitis, or leprosy or bubonic plague or something and couldn't come at the last minute and asked me to fill in for her—or I can just be as self-confident as Marion [Preminger] and not admit a thing![22]

The second half of their juggling act was to secure Miss Monterro a debut in Europe, initially in a country where Schuyler herself had never performed. They selected the small but impressive Kammermusiksaal in Zurich. The date would be April 25, 1963. The "biography," sent to a European agent, stated simply that Felipa Monterro came from North America, had recently performed in Central and South America, Spain, and Portugal, and had authored several books on Portugal and Latin America.

This biography was obviously quite different from the one sent to the John Birch Society, but mother and daughter hoped that in time they might meld into one person whose dossier as writer, musician, and lay missionary would be broad enough to encompass all of Philippa's interests and accomplishments.

❖ ❖ ❖

In the middle of March 1963, Philippa landed in Lisbon. Five days later, she opened a bank account under the name of Felipa Monterro y Schuyler. Two checks were immediately deposited—one from Italy and the other from Rhodesia—and some American money, bringing the opening balance to two hundred dollars.[23]

Felipa had no concerts scheduled in Lisbon, but through various connections she arranged two unofficial recitals for the director of the Music Conservatory; was interviewed on national TV; and met with the Portuguese foreign minister and others in the government. She also performed in the ancient university town of Coimbra. Toward the end of her stay, the foreign minister promised to engage Felipa as a soloist in Lisbon's Gulbenkian Music Festival the following year. Miss Monterro seemed to be holding her own.

She left Lisbon on April 21. Phil, however, was worried about her debut in Switzerland. Her tarot had been wary of the Mussorgsky (*Pictures at an Exhibition*), the Turina (*Tema y Variaciones*), and the Infante (*Variaciones Sevillanas*). But her mother had the final word and Philippa's program in Zurich included all three plus five smaller works by Casanovas, de Falla, Montenegro, da Fonseca, and Ravel. Except for the Russian war-horse, most of these pieces would have been virtually unknown to her audience.

The turnout for her Klavier-Abend, on April 25, was small but enthusiastic; Philippa felt she played well. The reviews, however, were mixed.

TAT wrote that in the Mussorgsky one could appreciate "the brilliant playing of Felipa Monterro, her natural presentation, her musicality. . . . The audience, captivated by her originality, became more and more enthusiastic."[24]

Tages-Anzeiger described her virtuosity as "remarkable," her temperament as "fiery." But the reviewer felt she took too many dynamic liberties in the Mussorgsky, giving the impression of an "exaggerated excitement. . . . Strength not poetry and delicacy was more her domain."[25]

Neue Züricher Nachrichten wrote that she possessed a "stupendous technique in which chords struck from an impressive height as well as running passages produced a brilliant effect. Other pieces also produced pianistic fireworks." The Mussorgsky, however, was not to the critic's liking. "This magnificently built work cannot be created only from a richly massive *fortissimo* and a *piano* as thin as breath."[26]

The most damning review came from the *Neue Züricher Zeitung*. The critic praised her technique but felt her "playing was not up to the exigencies of a concert career." Furthermore, the "harshness of her sound went far beyond the acceptable limits of artistry."[27] All adding up to the curious vagaries of the music-criticism metier.

To Josephine, Philippa sent some one-line translations, selecting such words as "extraordinary virtuosity," "fireworks," and "stupendous technique," adding in an out-and-out untruth that the reviews were "90% good." Going on the offensive, she then wrote:

> One said I should pattern my interpretation of the Moussorgsky on Horowitz's. I don't like his interpretation of it. I think it's mannered. . . . But you see, that is one of those pieces they reserve for the masters already accepted to play. I am much safer to play unfamiliar music in the big key metropolises. The familiar music can be played in the provinces where people want to like you not to destroy you. The critics in big places want to attack you. If the program is unfamiliar, they are somewhat restrained by this as they do not wish to appear ignorant. And if they want to be critical, they will criticize THE PIECES, NOT THE PERFORMER. . . . Of course I played the Moussorgsky wonderfully in Zurich—you should have heard the bravos and the six curtain calls I got from the audience.[28]

Uneven reviews notwithstanding, the Swiss experiment had been a success: It established beyond doubt that Philippa could assume the identity of her alter ego. Her African-American heritage had never been mentioned; all the reviewers referred to her as an "Iberian-American."

The Monterro persona continued to expand. "Felipa" moved to Rome. She gave over a dozen recitals in Italy that spring and summer;[29] published several articles in a small Catholic journal (in Italian); and was presented to Pope Paul VI—all as Monterro.

Occasionally, although not often, Philippa had her own problems trying to keep her two identities separate. In one unintentionally comic letter she asked Jody to send some money. But when her mother wired back asking which name she should use, Phil could not remember how she had registered at the hotel.

Top priority now was getting her Catholic mission book published. "I will need to have a book in hand if I'm going to *sell myself* as Felipa Monterro."[30]

She and Josephine crossed swords over this, Jody suggesting that the book be edited or introduced by "F. Monterro" but not authored by her. Phil disagreed.

> *NO* the book will *not* be by Philippa Schuyler—that dreadful name with which I certainly am not going to burden myself more than necessary. All my friends in Europe agree that Felipa Monterro is a better name. It's feminine, romantic . . . easy to remember. Nobody remembers Schuyler. I should have changed it 10 years ago. If even music critics say I am obviously an Iberian I must be! Not *one* person takes me for a (Negro!) and nobody in my whole life ever has unless it was written up in my publicity."[31]

Despite her vehemence, Philippa was concerned about publishing *Jungle Saints* as Monterro: What if her alter ego failed? She would have a book far more important to her than either *Adventures in Black and White* or *Who Killed the Congo?* under someone else's name. But for now, there were no publishers interested in the manuscript, anyway. Philippa had approached a number of Italian houses specializing in Catholic literature, but they had turned her down. Nor had any of her clerical friends made any real effort to help.

Finally, on August 3, she was able to give Josephine some good news: "God! What a struggle to get this Mission book published! I NEVER struggled so hard, so determinedly in my life for anything. Yet this book has more *meaning* than either of my others. It has more quality I think. That's why it was so hard to get it published. . . . And to think that it was ALBERTO MORAVIA [whom Phil met that spring in Rome's Cine Città], the King of Dolce Vita, the wicked man whose books are banned by the Church, that got me a publisher for my Mission Book."[32]

The publisher was Herder, an internationally respected 175-year old house, whose parent company was in Freiburg. One of its principals told Philippa that he would arrange a concert there. "This could be my entree into Germany!" she wrote Josephine.[33]

But after all the arguments the book was finally published under Philippa Schuyler, with an introduction by F. Monterro.[34] In a letter to Josephine she confessed that she used Schuyler "because it sounds German and they are Germans."[35]

It was shortly after coming to Rome that Philippa made the acquaintance of an Italian family, the Cosis, who became a source of comfort and support for her. Phil had met Dr. Renzo Cosi, director of the Commercial Bank in Rome, when opening an account at his bank. He was a great music lover, and he invited her to their home.

Dr. Cosi had described Philippa to his wife as a "very attractive and lively young woman,"[36] but when Signora Cosi first saw her—after Felipa had given a recital—she was stunned: "There she was standing in the doorway to our apartment wearing a long golden cloak down to her ankles, her hair pulled back in a tight chignon. She was like a vision, an incredible vision. There was something so surreal about her, I forgot to invite her into our living room."[37]

Hors d'oeuvres had been prepared. Philippa devoured most of them, and it wasn't until much later that evening that Mrs. Cosi understood the strange eating habits of her "surreal American": this had been Philippa's entire meal. "From then on, whenever she visited, I made sure to have plenty of cheese and crackers around. But I did not stock that brewer's yeast which she had all the time. I couldn't have stomached that!"[38]

A close friendship developed between Eschilia Cosi and Philippa. During her long stays in Rome, Phil would go to the Cosis' elegant and spacious apartment almost every day to practice on their vintage grand. In fact, to this day a small depression in the marble floor can be seen under the damper pedal—a hole worn by the high heels Philippa always wore while practicing.

Philippa ate many of her meals there, conversed endlessly with the Cosis, and often went out with Eschilia and her twin daughters (about ten or eleven then) who "adored Philippa. . . . But she did not display much affection around my children," Signora Cosi remembered.

> She never gave them any gifts. I think that was because she did not have much money. In a way she was very egocentric. She would play with people, she would

> even love them—the way she did me, my husband, and my children—but only if she could get something out of them. . . . Everything about her was out of the ordinary, including her personal relationships. People had to adapt themselves to her, not the other way around. Like all geniuses, she felt that she was owed something. . . . But we all adored her. . . . Although she was strong, she was also very afraid, very fragile. When things became too much for her, when she felt threatened by people, she would bring out her rosary almost like a defense mechanism. She was trying to protect herself against her past. We used to have a joke in the family. When she would come back to see us after a day out, my twins would run up to her and say, "Did you pull out your rosary today?"
>
> Sometimes she would get very shy and withdrawn around us, myself and the twins. She would clam up. And then we would have to solicit affection from her. I knew she was thinking about herself, about her problems, especially about her race and the problems between her parents which I never really talked about because it seemed so painful to her. [Mrs. Cosi knew that Philippa was "a Negro" but thought that her mother was the Schuyler and her father the Monterro.] Sometimes she seemed so starved for affection, so sad. When I would hear her play, there was such anguish there, such agony. . . . Sometimes I thought her playing was a need more than an art.

And it was at that point that Signora Cosi remarked about Philippa's eternal search—about her not being a complete whole. "But the one thing that *was* whole was her faith. She had an unshakable and unbreakable faith. You know, God gave her everything but she was very unfortunate."[39]

❖ ❖ ❖

Philippa landed in New York on August 7, 1963, five days after her thirty-second birthday. She had one major commitment: a contract with Middle-Tone Records, in New York, to tape "International Favorites."[40] Several days after the completion of the taping, Philippa Schuyler was honored at the Hotel Americana as part of the Fiftieth-Anniversary Celebration of the Delta Sigma Theta Sorority. She and twelve other black women, including Lena Horne and Leontyne Price, had been chosen to receive awards under "We Salute Women of Achievement, 1963."

Philippa left for Europe in the early fall. As Monterro she performed in Rome, Naples, and Brindisi; and in Paris she appeared on a television broadcast. She also performed in Freiburg, but as Philippa Schuyler. The concert had been arranged by her publisher, Herder, and for obvious reasons it would have been impossible to perform there as Monterro.

It was when she left the concert in Freiburg that her life took an unexpected turn.

31

L'Affaire Raymond

While changing trains in Frankfurt on her way back to Rome during the early fall of 1963, Philippa was robbed. A man watched her open her bag in the restaurant to pay for a glass of milk and noticed she was carrying a bundle of money. Phil had been paid in cash for her concert in Freiburg and it had been too late in the day to convert to traveler's checks.

The man followed her to the platform, and in plain view snatched her purse. She screamed, but the only person who came to her aid was a good-looking young Frenchman by the name of Maurice Raymond. Raymond helped her pick up her scattered belongings, including her passport and ticket, and accompanied her to the police. Philippa had no money left—not a cent—and Maurice paid for the cab and loaned her money for the hotel.

The encounter seemed providential. Raymond told her he lived in Lyons, was a graduate chemical engineer, and came from a "good family." Philippa revisited Frankfurt twice in the following three weeks, first to identify the robber (who was caught and convicted, although no money was recovered) and again in mid-October, to attend the International Book Fair, an important event in the publishing world. Each time, Maurice met her there and escorted her around.[1] On October 12, from Frankfurt's Hotel am Zoo, Philippa wrote her mother a glowing report.

> What do you think of the following man who has proposed marriage to me and MEANS it?
>
> NAME: Erik Maurice Raymond
> AGE: 37
> FAMILY: Paternal grandmother comes from one of France's great noble families. Father is an industrialist. No brothers or sisters.
> TASTES: Of course he is an intellectual and widely read and discusses ideas all the time in the French way. He plays chess well. Esoteric mysticism is one of his main interests. . . . He loves classical music. . . . He, like many Frenchman, has a "colorphelia"—a mystique about colored people. (These attenuated Frenchmen

> seek the "vital force" the darker races supposedly possess.) He thinks it's marvellous that I'm not white—like finding a golden pearl or something—a collector's item. . . . I . . . am the epitome of his preconceived romantic "mystique" about colored woman. He says I'm the most feminine woman he ever met—he thinks French women are not. Maurice appreciates my intellectual and cultural qualities. He says it's a privilege to know me. He loves my helplessness; believes I'm absolutely penniless and thinks that's wonderful. (I'm glad somebody does.) He is madly in love with me. He is not seeking adventure—he wants marriage and has said so a hundred times.[2]

Philippa was no less fascinated with Raymond's handsome looks: "He has green eyes and a sharp nose. From the front . . . when he is talking, Maurice looks typically French, but when his face is in repose . . . he looks the complete Aryan. Maurice says that when the Nazis occupied Lyon, they wanted him to join the Hitler Youth."[3]

It was a fast-paced romance; what Phil did not tell her mother is that before they left Frankfurt the last time, they made love.

Philippa's enthusiasm was contagious and her mother's response was quick and unusually positive although typically undermining:

> I'm sold on Maurice Raymond as a sincere and worthy friend and perhaps husband . . . so do nothing to upset the balance. Now that you have found Maurice, I think you should have done with Africa. . . . Now is the time to make a crucial change. You have experienced all the world has to teach you. . . . But now you need protection and beauty around you, and love and sympathy to develop your gifts to the fullest. For you have such tremendous gifts but like any flower, they need cultivation, time, patience, sympathy, not the hard daily knocking around one gets in the . . . ruthless untamed male world out there.[4]

❖ ❖ ❖

Before leaving Europe, Philippa visited Lyons briefly where she met some of Maurice's friends. She found them delightful and sophisticated.

The romance continued to flourish. He wrote passionate love letters to her in America; they made plans to meet on the first day of February 1964 when Phil expected to return to Brussels. Sensing her daughter getting very serious, Josephine asked Monsieur Monnerville, their longtime friend in Paris, to have Raymond "investigated." Phil and Jody waited anxiously for Monnerville's response, but it did not arrive until right before Phil's departure. The report was a shocker: Raymond had a criminal record and had been arrested in the past. Devastated, Philippa decided to have nothing further to do with him. Arriving in Brussels in a dejected mood, she asked her friends to shield her from him.

The first of February having come and gone, Maurice was beside himself. He tried to telephone Philippa at Roger Nonkel's in Brussels, only to be told that she had not arrived as expected. "Devoured by anguish," he attempted to telephone her in New York but Information was unable to give him the number. (The Schuyler number was always unlisted.) He then sent a telegram expressing his concern—a telegram with a pre-paid reply—but he received no answer.

On February 3, Maurice wrote a love-crazed letter to Philippa. He was "distraught with anxiety":

> I want you to understand that it is neither with the tip of my lips or a fraction of my heart that I love you. It is with my entire being: my heart, my mind, my soul and my body. . . . with that absoluteness, that totality of passion that only a Frenchman I believe, and I fear, is capable of. . . . There is no question of my imposing myself upon you like a tyrant! There is no question of my demanding that you sacrifice your career and your talent for my benefit.

He called her "La Dame de mes pensées," himself "a love-sick beggar."[5]

As a last resort, Maurice traveled to Brussels. But Phil's friends insisted they had no idea where she was. He persisted, and in the process made a very favorable impression on everyone: He was highly intelligent, attractive, and had all the social graces. Ultimately, neither Philippa nor her friends could bring themselves to believe that this man was a criminal.

And indeed, ten days later, Phil sat on the train to Lyons with Maurice, to meet his mother. She had not told him about Monnerville's investigation, but now decided to explain that Josephine had received an anonymous letter with allegations leveled against him. Her mother proceeded to have him checked out and these were the results. She showed him part of the letter. Maurice was shocked, and when they arrived in Lyons he rushed to the police and obtained a document stating that he had never been arrested, that it had been a different "Raymond." Philippa, overjoyed, reported promptly to her mother: "You see, he is not and never has been a desperado, a criminal, a highway man, or a member of the Mafia. . . . There is nothing he wants more than to help me achieve the success I merit. I have never met a young man who was so considerate, understanding, sympathetic and cultured."[6]

Possibly in response to some specific questions from Jody, Phil wrote: "Maurice's father is a difficult man, and an old fashioned 'Jehovah' who is descended on his mother's side from the family of the *Bourbon* and looks down on everybody, and is in general a bear. He has never forgiven Maurice because his son did not make a career of being an Army officer. He is not a Jew—he hates Jews. His second wife, Maurice's step-mother, is the niece of the Bishop of St. Etienne."[7]

Philippa discovered that Maurice's history included an ex-wife and a daughter. "I have met [the daughter], and she looks just like him. She is a sweet and very timid girl, weighted down by her Latin studies. . . . I'm glad he had a wife once. That gave him experience in human relationships. . . . He knows music thoroughly—and loves to sit for hours listening to me. He is not a communist or a leftist—neither is he an extreme rightist. His family and friends are lovely!"[8]

According to Maurice he had graduated from the University of Lyons in macromolecular chemistry, with postgraduate work in electronics, had worked in France and Sweden, where he felt more at home than in Lyons, and between 1957 and 1961 had been employed by photographic and chemical companies, including Poulenc (Rhône-Poulenc). Currently he worked for Dubois Electronics Corporation making 375 francs per week.

Philippa continued to sing Maurice's praises—in all areas:

> The other day some Russian music came over the radio, and he did a Russian dance, the one where they squat on the floor kicking their legs in and out. I couldn't do that! And one evening we were visiting a lady and I left something behind. When we were down on the street she found it and dropped it from the 12th story window to us. But the wind swerved it, and it landed on the second floor balcony. So he, without any hesitation, climbed up the wall to the second floor, retrieved it, jumped down, and landed on his feet! . . . He has perfect eyesight—and no physical handicaps that I can see.[9]

According to Phil, he also cooked beautifully, did housework, ate vitamins, had grown a luscious beard, and adored her. She reported that Maurice spent one entire evening washing and then straightening her hair, "and did it as well as Mrs. Smith," adding rather poignantly, "and he still loves me."[10]

"Imagine if we had a child, it would be perfect: Have his eyesight; his green eyes, his health and stamina; his Aryan appearance; my talents; AND NO NEGRITUDE.

"I hope you are happy at all this good news," she summed up.[11]

❖ ❖ ❖

On February 24, after Phil had been in Lyons about two weeks, she invited Maurice's mother to dinner:

> As I told you on the phone, Maurice offered to fix it but I refused because at my first encounter with her I wanted to impress her with my housekeeping ability!!! I told Maurice that it would be my first and last housekeeping venture for the next ten years. . . . Even if I never do it again, it was psychologically good to have the first meeting that way.
>
> Maurice's mother was *very* nice—not a bit like the prototype of the horrid French women with eyes like insect's antennae, ears like an elephant, and claws like a crab's for clutching sons.
>
> To get back to the dinner—I realized I must be conventional (Maurice is large about food since he has lived some years in Sweden. A bowl of wheat germ and yeast is fine for him). But for HER I decided it must be the best kind of utterly conventional dinner.
>
> I made two salads: The first was of lettuce leaves cut up in little pieces, pieces of black olives, and bits of cheese with shredded carrots, beets, turnips and cabbage, and bits of onions, and sardines with a dressing of lemon juice, olive oil and melted butter, all mixed together.
>
> The second was a fruit salad of cut up pieces of oranges and apples plus peanuts with a dressing of coagulated honey diluted in green muscat juice. For hors d'oeuvres I had black olives and smoked herring.
>
> For entree, I had broiled liver for them—as I decided that any moron could fix a piece of liver. (However, I ran into difficulty when the gas would not work, so I

> cooked it on the radiator). I also had asparagus soup. I had grape juice and wine to drink, a stick a yard long of French bread, goat cheese, a basket of fruit and a chocolate cake for her. They both said it was a magnificent dinner.[12]

At about the same time, Maurice organized a press conference to promote Philippa's career. As a result, an article appeared in *L'Echo*, Lyons's daily with the second-largest circulation in France. Essentially it recapped Philippa's musical and journalist accomplishments, and talked about the work of the missionaries in Africa whom she had eulogized in her most recent book, *Jungle Saints*. Curiously, she identified herself as Felipa Monterro, casting aside her previous precaution of keeping the two personae separate.[13]

❖ ❖ ❖

Although Philippa had known Maurice less than four months, she decided to marry him. "I have a chance to put my roots down in an important European country with a white husband who wants to boost my career," she wrote Jody.[14] Having consulted the cards, she picked March 29. She wrote her mother, urgently requesting that Josephine forward her birth certificate, adding: "If you can, deftly change my birth date from 1931 to 1934. . . . Don't tell anyone about my marriage. You know how people try to destroy the private life of stars with their malice. I don't want any tasteless speculation in newspapers. I want this happiness to be just for me. I've waited long enough for it. I hope you realize that."[15]

Having finally received the birth certificate, Phil and Raymond went to the marriage bureau to get the license. But when Maurice read the requirements of waiting and investigating that would have to take place, he became very angry and stormed out, saying they did not have time to wait. He then insisted they go to Scotland, where he had read it was easy to get married.

Madame Blanche Raymond, Maurice's mother, also appeared to be very anxious for her son to marry Philippa, and it was she who provided the money for the trip—and the wedding ring.

But in Scotland, they discovered that the laws that had once made the smithy of Gretna Green the legendary place of elopement had long since been changed and that it was now necessary to wait three weeks.

The details of the Scottish Fiasco are not clear. Her passport indicates that she entered Britain via Dover on Wednesday, March 25, and left via Folkestone a week later. On the first of April, after sending Maurice back to France and spending three lonely and sad days with her friend John MacKenzie in London, her wedding ring on her finger, Philippa was still far from giving up the idea of marrying Maurice: "He has a good heredity from the breeding viewpoint, don't you think?" she wrote her mother. "But imagine going to Scotland to get married without finding out the full details beforehand. . . . He was so keen on it, I couldn't stop him any more than you could stop George's 1,500 page pocket book."[16]

By April 8, Philippa was back in America beginning a busy concert schedule in the Caribbean bracketed by a few recitals in the States (all performed as Schuyler). Maurice continued to bombard her with love letters. He wanted to meet her in Iceland or Mexico and marry at once. And as if the idea of "hybrid vigor" had rubbed off, he began to imagine that if they could have a child, it would be a great genius. He called it a "love kitten" and fantasized that she was already carrying it.

Philippa had no immediate plans to go to Europe; it was mostly Josephine's idea to suggest that Maurice move to the States instead. Phil expressed mixed feelings about this. Nonetheless, she inquired about employment opportunities in his field and asked Maurice to send her his employment record.

But she knew, in the final analysis, that the idea made little sense. She had long since decided to become a European and live in Europe, and had basically written off America as a home base. As for Maurice, she was fully aware that he dreaded the idea of starting life over in a country whose language he barely knew, and whose work environment he would find alien; he would also be leaving behind his friends, his family, his child, and his cultural roots.

The truth is that as much as Josephine loved her daughter and wanted her to be happy, the idea of Philippa permanently moving to Europe distressed her. Forever machiavellian, she kept up a steady barrage of doubts about Maurice's integrity and intentions.

To make matters worse, Maurice delayed sending his work record. Instead he wrote about the race riots in America he had seen on French TV, whites hosing down blacks and attacking them with dogs; how it would be better for them to live in France where there was no racial friction. On another occasion, he wrote about the difficulties of obtaining an American visa, even a tourist visa; the need to prove that one had the intention of returning; and the discretionary powers of immigration officials at the points of entry who might refuse admission, even with a valid visa. Finally he confided that while he could scrape together the money for a one-way fare, he could not afford a round-trip ticket. In his appalling English, he wrote to Josephine on May 21:

> Please you don't see in my step, not any intention to recoil. I am writing at the same time to Philippa to inform her of the difficulties that one French try to travel in the United States. . . . I'm always loving Philippa. Today same yesterday and perhaps more again. My greatest desire is to marry her. But I think to have revoked my travel to the U.S.A. Philippa and me, at the same time of her next stay in Europe, can remain three weeks in Switzerland and to marry us without loss of our nationality.[17]

Josephine was becoming more and more suspicious about Maurice. But Philippa held out—although her mother's inferences dug deep—and suggested to Jody that perhaps Maurice's statements that he did not have the money should be taken at face value, and that there was nothing "sinister" about his inability to come to the States.

Unable, however, to get Maurice's irrational behavior at the French marriage bureau out of her mind, she decided to lay low for a while. She refused to answer Maurice's pleading letters. He tried to call her, once, twice, three times, but her mother told him she had already gone on tour.

Philippa had left some of her belongings in Lyons—a bag and her typewriter—and now in a fit of anger, Maurice wrote Phil that he was sending them to her friends in Brussels. For all practical purposes, he had broken off their relationship.

❖ ❖ ❖

Then, toward the end of June of 1964, something dreadful happened.

The Schuylers received from Raymond three poison-pen letters, and then two or three obscene postcards, mailed from Turin, Italy. On the picture side, they showed a man and a woman strongly resembling Philippa in a sexual act, while on the address side was a defamatory statement about Phil having been pregnant and having had an abortion. Josephine and Philippa were shocked beyond words.

There was also an indication that he had sent similar missives to some of her friends in Europe, Africa, and America (and to someone at the *New Yorker*). Not satisfied with that, he had tried to defame and ruin Felipa Monterro by sending similar letters to people in America.

The derogatory accusations were patently false. Phil had not been pregnant and thus had not undergone an abortion. The photographs presented a different problem. As soon as Philippa recovered sufficiently from the shock, she insisted that the pictures were fakes—montages.

As usual, her mother took charge of the situation. In an attempt to put a stop to Maurice's outrageous actions, she wrote the U.S. postal authorities, the chief of police in Lyons, and also the American consul general there.

On July 26, Josephine submitted "several postcards of defamatory nature" addressed to Philippa and herself, to the chief postal inspector in Washington, D.C., who advised her that while the Postal Office "had no investigative jurisdiction over material such as this entering the country from a foreign source," the matter had been referred to the French postal authorities.

In a brief communiqué from Paris, dated August 7, the French Postal, Telephone and Telegraph Service replied that they could not intervene in this matter, and that it was up to Josephine to refer it to the public prosecutor.

Jody had already done so, and the Lyons police took an active interest in the case. On September 11, Rupert A. Lloyd, Jr., American consul in Lyons, was able to report in detail to Mrs. Schuyler: "As you know, I have been deeply pained by the absolutely inhuman plot that has been woven against Miss Schuyler, and have been doing all I could to assist her. Please believe me also when I assure you that the French police are neither heartless nor indifferent [which Josephine, typically, must have accused them of] and are very keen on seeing to it that an end is put to Raymond's activities."[18]

Maurice must have fled the country as soon as the police started investigating him, for the consul's letter continued:

> I recently had a further talk with Chief Commissioner Julliard on the problem. He reiterated that as long as Raymond remains outside of France, the hands of the police are tied, much to their regret. Extradition to stand charges is regulated by bilateral treaties in force which specify the offenses for which extradition may be demanded. As far as the police here can see, and at the present state of affairs, the only solid charge that could be brought against Raymond is "defamation of character," which, while a criminal offense, is not an extraditable one. As a French police official remarked to me: "It is sometimes possible to do immense harm while exposing oneself to only slight legal risk."[19]

The consul advised Philippa to retain the services of a Lyons lawyer and to file charges in her own name. He also recommended that charges be filed by the New York police authorities in order that Interpol might be alerted.

32

Dennis Gray Stoll

Distraught by the Raymond affair, Philippa's practicing turned erratic during the summer of 1964. As a result, her fourth Town Hall concert on September 13 received mediocre reviews.[1] Her Beethoven, the *New York Times* wrote, had no line or cohesion. The critics complained of "too many missed notes," though granting that "she muddled through with aplomb." On the other hand, they uniformly liked her African compositions with their "vigor, color and authenticity,"[2] in particular her *White Nile Suite*, premiered on that occasion.

One week later Philippa, in a subdued mood, escaped to Europe intending to remain two months. Even though few concerts were lined up, for once Josephine thought this trip a necessary respite from the terror she and her daughter had gone through.

Her first stay was with John MacKenzie, now widowed and managing a boys' boarding school in Abbotsholme, England. He arranged a concert for her at the school (which she played as Monterro), and for a room in the master's house. MacKenzie and Philippa spent long evenings talking or listening to recordings from his extensive music library. John possessed a good deal of insight into the Schuyler family dynamics, and in the nurturing atmosphere of MacKenzie's home, Philippa began to recover some of her self-confidence.

With it came a rebirth of musical creativity. She worked on her *African Rhapsody*, and she began an ambitious piano concerto entitled *Nile Fantasy*, commissioned five years earlier by the United Arab Republic government.

She also started drafting her unusual work for piano and speaker, *Seven Pillars of Wisdom*, based on the book by T. E. Lawrence. Philippa felt a deep kinship with him. To her, Lawrence combined the intellectual life with a life of action. "He too had felt placeless and torn in his own country because of his birth—and he too went forth to seek the unknown in foreign lands," she wrote Josephine. "Like him I have known the terror of the tropics . . . the fever of the blood, the feeling of disintegration under the broiling sun."[3] Philippa was further intrigued by the fact that Lawrence had twice changed his name—after he had achieved notoriety—first to J. H. Ross at the time he enlisted as a mechanic in the R.A.F., and then to T. E. Shaw.

Late in October, Philippa flew to London for a few days. On her second night there she was invited to an informal cocktail party. Asked by her host to play, she chose a Casanovas, a Liszt, and three of her own African compositions. Particularly taken with Philippa's performance was a bespectacled, ruddy-cheeked man with an almost boyish face, although easily in his fifties, and a Pickwickian sense about him. His name was Dennis Gray Stoll and he invited her to tea the following day.

Youngest son to Sir Oswald, impresario and founder of the Coliseum Theatre, Dennis was born in 1912 and educated at Cambridge. He had begun his musical career as a composer, but soon took up conducting. By the age of twenty-one, he had already conducted for the Ballet Russe of Monte Carlo, and for a spate deputized for Sir Thomas Beecham at the London Philharmonic. His career was interrupted during World War II when he was jailed as a conscientious objector. After his release, Dennis moved to India to study philosophy and religion with Gandhi and Krishnamurti. In 1947, however, he was felled by polio and returned to England for treatment. Unable to conduct for the next ten years, he concentrated on composing and also began writing fiction.

Dennis's interests were as rich and varied as Philippa's. He was deeply spiritual—something he claimed was responsible for his almost complete recovery from polio—and like Philippa, he believed in reincarnation.

He also shared with her a precocious childhood in both music and letters. Distinct from Philippa, however, Dennis was "to the manor born" and at barely twenty-four, had already published, at his own expense, his autobiography, a slightly self-aggrandizing saga about his formative years.[4] As he matured, Dennis became a well-respected and prolific writer (he authored close to a dozen titles). In particular, his 1950 novel *Men in Ebony*—about an African who leaves his village to become a Roman Catholic priest and returns to care for their souls only to lose his faith—was hailed by C. P. Snow, Somerset Maugham, and Joyce Cary. Expressions such as "well written, brilliantly intuitive," "the richness of detail of a Balzac," and "the value of an Upton Sinclair sociological novel" dot the reviews.[5] And everyone commented on his ability to tell a story, strongly, economically, and compassionately.

Musically, Dennis and Philippa had remarkably parallel sensitivities. Both believed that the only possible future for Western music lay in Asian and Arab inspiration. "Now there are Arab composers who have tried to put Near-East, North-African idioms into Western garb," she wrote her mother, "but their efforts have seemed to not quite click. . . . Mr. Stoll says my *White Nile Suite* is magnificent. We can form the nucleus of a movement together! Like the 'Five' in Russia. He has great experience as an orchestral and ballet composer and opera too . . . but he needs a pianist who understands his . . . idiom to play his works, and wants me to be his protegee!! He is going to write works just for me and find me concerts to play them."[6]

Dennis was extraordinarily attracted to Philippa from the moment he heard her play—not only emotionally, spiritually, and intellectually, but also physically.

They saw each other a great deal during their first weeks. He introduced her to prominent personalities such as Dame Sybil Thorndike; planted seeds for future recitals in England; reviewed her one London performance (played as Philippa Schuyler); and decided to write a piano concerto in her honor, which she would premier. Rather surprisingly, given her previous feelings about England (she once said, "I can't pass here as I can elsewhere. In Southern Europe I look like one of *them*, for so many of them are dark. In America, I can pretend I'm Oriental or a Latin. . . . In England they're prejudiced against Orientals *and* Latins—so there's no hiding place for me."),[7] Phil wrote Jody, "I actually foresee the possibility of my becoming a figure in British musical life."[8]

Philippa was initially captivated by Dennis. He seemed kind, considerate, and highly accomplished. The fact that their musical interests were not only similar but complementary fed their mutual attraction. And she quickly realized that he could promote her career more effectively than anyone she had previously met.

Still lacerated by the Raymond experience, however, she harbored no intentions of getting emotionally involved with any man—just yet. Nor did she feel any sexual attraction to Stoll. Besides, Dennis was married to Patricia, his third wife. All this made Dennis "safe" in Philippa's eyes.

They both referred to their relationship as a "partnership," perhaps in an effort to protect it—he to isolate it from his passion and she to shield it from her emotional indecision.

"I believe we are on the threshold of great things," he wrote her on November 14.[9] "I think the Stoll-Schuyler partnership is going to work out wonderfully. I do feel that it is some part of God's will that we should work together."[10]

"You are a great artist," he added on the eve of 1965, "and I want to see you make a big name here. . . . I want your name built up as a composer, too. . . . I look forward one day to meeting your parents, who must be wonderful people to have produced you. . . . It must be a thrill to produce a second Mozart."[11]

❖ ❖ ❖

Early in 1965, a nagging anxiety was lifted from Philippa's shoulders. She learned that Maurice had returned to Lyons on November 29, 1964, and was arrested the following day. The examining magistrate reviewed the evidence that Josephine had submitted to the police and obtained a written statement from Raymond. Prosecution was to be considered under Article 283 of the French penal code (offense against public morality), which provided for imprisonment from one month to two years and/or a fine of from 300 to 18,000 francs. Raymond was released a month later upon posting a bond or 400 francs ($82)—which he had great difficulty raising—and accepting an obligation to desist from further harassing Philippa, on pain of having his provisional liberty revoked.

Philippa did not return to Europe until March 11, 1965. In the intervening months, Dennis had taken on the unofficial role of her British impresario (although Phil was already represented in England by a "Miss Jennings").

As music director of ICE (International Cultural Exchange), Dennis was well respected in artistic circles and had extensive connections. To promote her career, he was arranging dinners for her at the House of Commons and the House of Lords as his special guest. He contacted Lord Sorenson, Sir Michael Redgrave, and others for whatever help they could offer. When news of a concert had gelled, he would write Philippa in his typically overblown style: "GOOD NEWS FROM LONDON! Hurrah! Hurrah! Yip-peeeee! . . . You are going to be so famous and in demand we shan't know which way to turn!!!"[12]

Dennis, however, did not neglect his own creative muse. Busy finishing the piano concerto inspired by and dedicted to Philippa, he rushed her weekly drafts. He was entering it in the Grenada Festival, which awarded a handsome purse as well as a debut performance in the exotic palace at the Alhambra. Phil would be the featured soloist. "That you loved the CONCERTO I am creating for you is most encouraging," he wrote, "for I would indeed be thrown into the depths if you hated it!! The events of the past few days have introduced an agitato movement with pont.[icello] strings, and muted trumpets in the last movement. The biography of PHILIPPA SCHUYLER continues to be unfolded in musical terms."[13]

The "events of the past few days" referred to unexpected problems concerning Philippa's labor permit. As required by law, Miss Jennings had applied to the Incorporated Society of Musicians (ISM) asking that Phil be allowed to perform twenty concerts a year in England. But they turned her down, on the grounds that Philippa was not an "internationally known artist." Dennis intervened, requesting an appeal. He encouraged Philippa to get well-known figures to write letters of support. Aaron Copland and Deems Taylor were among those who replied promptly. "Her talent and career are truly remarkable," Taylor's letter read.[14]

While waiting for proof, ISM decided to postpone any final decision. But Dennis had already taken the problem to a higher level—to Lord Sorenson. Philippa was upset, convinced that the refusal of a permit was a racial issue, and she shared her fears openly with Dennis. He responded immediately.

> Your letter of Feb. 5th got here today. What a cry of pain from New York!! My dear, I don't think this is colour prejudice. There are some people who are just stuffed shirts or pompous puddings who like to exert their authority. If they haven't heard of someone, that someone doesn't exist! . . . When you come again to Britain to play for us in the Spring, put that RIGHT OUT OF YOUR MIND. Go in and conquer them with your GREAT GIFTS, YOUR CHARM, YOUR ABILITY TO BE PROUD of being the daughter of a great negro writer and editor. I think the mixture of races in you is a TREMENDOUS ASSET, and if some very LITTLE people

> have made you SUFFER (and I expect they have), realize that without being WHAT YOU ARE you could never have been such a fine composer, writer and pianist.[15]

By February 12, Lord Sorenson had intervened on Philippa's behalf. Dennis wrote Phil the good news and asked her to forward a brief biography. Always a promoter, he shared with her an idea he thought would be terrific: "Why not do your hair Nefertiti style [One of her planned recitals was entirely Afro-Arab music]. . . . It would show off your beautiful face marvelously. Patricia has jewelry which is an exact copy of Queen Nefertiti's, and she would lend it to you."[16]

In a curt response, Philippa sent Dennis her biography, emphatically stating, "My father is not a Negro."[17]

"The biographical notes on parentage are most interesting," Dennis wrote back. "I'll use some of them. . . . So you're EUR-ASIAN which as far as I and ICE and MRA are concerned is socially marvellous!!!!!" But again putting his foot in his mouth, Dennis continued. "What a pity you couldn't have some connection with the African mainland. However, wearing the Nefertiti jewelry should make up for this."[18]

"Make up for this?" Phil was beside herself. Dennis was going to advertise her as a "Negro" no matter what she wanted. He was going to make her dress up like an African. Dennis seemed determined to ruin her. Completely undone, she sent him a woeful note. Having received it, he tried to reach her by phone but there was no answer; he immediatley posted a special-delivery letter:

"My dear Philippa," it began.

> Put your great heart at rest. Ken [their public relations agent] has not, and never would have, advertised you as a Negro pianist. . . . Most MRA people knew your father was editor of a big Negro newspaper anyway. Now, we know you are Eurasian, and the matter can rest.
>
> But please, my dear sweet Philippa, try to let it rest in you. . . . Don't think I am unaware of the vast pain and hurt that can be inflicted, and obviously has been on you. . . . I hope you know that I accept you completely as a person, just as you are. When I wrote of being "disappointed" you hadn't connections with the African mainland, it didn't mean anymore than that it would have been a link with your great knowledge of African and Arab music. It didn't mean disappointed in *you* in any way. Obviously, I realize that a racial link is not necessary to understanding of Afro-Arab music, otherwise I'd be in a bad position myself with my Irish-English ancestry![19]

Dennis and Philippa also shared views on intermarriage. He believed that mixture was a good thing, "and the more contrasted the better. It leads to beauty and intelligence."[20] Dennis had referred to intermarriage as "beneficial to the human stock," in his 1944 novel *Comedy in Chains.*[21] Yet he was keenly aware of the often unhappy lot of the racially mixed child. His experience had been firsthand: a daughter by his second wife, an Indian, had struggled with identity problems similar to Philippa's. "We Eurasians are lonely shut-in creatures," he had the protagonist of the same novel say. "We can turn for sympathy and justice to no community. We, who are born of no-nation"[22]—lines which could have been written by Philippa about herself.

❖ ❖ ❖

Dennis had arranged seven concerts for her in England.[23] They were prestigious recitals, some at world-class halls, with top professional orchestras, but first she wanted to go to Brussels and Paris.

Prior to departure, the Raymond affair had again resurfaced. Philippa received a letter from Marcel Gros, the Lyons lawyer retained at year's end by Josephine for the case, explaining that he had acquainted the examining magistrate with Philippa's contention that the photographs were fakes, representing a person other than herself. To date, however, he had not received any formal complaint from *her*, only from her mother. He continued:

> As no offense has been commited in France (since the photographs and postcards were mailed from Turin in Italy and addressed to New York, Brussels, Holland and Africa), it is necessary, unless governments of those other countries lodge a complaint with the French Government, that you yourself bring the charges.
>
> Otherwise, the Judge cannot act officially in this affair, and he finds the intervention of your mother, on your behalf, surprising. He desires to receive a letter from you which could be worded as follows: "I bring a complaint against the defendant RAYMOND by reason of deeds felonious and gravely prejudicial committed by him against me during 1964, notable for having sent to numerous persons, photographs of pornographic nature, and of postcards circulated in which he imputed me of an alleged abortion. I refer for details to letters sent by my mother to the police. I reserve all rights of civil recourse and demand that the prosecution take place . . . .

Maître Gros, furthermore, advised her to go to Lyons to see the Judge.

> He would like to know in particular whether you contest the authenticity of the photographs, which according to Raymond represent you. He has subpoenaed the negatives. In addition, Raymond pretends to have your correspondence demanding the return of these photographs and alluding to your state of pregnancy. The crime Raymond is accused of is independent of the attribution of the photographs to a specific person—but his responsibility would evidently be graver if he had fabricated these photographs to harm you.[24]

Philippa decided to consult the one person in Europe who could give her sound legal advice, her friend from Togo, Georges Apedo-Amah, who at the time was staying in Paris. She had kept him abreast of l'affaire Raymond.

Georges strongly advised against her going to Lyons, or even telephoning the lawyer from French soil. He felt that the judge wanted to arrange a confrontation between her and Maurice. Apedo-Amah suggested she write Gros asking him for a list of questions the judge wanted to raise, write out the answers, and have the lawyer submit them instead of her confronting Maurice.

Philippa followed his advice. She phoned Gros from London only to discover that indeed the judge had wanted to arrange a confrontation.

Somewhat disconcerted, Philippa nonetheless played to excellent reviews in the first two recitals arranged by Dennis. As promised, he had set up dinners and parties with London's finest: Lord Sorenson, Sir Michael Redgrave, Dame Rebecca West, and others. Philippa was genuinely grateful to Dennis for his efforts.

But on April 1, she received even more disturbing information about the Raymond case. The judge had seen the negatives: In his mind there was no evidence whatsoever they were montages or fakes as she contended. As her house of cards collapsed, Philippa folded.

Appearing hypnotized, she performed two concerts in Belgium and returned to London to give another one on April 8 at the U.S. embassy in Grosevnor Square. There she gave her first London performance of her *Seven Pillars of Wisdom*, dedicated to Dennis.

Although her reviews were good, Dennis noticed she was distracted. Philippa decided to confide in him. She had already hinted about her unhappy relationship with Maurice. But now when he asked her what was wrong, she told him the whole truth.

Later, Dennis would reminisce about that evening: "The small crushed rose who had been trampled in the mire sat in my car opening her heart and soul to me, and saying over and over: 'Doesn't this make you hate me? Can you still like and respect me now you know this?' This answer is yes, and still is. Please remember this of me. . . ."[25]

Caught in the tide of the moment, and her own needs, Philippa told Dennis she loved him. But she was unprepared for his reaction: Dennis went to Patricia asking for a divorce. In a state of utter confusion, Phil abruptly left London for Paris on April 11 to seek shelter in the arms of Georges Apedo-Amah.

33

Dennis Redux

It was a typical April day in Paris. Elegant women walked the boulevards. The *taxiste* on his way to the Palais Luxembourg leaned out the window to call a fellow motorist "espèce d'idiot."

Philippa's hosts, the Monnervilles, were glad to see her, and she was equally pleased to sit down at their grand and practice.

At 8 p.m., Georges Apedo-Amah arrived. They dined at their favorite restaurant and talked. Georges was comforting—assuring her that Maurice would draw a stiff sentence, and never again be able to harass her. Sometime during the evening he said quietly, "I have been in love with you for 12 years. You must have noticed . . . ."[1] Finally, they went to his place and made love.

The next day Georges was off to Madagascar "to solve the Congo problem," and a letter arrived from Dennis, an unhappy letter. He had not won the Granada Prize for the concerto he had written for and dedicated to Philippa, and in some oblique way he was blaming her.

This prompted Philippa to dash off a six-page tirade to her mother finding fault with Dennis on every score: She criticized his music and his writings, and mercilessly slammed his manliness.

> No wonder they turned his Concerto down. It is 2½ times as long as a Concerto should be. It doesn't follow the accepted form which alternates piano and orchestra. Since the piece is just one long orgasm, naturally he had to have the piano in all the time. . . . Also, he does not show enough different types of piano technique.
>
> Another fault in his music is that his childish underdeveloped sexuality gives an effeminate young-lady quality to it. He cannot express a variety of emotions. I noticed in his novels that he could write wit, and sugary descriptions of scenery, but anything on the man-woman level was utterly unconvincing—like Peter O'Toole and the Eurasian girl in Lord Jim.[2]

Continuing the litany of his shortcomings, she wrote:

> He was a coward in the last war preferring to go to jail . . . rather than fighting like a man. . . . He *never* had sex with his first wife. He is always accusing men of

> being homos and women of being BAD. That Hindu girl seduced him when he was 34—the first time he'd had a woman. But their marriage broke up because she "treated him with contempt" and took to drink. Patricia [his current wife] was unfaithful to him for two years with a Welsh workingman—to whom she wrote a lot of poetry about how he was more manly than Dennis. Dennis says he found the poems "accidentally." Since then, 18 months ago or more, he has not slept with Patricia or any other women. He can't. But with me has been able to create a fantasy and in this fantasy he CAN. . . . He's about 6′ 4″, enormous and bulky, no shape at all. I haven't the LEAST desire to jump in bed with him.[3] [In his pictures as a young man, one might call Dennis ascetic looking, and even now, approaching fifty-four, he could be viewed as handsome.]

Yet at the same time she expresses her fears: "He's now going to insist flatly that I have to sleep with him or he will never arrange anything else in England for me and I'll never have another Labor Permit. No sex, no concerts; it's just that simple," she writes, adding painfully, "Oh, WHY can't Dennis have Georges' personality, and WHY can't Georges have Dennis' nationality?"[4]

It is a brutal letter. She viciously attacks a man who had turned himself inside out for her, who wanted desperately to promote her, to be her "guide and philosopher" and lover, for life, who had poured out his soul on a piano concerto dedicated to her.

Why was Philippa unable to accept the love that Dennis offered? He could open the doors to the white world of classical music, first in England and then elsewhere, including America. And he had a burning desire to help and protect her.

Was it that Philippa was no longer capable of loving and trusting anyone—least of all herself? Had her constantly nurtured and almost never challenged sense of superiority in all things made her incapable of acknowledging intellectual accomplishments in others? Or was the tragedy that her entire life had failed to prepare her for a meaningful encounter with a man like Dennis Gray Stoll?

Philippa's insulation, as a child and adolescent, went far beyond the loneliness of a prodigy. John MacKenzie wrote insightfully:

> I felt much compassion for Philippa who I always felt had an unnatural childhood and youth and was rather a forced product and desperately lonely and isolated. . . . Poor Philippa on more than one occasion shed tears and confided in me her almost dread of her Mother's strong urges . . . that she was not measuring up to [Josephine's] driving ambition. . . . [Her Mother] seemed to be the great . . . force behind Philippa's precocious career development. In fact, I felt a parallel in her to Leopold Mozart in his relationship to his genius son Wolfgang."[5]

But most of all it was Philippa's attraction to powerful masculinity that turned her away from a man like Dennis. Theo van Eyk may have been the most accurate when he volunteered that sex "was the most important thing in her life." It could be cruel, like Rudolph's in *Appassionata*, or soft and dreamy, as she fantasized it would be with André Gascht—but it could not be unmanly. Philippa had come by this predilection rightfully; after all Josephine never hid the fact that her attraction to George had been due in large part to his sexual prowess.

In Philippa's mind Dennis's masculinity was highly suspect. His history of repeated failed marriages, his references to not sleeping with his first wife or with Patricia in recent months, all heightened the suspicion that he was partially impotent—possibly as a result of his bout with polio—or at least seriously concerned about his potency. Philippa and Dennis had not made love. Strangely, on June 7, 1965, he would write her: "Do you remember a night when you suggested we should draw the car off the road where there was no light? You told me of someone who always carried the necessary things in his car. My God, I wanted you, but NOT LIKE THAT. In the end nothing happened except that you laid your head in my lap like a tired child."[6] Possibly Philippa had felt rejected, if not outright offended.

❖ ❖ ❖

Returning to England, Philippa decided to play it cool. She delivered four concerts in a row and met Dennis every day. As far as he was concerned, her one expression of love to him a week earlier held.

They must have continued to discuss marriage, for on arriving in New York, on May 10, she found that Josephine had already received a letter from Dennis. It confirmed his deep feelings for her daughter and their desire to marry. "Because of the particular and wonderful part you play in her life, I would very much wish for your approval," he had written.[7] His marriage to Patricia had gone very wrong some years ago, and he was about to initiate divorce proceedings. "Your sweet daughter has had a lot of unhappiness, and she needs the love and stability from a husband who understands her. I think we could help and give solace to one another. . . . Please believe that I seek Philippa's whole good in this as well as needing her love, and her partnership in music. Her character—a blend of intelligence, sophistication, the primitive and a rare innocent simplicity—is one which I can understand, though she has certainly puzzled me sometimes."[8]

To Philippa he wrote impassioned letters, but she remained aloof, neither committing herself nor abandoning the idea of marriage. Finally Dennis called her in New York. He wanted to hear her voice, to tell her how the divorce was progressing, and to express his love. She said she was "thrilled" about the news of his divorce, but nothing more.

"Felipa" went on tour lecturing for the John Birch Society in mid-May, and although Dennis wrote practically every day, he heard almost nothing from the woman he intended to marry. In great distress he turned to Jody again.

> The Egyptians have invited me to see their hour-long travel film with the proposition that I should compose the incidental music. I at once thought of Philippa, and how nice it would be if she could play in it. Then came the stab—the feeling that I am gradually finding kills everything I am trying to do for her—the thought that she has not written, and has perhaps changed her mind. . . . I asked her a question to confirm her love now that I am getting my divorce. She has *not* done

> this, and I feel emotionally shattered. . . . This silence of hers is even poisoning my negotiations with the Royal Philharmonic [Dennis was planning future concerts in October with her as soloist]. Someone has suggested that Philippa is unreliable.[9]

Finally, on May 27, Dennis called New York to talk to Josephine. She explained her daughter's silence: Phil did not want to be blamed for the divorce. Of course, she still loved him; she was just being cautious.

Dennis was relieved, but he now asked Philippa to consult with her priest about marrying a divorced man. He thought that there were special mediating circumstances. His first marriage had been annulled, his second, not recognized by the Catholic Church since it was the union of a Hindu and a Christian. His third, to Patricia, he hoped would also not be recognized, as it was only a Registry Office marriage. But he did not get an immediate reply because Philippa was away, lecturing in some western states as Miss Monterro.

Aside from the fact that the lecture tour was most remunerative (bringing in $1,500 a week), she was staying with affluent, and likely bigoted, white Republicans, seeing an aspect of America new to her. "How horrid that I have been kept from knowing this side. So this is the way most Americans live!!! Aren't they having the good life."[10]

Dennis had a different view of the Monterro scheme. "My lovely One," he wrote her from England, "One day, I hope, these two people will be merged because Philippa will realize she is deeply loved for her own sake, neither because she is white or brown, nor because she is intelligent or a great pianist, nor because she is beautiful and sophisticated and a sweet frightened child—though she is all of these. Philippa will know that she is loved just because she is PHILIPPA, PHILIPPA, PHILIPPA."[11]

Although Phil was thoroughly enjoying her tour, she complained of chronic stomach trouble, which she thought was too severe and drawn out to be accidental. She even fantasized that Patricia Stoll, who had taken an intense dislike to her, might have poisoned her.[12] But toward the end of her tour the condition improved. "Now that I am convalescing I see how ill I was. I have to have a complete change of diet now because some enzyme necessary for digestion is missing from my stomach. So when I come back to N.Y. I must eat: No fat fish like mackerel, cod, etc. No beans, or greasy or heavy things. I can eat: All fruits except avocado and banana. All greens. All fruit juices. Shrimp, crab or other non-fat sea foods. YOU'D THINK I'D LOSE WEIGHT!!"[13]

Philippa returned from her tour on June 18 to find a letter from Marcel Gros reporting that the Raymond case would be heard on October 1, 1965, before the Sixième Chambre Correctionnelle du Tribunale de Grande Instance in Lyons. Following Phil's instructions from London, Gros would let the prosecutor handle the case without intervention, to avoid unpleasant publicity. In any case, Raymond had been revealed as incapable of paying damages.

Dennis was now in the Middle East as the guest of the Egyptian government. He had secured several recitals in England and two in Cairo with their orchestra, where Phil would perform Dennis's *Concerto Arabo* on December 4 and her own *Nile Fantasy* on December 10.

Although their plans for musical collaboration were progressing, their future as husband and wife was at a standstill. Philippa kept delaying, finding one reason or another. To her, it had become a chess game, each move carefully considered to confuse or frustrate.

By midyear, though, Philippa began seriously to consider a marriage purely for the sake of her career. She turned to her old friend, John MacKenzie, seeking advice. Should she marry Dennis for the sake of her "art"? She asked if he might "investigate" Stoll—to find out if there were any skeletons in his closet. "Your request . . . is I fear rather too delicate for me to handle at this end," John wrote back. "In fact, it seems closer to the Italia of the Borgias! Of course you must take all steps possible to make . . . doubly sure in a career marriage. . . . Anyhow my dear Philippa, I wish you all success and above all much happiness."[14]

So the chess game played on. Some of Dennis's letters were acquiring an undertone of disgust.

> If what you say is true about the R.C. Church, THEN HOW ARE WE EVER GOING TO GET MARRIED? As you know, to get this divorce I am pretending to be the guilty party [the charge was mental cruelty]. So your Church WON'T MARRY US. . . . PLEASE ANSWER THIS ONE AT ONCE. My lawyers, as I advised you earlier, suggest that we should look into this position, which is a religious one. You are the Roman Catholic, so it is your decision. We could have a legal ceremony of marriage after my divorce, EVEN IF YOUR CHURCH SAYS NO. But then, I take it, they would excommunicate you? . . . If I were the Roman Catholic, my view would be that I loved God and my Beloved more than any priest. . . . Does truth, beauty, and sincerity not touch you at all? Do only COLD IDEAS and LAWS govern your existence?[15]

Dennis took off the gloves. He wrote her that if she were not always so anti-anti, especially about communism, he could get her some wonderful things in Iron Curtain countries for the spring. And finally Dennis told her the true story about her labor permit, something only unbridled anger would have dragged out of him.

> I talked to Hungary about you, and I fear they won't have you there. I don't think you realize [how much] prejudice creates prejudice. . . . A lot of the trouble with your Labour Permit here was this too [her reputation as an ardent anticommunist]. . . . It is ironic, you may think, that it was Lord Sorenson, a man of the Left who got your Labour Permit through. But in this country the Left wingers are far more against Communism than the Right wingers. The Right wingers like to tread what they call a sensible and cool Middle Path.[16]

But as hard as Dennis tried, he could not turn off his love for Philippa. Slowly his anger subsided and his letters turned warmer again. "At the moment we are free," he wrote on August 22, "and without any obligation except to be kind to one another, and not cause unnecessary pain. My whole wish is to understand you, your needs, your character. My fulfillment is in seeing you fulfilled. One day I hope you may feel that way about me. Then the time will be ripe for us to marry but not before."[17]

Rather unexpectedly, toward the end of August, Josephine and Philippa radically changed their minds—marriage, they wrote Dennis, was a wonderful idea, and the sooner the better. They proposed a quick divorce and a quick marriage. Dennis was overwhelmed, but he was alarmed at the method that the Schuyler women were proposing. He wrote a measured response:

> You both mention the possibility of a divorce, and a quick marriage to follow, under the wing of Texas or Mexico. Attractive as this idea is, it does not seem to me very practical. . . . It would not be valid in England, which would be a disaster for me, since presumably I could be arrested for bigamy on British soil! I doubt if Patricia would agree; and I must say that the idea of spending the first five years of my marriage behind bars, and then having to start over again getting a divorce in England, is not an attractive one—even to an impetuous Irishman, with or without a red beard![18]

❖ ❖ ❖

Dennis did not know that in mid-August, Philippa suddenly realized she was pregnant. She had been menstruating all along and had skipped only one cycle, but now she was missing a second. Philippa was shocked, and then it dawned on her that the stomach problems she had been experiencing were not the Asian flu, an enzyme missing from her digestive tract, or even the fanciful idea that Patricia had tried to poison her. By now she was more than four months pregnant. The father of her unborn child, conceived on April 14 in Paris, was Georges Apedo-Amah.

Philippa was frantic. Obviously, she thought, she could not cope with the idea of having an illegitimate child. It would ruin her career. It was at that point that mother and daughter had hatched the plot of a quick marriage to Dennis. But perhaps more distressing to the Schuyler women was the fact that the child's father was black.

A woman friend, a young physician working in a New York clinic, saw Philippa (without her mother's knowledge), and confirmed her pregnancy. "Philippa was beside herself," the woman reported to me over twenty-five years later. "She told me the father of the child was black. 'Having a child is the greatest miracle in life,' I told her. 'But I can't have this child.' 'Why?' 'Because I'd disappoint my mother.'"[19]

Philippa's search for a mate was greatly influenced by her eugenic agenda. At heart, she considered her parents' marriage, because of her own insurmountable difficulties, a mistake, even though it had produced a genius child. She wanted to correct this by marrying an Aryan type—a Maurice Raymond or an André Gascht—so that her children might not suffer as she had.

Abortion seemed the only solution now. It would be illegal in America for another eight years. Certain European and Latin American countries allowed abortions, but even there it was not easily arranged at the end of the second trimester, and they were costly. At first, Philippa tried holistic and homeopathic medicines. But nothing worked. She decided in late August to fly to Europe.

Phil arrived in Holland on August 24. She was staying with Lineke and Theo, in whom she confided. "Abortion was illegal in Holland then. I did not know where to send her," Lineke said to me many years later, adding almost as an apologia, "I had several children."[20] Somewhat reluctantly, she also admitted that from that point on their relationship changed. Most likely, Philippa never forgave her good friend for not trying.

From Holland, Phil telephoned Gerd, who advised her to come to Norway, to stay with her; she would do whatever she could. Abortion was legal, but at her advanced stage only on the advice of a psychiatrist. Philippa would have to be "interviewed" and sign a document that she was not mentally fit to carry the child to term.

Phil did not go to Norway. That the abortion would not be anonymous troubled her—she would have to sign documents and verbally confront her feelings about the unborn child and her own state of mind.

Gerd suggested Denmark. Phil flew to Copenhagen and checked in at the Grand Hotel. She stayed a week, waiting, searching, throwing her tarot cards. Then suddenly on the third of September she called Dennis in London, asking him to fly over and meet her. Dennis, busy with preparations for the imminent, prestigious Edinburgh Music Festival, dropped everything and flew over.

What happened in Copenhagen is largely shrouded in mystery. By the time I had begun research on this part of Philippa's life, Dennis had moved to Australia, and he died in 1987. I regret that I did not have any opportunity to hear his side of the story. Apparently, their first day together after the long separation was quite idyllic. On their second day, however, Philippa, despairing of being able to have an abortion, raised urgently the question of the quick divorce and immediate marriage. When Dennis patiently explained to her again the irrationality of the scheme, everything went awry.

Shortly after they parted, Dennis would write Josephine:

> The first day [together] was joyful but the last three were ruined by those very strange elements you mention in [one of] your wise letters. So often Philippa seemed just like a little jungle creature peering at me angrily through the undergrowth. She has so many psychological chips on her shoulder that for a free-thinking Irishman like myself it is very difficult to steer clear of the taboos unless one confines conversation to music and the weather. . . . I could not live in this stifling dungeon she seems to occupy. . . . The only thing that seems to draw her out is music, or some chance acquaintance or stranger on whom she can lavish natural human affection for a while.[21]

Philippa wrote her own version of their time together:

> Dennis was lovely yesterday but today he made long tirades against religion, government and most of the world's important political leaders . . . I think he hates underneath the fact that I am MAKING my price marriage—and he is having to

bend to that. He is against marriage as an institution. He thinks it is the most immoral crime after murder if there is no romance in it. . . . He says it's "nasty" of me to say that I would not want to live with him without marriage. I said that a man is not keen for marriage if he can live with a girl. . . . Then he accused me of not loving him enough—and I said that if one wishes to be loved one must act lovable—that attacking everything another person stands for is not appealing behavior. Well, I'm holding out for MEXICO. If he will go through the process of a Mexican divorce with all it entails NOW, I will marry him.[22]

The following day, in an attempt to explain her failure, Philippa wrote her mother an even more unbalanced letter:

On Sunday afternoon he made the position clear to me. Oh please don't think he does not want to marry me. He does, but the price is that I turn RED. I must abandon my "fanatic religious and political opinions" and come over to his. He spouts the REDDEST political statements I've heard this side of Peking. He said America under the dread terror of McCarthyism was more totalitarian than Russia. I said I found the USA just fine in [the 1950s]. He said that's because I was one of those who was doing the persecution of the innocent.[23]

One thing is certain: they parted on inconclusive terms, and their personal relationship hung in the balance.

Their professional relationship, however, continued. Dennis was planning to come to New York in two weeks and Philippa scheduled to return to England in mid-November for her Wimbledon recital. But Dennis caught the flu and his doctor advised him against flying.

The subject of Copenhagen was studiously avoided until shortly before her trip to England:

I think maybe sufficient time has passed for me to write you something on the personal level, something to help us understand what happened in Copenhagen. . . . I found that you were much more identified with ideas than I had imagined, and you found that I was not so ready to please you by doing exactly what you wanted. . . .

What came out . . . for me was another view of you. I had realized that you had problems, but I was not aware of how deep [they went] . . . nor how bitter were your resentments. . . . My dear sweet one, when you said that I was trying to "take advantage of a coloured girl" I was absolutely flabbergasted. For one thing, I have never thought of any human being, least of all you, in those terms. Colour means nothing to me at all on the human skin, and while I realize that you live in a society where too many are colour conscious, I had hoped that this small facet of my personality had filtered through to you.

Your cries on the swings and roundabouts were sweet and pathetic to hear, just like a child "deprived," as you said "of childish joys like these" because you were an infant prodigy. Here, I confess, was a facet of you I had not suspected. It was nice to do the things you had not done since you were six, but somehow that experience did not make you deeply happy, perhaps because what you needed even more was to have the experience of an adult. . . .

You said in Copenhagen . . . with such blazing bitterness in your eyes: "You want to save my soul!" Only you know what deep self-accusation underlies that cry.

> My darling I would save you from hell and high water if I could and there was the need. But I never thought your soul needed saving. Is it that you need a Father figure, and at the same time resent one?
>
> I have been playing over some of your tapes and recordings, wondering how far these interpretations reveal you—the real you. . . . Here, in almost every piece you have an emotional approach. I find myself in complete agreement with you in this music. In life, too, it is the emotion . . . that counts so much more than the intellectual. I, therefore wonder why you allow this to predominate in the closest of all human relationships—that between a woman and the man who loves her. The answer could be that you simply do not love the man. Or is there another reason? . . . Sometimes I think there is a strange neurosis of fear and doubt in you, as there was in [T. E.] Lawrence. I never knew him, but close friends of his have told me that he was a man who was afraid to love. He was a desperately unhappy man, and ended his days isolated among "the common man" in the RAF. I think that Lawrence only had to care for somebody more than himself to find himself. Please do not resent my speculations, which are intended in tenderness and friendship, and in search of true understanding. If there is one thing I would like to see brought into your life, it is greater fulfillment and happiness, and freedom from torment of doubt, and the serenity that comes from loving and being loved.[24]

It was an extraordinarily perceptive analysis. An important piece was missing, though, from the jigsaw puzzle: Dennis did not know that Philippa was pregnant. And desperate.

34

Tijuana

Unsuccessful either in obtaining an abortion or in convincing Dennis to embark on the Mexican adventure, a very anguished Philippa left Europe, landing in New York on September 7.

She called her mother's nephew, a physician in New Orleans; he reluctantly gave her the name of someone in California who could direct her to a doctor in Mexico. By now, Philippa was on the cusp of her fifth month, and frantic.[1]

On September 14, she flew to the West Coast and spent a restless night in San Diego. The following morning she awoke gripped with apprehension that the Mexican doctor might not take her. Her contact was a Señora Guinar, in Tijuana. And indeed, when Phil called her at 9 a.m. she was advised that the doctor would not perform the operation. She was too far along.

She begged and pleaded, and finally Señora Guinar agreed to keep trying.

At 1:15 p.m., Philippa got through to Tijuana again. The answer was still a firm no, but she might try another clinic. With great trepidation, she dialed the new number. The woman on the other end was more encouraging: Be in Tijuana at 2 p.m. A young man in a dark blue Mercury will meet you at the benches right after you cross the border. But she promised nothing.

Changing her money into traveler's check's delayed her, and even though a passerby gave Philippa a lift to the border, she was fifteen minutes late. As she looked around apprehensively, a young man came up to her. It was the driver. He had recognized her from the description given over the phone: tall, thin, a red kerchief tied around her dark hair.

They drove around Tijuana picking up women at various prearranged locations—in front of Woolworth's; at the jai alai fronton; in a parking lot. Then he took everyone to a small building with a sign, CLINICA CENTRAL.

Philippa waited a long time. Finally the doctor examined her; he recognized immediately that she was far into her second trimester. But all he said was, "You are on the threshold of your twenty-first week." Again panic gripped Philippa. Was he trying to tell her that he would not take care of her? After another agonizing wait, an attendant came out to say they would accept her.

It would be a two-stage operation performed on consecutive days.

The first step was done immediately. They put a short white hospital gown on her, strapped her legs down, and the doctor began to probe. The pain was excruciating. They held a mask over her nose. It seemed ages to her before unconsciousness came. The doctor injected some concentrated saline solution into the uterine sac in order to initiate labor within twelve to fifteen hours.

When Philippa came to, the doctor was just finishing. She now noticed he had a "strong, rather noble face, kind but ironical"—a face, she thought, that spoke of an interesting life. "I think you have been to Cuba," he said to her, interrupting her thoughts. "You tied the red scarf around your head the way they do it there."

"Are you from Cuba?" she asked.

"I am from Heaven. De los Cielos."

Philippa awoke the next morning with blood all over her white robe and the bed sheets. Frightened, she looked in vain for the young nurse who had spent part of the night with her. Philippa groped her way to the bathroom and tried to urinate but couldn't. She went back to bed but the need to void was becoming more and more urgent. She rang the bell on the little table many times before the nurse came and helped her by pulling out the tightly wedged tampon.

Soon, two other women were brought into the room, accompanied by "big, important-looking men." Philippa felt suddenly bitter that they had people to come with them, while she was alone.

She expected the doctor to do the second operation that morning, but for some reason it was delayed. They moved her to a little room with no real bed, just a hard blue couch with two blankets on it. Lying on it made her back ache. The only window was covered with opaque glass and could not be opened. She wanted to pray but could not find her rosary. "Has God forsaken me?" she thought.

In the afternoon, Philippa was advised that the major operation might not take place until the following day. But at 7 p.m., she was suddenly told that the doctor would operate now.

On the table, she felt as if some awful machine were being screwed into her. The doctor kept saying, "Relax, Relax." He told her to hold the mask to her face. She asked the nurse to help, and began breathing hastily, hoping for swift release. "And then things became a black hole in which nameless horrors were happening."

After an undetermined period of time, the "swirling darkness became grayer and lighter." She was regaining consciousness only to realize, with great pain, that the operation was over.

The young attendant lifted her and carried her back to the little room. They must have administered prostaglandin to accelerate her labor, for she felt freezing cold, "shivering like a leaf in the deathly autumn sun." Piling five blankets on her, the attendant said it was always like this afterwards. Philippa prayed to Christ and all the saints. She felt sore and ripped inside. It was hard to find any position to lie in that was not sheer agony. She wanted desperately to ask for a sleeping pill, but decided not to. At last, sleep mercifully came.

Sometime during the night—she couldn't remember when—they moved her to a cubicle adjacent to the operating room, and there she awoke the next morning feeling strangely happy, even though an old and wretched face looked back at her from the bathroom mirror.

She was told she would have to leave after the doctor checked her out. "I can't possibly leave today," she thought. "How could I walk across the border under the searching eyes of those immigration officials?"

At 4 p.m. the doctor arrived. He operated on two other women. Despite the loud radio, she could hear the cries of pain through the wall. "I was braver than they," she thought. One operation must not have gone well. "More terramycin! Try more terramycin!" someone was yelling. Then complete silence.

Philippa heard her name and walked slowly into the examining room. She lay taut and rigid on the operating table. The doctor started to probe. It was so painful she gripped the table, hard, with both hands to keep from crying. The doctor was repeating, "Relax, Relax," and then, "Put your hips down, *down!*" The probing and prying, the washing off seemed to continue forever.

"Shall I tell you a funny story?" he finally said in his Cuban accent, feeling obviously satisfied with the examination. "For the life of me I don't see how you girls get pregnant. Here I am trying to help you, and you squeeze up so tight that I can't even get an instrument in—so how do you ever get pregnant?" It took her a little while to take in the humor.

The doctor said she could leave right away. Philippa countered weakly that she did not think she could make it across the border. "Oh, a woman is a powerful thing," he said. "What they can do, I couldn't do."

She threw on her clothes in haste, and sat on the bed forlorn and scared. Then unexpectedly she was told she could stay another night.

When Philippa awoke the next morning, it was Saturday, September 18, her fourth day at the clinic. At 9 a.m., the Mexican nurse brought her two slices of white bread with strawberry jam and a glass of milk. An hour later one of the young attendants performed the final cleansing. He pushed a tampon in hard. "It hurt like fire." Philippa dressed absentmindedly. The young man who had picked her up now drove her to the border.

"I hated to leave," reads the last line of her little journal "—the dark room with its light brown panelled walls that shut out the world was such a safe little box. I did not want to leave Tijuana behind, face the cold awful world with its terrifying decisions again. I wanted to cry and cry. The young man—"

Abruptly, her journal stops.

❖ ❖ ❖

Philippa returned to New York to pick up the pieces of her life. Ironically, Georges Apedo-Amah was there for a U.N. session, and he stopped by to see the Schuylers. Philippa told him about her recent abortion and that he was the father of her unborn child. At first he was shocked, then highly skeptical, but he offered to reimburse her for the cost of the operation. Despite the trauma—for both of them—Georges and Philippa parted on warm, even tender terms, and over the course of time he became convinced Philippa was telling the truth. "I want you to be mine forever," he would later write her, "and I view with great nostalgia him who did not arrive . . . he or she would have been so EXTRAORDINARY."[2]

Two weeks after her abortion, on October 1, 1965, the trial of Maurice Raymond took place. At the request of the prosecutor, the accused was interrogated behind closed doors. The court then ordered the doors open and gave a public declaration: "RAYMOND, Maurice Pierre Alfred, born 17 September, 1926, in Lyon, the son of Pierre Antoine RAYMOND and Blanche Hélène DIEU; superintendent of manufacturing; divorced, with one child; stood accused of an outrage against good morals having during the month of June 1964 at Lyon, on French territory, and in Turin. . . . knowingly transported and distributed by some means numerous photographs contrary to good morals, representing a man and a woman, nude, and in obscene poses."

Maurice was found guilty and condemned to forty days imprisonment and a fine of 800 francs ($163), plus court costs of 255 francs and 15 centimes ($52). The court further ordered the destruction of the six seized negatives.

❖ ❖ ❖

On a balmy day in June of 1991, I was wandering through the narrow streets of Lyons's old Villeurbanne district. The blue-collar area had seen some change in recent decades. Many of the small industrial establishments were gone, and an increasing number of foreign workers, mostly Algerians, had moved in. A sprinkling of high-rises dotted the arrondisement.

I had just obtained a confidential report on Maurice Raymond. "Monsieur RAYMOND," it began, " . . . died on 12 April 1984 at his residence, 23 rue Charles Robin, in Villeurbanne. M. Raymond, who lived in the left ground-floor apartment of the [four-story] building, had been discovered dead by his neighbors. Taken to the [morgue], the coroner had concluded that he died of natural causes.

"We have further learned," the report continued, "that in addition to his 1965 sentence, he had been sentenced in 1957 to four years in prison for aggravated robbery [these were the four years Maurice told Phil he lived in Sweden], and again in 1979, to one year, for assault and battery."

I rang the bell of the ground-floor apartment formerly occupied by the Raymonds—Maurice and his mother—but a frightened woman refused to open the door. After wandering around the neighborhood, fruitlessly trying to find someone who remembered the Raymonds, I noticed that the shutters were open on the ground floor of the apartment building catercornered from the Raymonds'. M. and Mme. Bonnet were having their lunch inside.

Yes, they had lived in their house for about twenty years, and oh yes, they knew Mme. Blanche Raymond, Maurice's mother, well. She was a lovely woman, kind and refined. Mme. Bonnet went back to another room and produced a small color photograph of a nice-looking gray-haired lady seated close to a young girl.

Was the girl in the photograph her granddaughter, I asked? No, it was a neighbor's child. The granddaughter, Marie, was not a nice person. She showed up only when she wanted to worm some money out of her grandmother or her father. They had never met Maurice's father.

Blanche Raymond had died in June 1981 and her son kept the apartment. Maurice, they told me, was an exceptionally handsome and well-educated man. But he had been in one or another kind of trouble with the law all of his life. His mother had shed many a tear over him.

They had not met Philippa, since they moved into the neighborhood only in 1968. But they did remember the 1979 incident. Maurice had taken to drinking quite heavily. One night, obviously drunk, he had become embroiled in a brutal argument with a friend and bludgeoned him with a hatchet. The friend staggered, wounded, down the street before he collapsed and was found barely alive lying on the ground at dawn.

The circumstances under which the obscene photographs of Philippa had been taken puzzled me for years. They clearly depicted her in an explicit sexual act with a man. I had known of their existence for some time. André had received them. His mother, who opened all his mail, was the first to see them and was shocked. It was then she had destroyed all of Phil's letters, which André had been saving. Albert Maurice may also have received at least one, and that may be part of the reason he ultimately ceased communication with Philippa. (When André ran into Albert in 1988, in the south of France, he reported that I was working on a biography of Philippa and that I would like to talk with him. "I never want to hear the name Philippa Schuyler mentioned to me ever again, as long as I live," was his response to André.[3]) The director of Herder (her publisher in Rome of *Jungle Saints*) had also received several. He refused to believe they were of Philippa—although it looked so much like her. But he must have been sufficiently concerned because he later admitted to taking them to a photo expert—who said they were, in all probability, fakes.[4]

There is no doubt, however, that the photographs, as the courts decided, were genuine. But why had Philippa done this? Was she carried away with the passion of the moment? Was she imitating her mother who, in her salad days, had been a pin-up girl? Now listening to the Bonnets' account of Maurice's lifelong association with crime, the idea occurred to me that these photos had been taken with Phil's full consent, as a scheme to make money. The professional quality of the prints suggests that they were intended for submission to hard-core pornographic publications, which were blossoming in Denmark at the time. Both Maurice and Philippa were hard up for money.

But why had Maurice mailed the photographs after their relationship had collapsed? The obvious motive of revenge was only part of the answer. A con artist par excellence, Maurice must have felt that money could be gained through blackmail or some other scheme of intimidation.

And another light turned on: Raymond could not have come to visit her in America—because of his criminal record.

Yet despite these insights, and despite the declaration of the court, I was never able to declare this case closed.

35

Sic Transit

By 1966, except for the remunerative lecture tour with the John Birch Society, the Felipa Monterro charade had run its course, proven an unsuccessful gambit into the white classical-music milieu of America. Her relationship with Dennis had also come to an end. After honoring her engagements with him in London and with the Cairo Symphony Orchestra—where her piano concerto *Nile Fantasy* received rave reviews but Dennis's conducting did not, and which she called a "season in hell"—Philippa left for a two-month tour of sub-Saharan Africa. She was never to see Dennis again, although occasionally they corresponded about music matters.[1]

Phil's tour of Africa was particularly difficult. Caught in the crossfire of three coups, Nigeria, Ghana, and Dahomey, Philippa decided that things were too hot, even for her, and she left abruptly in the middle of January to go home, via Togo, for some rest.

In Lomé, she stayed with Georges Apedo-Amah, whom she had not seen since the abortion. Even with Georges, her time did not go well, but it is not clear why. Long after Philippa left, he wrote her: "I was not pleased with certain things that happened during your stay here, but this does not change any of my feelings toward you. I feel that I must help you even more."[2]

While in Togo, Philippa had begged her mother not to book any concerts for at least a week after she returned from Africa. But Jody disregarded her wishes, scheduling several recitals that started less than three days after her arrival. The first one was set for North Carolina.

Although Philippa's flight had been delayed by thirty-six hours, Jody insisted that her daughter proceed to North Carolina almost immediately. The scene in the international-arrivals building engraved itself in the memory of a young physician friend, the same woman who had counseled Philippa about her pregnancy: "I had gone out there to pick up Philippa with George and Josephine. She came off the plane obviously very tired and not well. Josephine didn't even ask her how she was. She just said, 'You have concerts to give tomorrow and you had better be ready.' I was shocked, simply shocked, by her mother's behavior."[3]

Philippa flew to North Carolina that evening for a concert the following afternoon. A white couple, Joseph and Mary Myers, were in the audience. Mary clearly remembered her first encounter with the American pianist: "Philippa walked center stage, her white dress wrinkled and stained, her hair a mess. She played terribly, like an automaton. But there was something there that just broke my heart, and I wanted to help her."[4]

Photographs of Philippa taken during this period, in fact, corroborate Mary's perceptions. There is something elusive and disquieting about them. Philippa appears to be staring out into the middle distance. And it is difficult to size up Philippa, as if her constant search for an identity had indeed fragmented her *force vital*. In some photos, Phil appears African American, in others, white; in still others she looks Hispanic. In Vietnam, a long, straight wig flowing over her *ao-dai*, she could easily be mistaken for an Asian. A reporter tried to catch Philippa in a word painting:

> She spoke fast and never faltered. Dates, names of people and places, swatches of history, musicology, anthropology and self-revelation all poured out in an almost unbroken stream, like a massive but random recitation. Her tapering short-nailed fingers wove together, twisting and flexing as she talked, suggesting twanging tension and threshing compulsion for constant activity. Yet an atmosphere of child-like emotional fragility hung about her, combined with the electricity of awesome energy.
>
> She wore sunglasses, even in the dim room. Her skin is lighter than an East Indian's. . . . Her hair was shoulder length and dyed coppery red. She wore no jewelry except two medals around her neck and very little makeup. . . . "I'm not sure what I am," she told me.[5]

The Myers developed a close relationship with Philippa. They began to arrange concerts for her in the South, for white audiences. This had never happened to Philippa before."You can't imagine how grateful I am for your kindness," she wrote Joseph and Mary, shortly after having met them.

> For years I have deplored the way I have been effectively segregated as if I had a wall around me. So I spent most of my time abroad. . . . I am very anxious to appear before white groups in America. White Americans know me more as a writer and a journalist than as an artist. This segregation has been so effective that though I have been a successful artist abroad, and almost all American Negroes know me in that capacity—many whites think of me as the *Who Killed the Congo?* girl who wrote on Katanga for UPI.
>
> My father's way-out extreme right conservatism has been an extra factor in segregating me. For some reason, he has chosen to be politically on the same side of the fence as the most prejudicial whites in this country. The pro-Rhodesia people, etc., the anti–civil rights people. You can see how helpful this is to me! He attacks in his columns the people who might help me and is friendly with the people who won't have me.[6]

In fact, George's obsessive conservatism, his anticommunism, had intensified over the years. He viewed almost everything coming out of the 1960s as a communist conspiracy. And although he had his own private agenda for the black race, whose welfare and survival were uppermost in his thoughts, George did not support the black revolution. Carl Rowan pigeonholed his rival's political leanings when he wrote that Schuyler was the type of man who would have thought the Renaissance subversive.[7]

The civil rights movement was anathema to George; the freedom riders, "brainchildren of crackpots and conspirators . . . no more than hitchhikers pressing their luck."[8] He believed almost every leader coming out of the movement invited racial polarization and contributed toward a breakdown in interracial communication. He denounced the marches on Washington and other demonstrations as "part of the Red techniques of agitation, infiltration, and subversion."[9]

In 1964, George wrote an article for the *Pittsburgh Courier* criticizing the award of the Nobel Peace Prize to Martin Luther King, Jr. The *Courier* refused to publish it, but William Loeb's *Manchester Union Leader* placed it on the front page. It was a scathing attack: "Neither directly or indirectly has Dr. King made any contribution to world (or even domestic) peace. Methinks the Lenin Prize would have been more appropriate. . . . Dr. King's principal contribution to world peace has been to roam the country like some sable Typhoid Mary, infecting the mentally disturbed with perversion of Christian doctrine, and grabbing fat lecture fees from the shallow-pated."[10]

George's disrespect toward the honored Dr. King was his final undoing. He ceased most of his activities with the *Courier* during November 1964 and was relegated to writing book reviews. By 1966 he had no relation to the newspaper he had been intimately connected with for almost forty years. His daughter was now the family's primary source of income.

❖ ❖ ❖

In the spring of 1966, Philippa was invited by Henry Cabot Lodge, then U.S. ambassador to South Vietnam, to perform for wounded soldiers there. Vietnam was a country she had never before visited. William Loeb also engaged her as his paper's foreign correspondent.

She left that summer, and did a sweep of the Far East. In Hong Kong, she met Ernie Pereira again. Although they had continued to correspond, Philippa had not seen him in almost six years. Had he changed? she wrote Ernie, anticipating her arrival. "Time," he answered, "like the draftsman, rubs out the weak lines in the drawing of life."[11] When he posed her the same questions, she wrote back a little flippantly that her hair was now red.

Philippa decided to reconsider Ernie as a mate. "If he is in a serious mood, would you consent to him?" she wrote her mother like an adolescent. "He is nearer my age and my physical type than ANY of these other men. As regards looks, we would produce a GOOD LOOKING child. A marriage does not have to last an eternity. But it should produce some positive thing. A pretty olive-skinned child, with straight hair and elegant features."[12]

On Tuesday, July 26, after playing two concerts in Taiwan, Philippa met Ernie at the Hong Kong airport. "He has charming manners. . . . I have not assessed him yet," she wrote home elliptically.[13] Ernie, on the other hand, was totally unreserved: she was as beautiful and exciting as he remembered.

"This encounter we had was memorable, to say the least," Ernie wrote Philippa after she had left for Saigon. "This is the first time that you allowed your mask to drop and I see you for the first time in all your 'nakedness.' It was indeed a happy, brief encounter and the evening was delightful in every sense of the word. And if Venus had a hand in it, I am quite thankful to her. I hope such nights can be repeated daily. . . . I think you will agree that the unexpectedness of it enhanced the occasion, making both of us prurient for more. . . ."[14]

36

Good Men Die

From the first, Philippa was both intrigued and appalled by Vietnam. She found its people the most beautiful she had ever seen – even more so than the Chinese or Japanese. Their interest in the occult resonated with her, as well as their approach to healing and medicine. Their cunning was like nothing she had ever encountered. She was particularly attracted to their children, especially their "half-castes" – products of the American GIs and Vietnamese women – and she co-founded an organization called the Amerasian Foundation to help these families. Philippa wrote her mother that she felt an affinity bordering on kinship with these graceful Asians, with their large limpid eyes and pale skins, adrift in a country that had known no peace for almost thirty years.

Invited by Henry Cabot Lodge to perform at the National Conservatory of Music, Phil arrived in Saigon on September 2, 1966, and was met at the airport by a Mrs. Dorian, an employee of JUSPAO, the Joint United States Public Affairs Office. Mrs. Dorian drove her to the embassy (where Phil was staying) and proceeded to attach herself to the young American, screening her appointments, mailing her letters, and generally keeping her away from taxis, pagodas, and "undesirables," which apparently included all Vietnamese people.[1] Mrs. Dorian had a thick French accent but was vague about her background. There could be little doubt that she was a CIA agent.

The Lodges knew Philippa's parents and immediately after the concerts, the ambassador, undoubtedly concerned about her safety, forcefully encouraged Philippa to leave the country. She reminded him that she was also a journalist and wanted to learn more about the situation, but to no avail.

Philippa decided to move out of the embassy. As for Mrs. Dorian, she was also dedicated to the idea of getting Miss Schuyler out of the country as quickly as possible. There ensued an elaborate game of cat and mouse between the two women, Philippa repeatedly booking a flight for her next destination and then canceling at the last moment.

Thus dodging Dorian, she had played two additional recitals for wounded American soldiers at the Third and Seventeenth Military Field Hospitals, and met an impressive array of people, even before her first week in Vietnam was over. "I have visited the Vietnam People's Hospital, asylums, orphanages, refugee camps," she wrote her mother. "I interviewed ex-Viet Cong terrorists . . . have talked with student groups, priests, social workers . . . American officials. I have gone into poor Vietnamese homes . . . and have talked to 'the man in the street' and . . . I have interviewed Negro and white civilians, and military personnel."[2]

Philippa found the political theater of Vietnam difficult to comprehend. She attended an official press briefing, a morass of misinformation and nonspeak. "Vietnam beats Katanga any day," she wrote home after ten days in the country. "One hears so many conflicting opinions here. It will take a long time to straighten this mess out. . . . What an involved situation. Corruption plus idealism plus brutality plus deceit plus incompetence plus everything else confusing. . . . But I have run around here . . . digging up hidden facts like a little pig digging roots."[3]

Philippa had several advantages over many American journalists in Vietnam. First, she was fluent in French and even knew some Vietnamese. Second, although the Catholics represented only a small percentage of the population, innumerable missions dotted the countryside; and as a well-known and devout Catholic lay worker, Philippa could always be sure of a place to stay and a meal, sure to meet Vietnamese from all walks of life, and sure to hitch a ride into areas normally closed to correspondents.

Phil had another advantage as an investigative reporter. With her golden skin, and dressed in a conical hat, a long, straight black wig, a blue *ao-dai* gown, high-necked and slit from ankle to waist, a small jade amulet of Buddha around her neck, black satin trousers, and high-heeled sandals, she could assume a disguise that would fool 95 percent of Americans and most Vietnamese.

This irked the U.S. intelligence community no end. Despite her association with an accredited ultraconservative newspaper (the *Manchester Union Leader*) and her endorsement of the American presence in Vietnam, she soon discovered that the information and security arms of the U.S. military would go to great lengths to restrict her movements and access to independent facts. She was hardly alone in this harassment; other correspondents had been equally discouraged from venturing outside the protocol-ringed ivory towers of Saigon or beyond the bars of Cholon. But the military (and the CIA agents who swarmed the country in its service) considered Philippa a particular gadfly. At least the other journalists could be spotted; Philippa blended into the crowd.

In her disguise and unaided, she found her way to the understaffed and overcrowded civilian hospital in Saigon, where the bug-ridden wards stank in the sweltering heat and live patients might stay in the same bed with a corpse for a day or more. "This is the incarnation of a seventeenth-century charnel house," she wrote in her notes.[4] She visited the Vietnamese military Cong Hoa hospital where 2,500 wounded patients were often squeezed into 1,500 beds.

The doctors were doing a heroic job, but, surprisingly, there was an appalling lack of medical supplies despite massive shipments from many countries of the world. Corruption of homeric proportions had siphoned them off.

She met two African-American soldiers from the United States Agency for International Development, Leonard Holsey and William Mann, with whom she would form a close friendship. It was Holsey who prevailed on her to leave the embassy and found a place for her to stay, making it a little easier to shake Mrs. Dorian. And it was Holsey who took her to visit a detachment of Negro soldiers within a Viet Cong area.

Despite the many obstacles placed in her way, Philippa witnessed the internationally supervised election, interviewed both President Nguyen Van Thieu and Foreign Minister Tran Van Do, and took a whole day's trek by jeep, canoe, and pedicart into the interior of Long Xuyen Province.

"This is the funniest mixed up war I ever saw," she wrote Josephine on September 15 from Saigon. "Everyone around here must have done something wrong in their last reincarnation and is getting punished for it now. . . . The Viet Cong are AWFUL CRUEL; the Americans are AWFUL CRUDE; and the South Vietnamese are AWFUL CORRUPT. Take your choice."[5]

To Philippa, Vietnam had increasingly become a sea of despair. She condemned the brutality of the Viet Cong, the booby traps, the burning of villages, the attacks on women and children, and especially America's incompetence: innocent victims of our own "friendly fire, in a land where each hamlet was infiltrated by the enemy, where one had the greatest difficulty telling friend from foe. . . . A disfigured girl, the victim of one of our napalm bombings, looks straight through you, with a blank . . . stare. Should one look back at her? Sympathy would make her even more bitter, curiosity would be an insult. So should one ignore her, and become as calloused as many Westerners are, or should one feel part of a profound and collective guilt?"[6]

Or the man whose head was smashed during an interrogation because American soldiers mistook him for a Viet Cong guerrilla. A U.S. intelligence officer told her about the torture methods they used to extract information from captured Viet Cong. He had prefaced it with the remark, "We are not permitted to touch the bastards, of course, but there is always our Vietnamese counterpart who does not labor under the same idiotic restrictions."[7]

Philippa felt the American presence in Vietnam, however, was part of a grim and necessary crusade. In fact, her dispatches for the *Manchester Union Leader* became some of the most conservative coming out of Vietnam at the time.[8] And yet her political conservatism did not blind her to the mistakes being made by the Americans in Vietnam. Like the much more liberal Frances FitzGerald, Philippa could also see clearly that "the inability of the Americans to understand the Vietnamese and their culture was dooming them to failure."[9]

❖ ❖ ❖

By mid-September, Philippa had already decided to write a book about Vietnam. She bought American army fatigues in the Cholon black market and finagled a reservation on Military Flight 651 to Hue, to see "some action" beyond Saigon's nightly artillery barrages. On September 16, she left her room at 4 a.m. for the Tan Son Hut airport, and after going through numerous checkpoints, wading through a huge crowd of mostly Vietnamese soldiers—some pitifully young—and waiting endlessly, she finally boarded the transport at 8 a.m.

The hard, uncomfortable seats were lined up against the walls, leaving the center area for cargo. A man in civilian clothes sat down next to her, helped her adjust the seatbelt of unfamiliar design, and then proceeded to study her. She, in turn, noted that he was tall, six foot one or two, rangy, and about thirty-five years old. He had blue-green eyes and brown hair, a sharp nose, a strong chin, and high cheekbones.

There was something elusive, however, about the handsome stranger.

Philippa decided to play it cool. She was going to Da Nang, the stop before Hue, she said. He replied that he was going to Hue, and without much ado asked her to join him. He promised to show her around the ancient city, and wrote down his name, Captain James Leiter, and a telephone number. He added that he had forgotten his army uniform.

Philippa demurred, insisting that she was "expected at the Press Center" in Da Nang. To her surprise, Captain Leiter deplaned with her in Da Nang. He immediately spotted three Americans in civilian clothes, instructed one of them to commandeer a seat for Philippa on an Air America flight to Hue later that afternoon, and returned to the waiting plane.

Arriving in Hue that afternoon, she found an army phone. A sergeant whom she asked for help was puzzled: He had never heard of the telephone number 7332, and indeed it was not listed in the directory. They tried the number anyway and a faraway voice said, "Captain Leiter speaking."

Leiter had not really expected Philippa to respond to his offer, but one could not tell it from the way he said, "Stay right there. I'll pick you up in twenty minutes."

The clues started to fall into place: the civilian clothes, the unlisted phone number; the ability to commandeer a seat, all pointed to Leiter being high-up in military intelligence.[10]

"Jim" drove Philippa around in his jeep, showing her the sights: the university, the odd-looking Catholic cathedral, the American consulate, and the United States Information Service Library, wrecked and burned by a rioting mob only a few months earlier. They passed the large statue of a Chinese dog that stood outside the Imperial Citadel.

Their conversation had a peculiar flavor: He asked her questions that she evaded and she asked him questions that he evaded. He finally drove her to a villa, a large pastel-colored house with armed guards in front, where she spent the night.

Over the next several days, Jim introduced her to some of his colleagues. He took her to a birthday party in a "club" run by Chinese. She was the only woman there as well as the only sober person among a bunch of inebriated men. Furthermore, she was making an upsetting discovery—that personal cruelty was not the sole prerogative of the Viet Cong. "I get a kick out of killing now!" one officer said drunkenly. "When I see a wounded Cong lying on the ground, writhing and spewing out his guts, I feel good."[11] She found it hard to enter into the spirit of the occasion.

But Philippa was more relaxed when Leiter had her over for dinner the next night in his house, a large villa surrounded by a wall and high gate, quite a distance from the MACV compound. He shared it with a dozen other intelligence personnel. The occasion was quite convivial.

Philippa saw quite of bit of Jim. One evening as they were sitting in the jeep gazing at Hue's Perfume River, he told her she was the first good thing that had happened to him since he had come to Vietnam.

Philippa, for her part, "felt a genuine sympathy for the 'irrelation' suffered by the few people in this '[intelligence] business' who are basically sincere. This 'irrelation' may be the quixotic extension of the idealism that originally [brought] them into this unrewarding profession. There is a depth of loneliness . . . that isolates [them] from the world [they] are so closely observing. It is hard for the hunter to feel part of society, or for the eagle to feel at ease with the lambs."[12]

One story in particular that Leiter told Philippa touched her. While he never discussed the methods of interrogation used on the Viet Cong by their South Vietnamese counterparts, he confided that on a certain occasion, witnessing it became too much for him. Right before a break in the interrogations, Leiter took his service revolver, loaded it, placed it in the open drawer of his desk, and left the room—all in full view of the prisoner. Shortly thereafter Leiter heard a shot. The person being interrogated had grabbed the revolver and blown his brains out.[13]

At first, James Leiter seemed like a hero to Phil: a man valiantly and bravely fighting communism. Only gradually did she realize that he was no idealogue. To him, a professional soldier, the overriding objective was to win the war in Vietnam, not destroy communism. "She believed in the domino effect," he said many years later. "I thought it was balderdash."[14]

They argued some about communism. He felt that she had "too many misconceptions about it."[15] And he disagreed with her comparison of the struggle in Africa and the war in Vietnam. She had given him a copy of *Who Killed the Congo*? to read. He thought it a remarkable book by one so young. And when he asked her if she had written it by herself, "Philippa got insulted. . . . This is about the only time I saw her agitated.

"She was a very caring individual, a great humanitarian," Jim said when I interviewed him many years later, lost in memories of events he had not thought about in years.

> She could have been another Mother Theresa. . . . She felt the biggest problem with the war was its effect on children . . . and isn't it in an effort to save some of them that she perished? . . . Everything in her life was a "mission." . . . The

> things she championed were often lost causes, though. . . . Maybe this was a metaphor for her life, [adding pensively] with men in particular.[16]
>
> In a way she was a strange creature. She did not seem to have a base of operation that I was aware of. I'd run into her in Da Nang, even Quang Tri—don't know how the hell she ever got the permission to go there. . . . She was always looking for a ride on an airplane or some kind of transportation. And in a lot of instances she was looking for a place to sleep. . . . Occasionally her persistence grated on people.
>
> She traveled like a hobo—with bags and bags . . . things tied up in bundles, straw baskets . . . with tons of things falling out of them. One time we were talking and I said, "You must have been a very pretty child," and she said, "Let me show you a picture, if I can find it." And out of one of these falling-apart straw bags she pulled out a picture of herself at about eight or so, in a long concert gown with her parents behind her.[17]

By then, Jim knew that she was part black—although this had not been his first impression. They had talked "a little about it . . . but Philippa kind of pulled back when the subject came up."[18] Trained and accustomed to carefully analyzing facts and observations, Jim had suspected there were deep schisms in the Schuyler family. Among other things, he came to the interesting conclusion that Philippa was jealous of her mother's writing skills. He also sensed a resentment toward her parents for making her "neither fish nor fowl."[19]

❖ ❖ ❖

Philippa concluded her first trip to Vietnam in the late autumn of 1966, having spent almost six weeks there, and returned to America for some concerts. But her career seemed to be moving more and more toward writing—so much so that a week after a very successful recital in New York, on January 29, 1967, Josephine wrote to her daughter: "You must never let music become secondary to your other activities. Writing—as you recall—started just to fill in and to get wider knowledge of your name. It is only to help your music, not to replace it."[20]

By then, Philippa had gone to East Asia, to Taiwan, and was planning to go back to Vietnam. She even hoped to obtain permission to visit Hanoi. Josephine was deeply concerned: "Please don't take any risks that might injure your beautiful legs or your wonderful hands," she wrote. "A headline is not worth your being hurt. . . . You are too valuable to be used just to get some scrap of news for Mr. Loeb. . . . Don't risk your life. *Don't. Don't. Don't.* Don't go to Hanoi, and don't consider staying around Vietnam too long."[21]

Philippa arrived in Vietnam for her second stint in very early March 1967. No official was there to greet or obstruct her this time. She arrived quietly and stayed with Asian friends made on her earlier trip, having learned from her previous experience not to spend more than one or two nights in any one place.

Her travels were exhausting. She went with missionaries by jeep and bicycle to villages in the northernmost provinces of Quang Tri and Thau Thien, over insecure roads where Viet Cong landmine explosions were a daily occurrence. She climbed, in the company of a young priest and two villagers—past sleeping sentries—the spiral staircase to the top of a pagodalike watchtower, to gaze across the lazy Ben Hai River, demarcation line between the two Vietnams. And she thought sadly how much "the division of this verdant and fertile country was the tragedy of the whole world."[22] She may have been the only American correspondent to visit the DMZ that spring.

Unhappy events plagued her as soon as she arrived. In Saigon, a bicycle swooped out of an alley, knocking her down and injuring her leg. In Da Nang, her suitcase mysteriously disappeared and was returned hours later with clothes and valuables gone, and her papers in disarray.[23] In Hue, as she sat writing in her hotel room, soldiers—two Vietnamese and one American—broke down the door. They told her they thought she was harboring Viet Cong in her room. A fortune-teller warned that she was in danger of assassination and to beware of crowds.

But Philippa fearlessly remained in Hue despite a virtual siege of the city. The roads around Hue were heavily mined, and the sounds of gunfire and shelling echoed at night.

In a dispatch to the *Manchester Union Leader*, published April 21, 1967, Philippa provided a detailed view of the situation:

> One night I was visiting the MACV enlisted men's club when word came that Viet Cong were only a block away. . . . A great deal of machine-gun fire could be heard. The explosion of plastic bombs and grenades made a jungle rhythm around the area. Suddenly the lights of the compound were blacked out. We rushed downstairs . . . and the soldiers dashed to various posts. Soon, they stood with rifles poised at every window. No one knew exactly from which direction the Viet Cong would come. I recalled the horror pictures I had seen of Vietnamese villages half-destroyed by the Viet Cong, and of the mutilated corpses they left in their wake.
>
> I was told to hide in the latrine if the Viet Cong should break in . . . or if bullets should start coming through the window. . . . While I was not personally afraid, the noises . . . from the dark shadows of Hue made [me] tense and nervous.
>
> [The barrage] went on nearly all night. But miraculously, the compound was not attacked. About midnight, I walked alone out of the compound . . . through ominous Hue. . . . I passed groups of Vietnamese soldiers, half of whom were asleep. The other half were chattering amiably. They did not seem greatly disturbed by . . . [what had] taken place a few blocks away. Thinking I was Vietnamese, they greeted me cheerfully. They seemed to feel a bit of conversation with me would break up the monotony of the night.
>
> After twenty-seven years of military friction, war has become monotonous in Vietnam. The Vietnamese soldier, who has never known a state of peace, has become very casual about the whole affair.
>
> Can one blame them for their terrific desertion rate? Or their frequent lazy approach to combat? Vietnamese are affectionate, sensitive men, who love their families. They are very sentimental about their wives and childen. Seeing little purpose or progress in the war, they want to preserve their lives for their families. The Viet Cong are intense and ferocious fighters, because they have an ideology. Our vacillating strategies have kept these Vietnamese who are on our side from having a sense of direct commitment to the war.
>
> The next morning, I asked many people what had really happened in Hue the

> night before. As is typical in the Orient, everyone had a different story. . . . It is hard, under these circumstances, to feel sure that "we are winning the war." And even when you live in a city like Hue, it is difficult to know what is really going on, or to foresee what tomorrow will bring.[24]

There were two things that drew Philippa to Hue again and again: her mission of evacuating children to Da Nang where they could continue their education, and her relationship with Leiter.

Throughout 1966 and 1967, Philippa had maintained contact with Leiter, by now a major. When she performed in Taipei in late February 1966, she decided to invite Jim, who, coincidentally, was planning to be in Taiwan. He attended her recital,[25] and Philippa, in her typically unrealistic assessment of men in whom she was interested, wrote Josephine:

> I introduced him to scores of my distinguished friends, and they all loved him. . . . He made a better impression than any other friend I have ever known or introduced to people! As far as I am concerned he is the best Man I ever met. . . . The reason I never met anyone as nice before was because I tried to find single men, not married ones, and men who are still single at my age are impossible. Jim says his marriage (at 21) was a mistake . . . the obstacle is that he wants his children to think well of him . . . and unless I have the patience to just wait and see him in a relaxed way, and let him make up his own mind, I shall just ruin it.[26]

❖ ❖ ❖

Interviewed twenty-five years later, Leiter admitted that he was overwhelmed by Philippa's fame. She had taken him to a dinner reception in her honor at the Dominican Republic embassy in Taipei. Not easily impressed, he could not "get over how important people treated her like someone very special. You know how it is," he said. "You meet someone in another country and they give you their history, like being a renowned concert pianist and having been a child prodigy, and you say, 'Yeah. Right. Sure.'"[27]

He was equally impressed with her playing, even though he admitted that he did not know much about classical music at the time.

Philippa had been very cautious in her letters to Jim, "carefully screening out all culture."[28] The result was that Leiter was even more surprised than he might have been when he encountered her in Taiwan: "He [met] me in the middle of the jungle practically," she reminded her mother, "and I was GI Jane, the Angel of Dien Bien Phu, or something. Then he comes to Taipei, and instead of finding GI Jane, he meets an elegant young lady in luxurious surroundings."[29]

In many ways their relationship was a strange one. Keeping in mind that she was a journalist, he was careful about what he told her. She, in turn, had a knack of making herself appear "mysterious," which appealed to a man embroiled in a business that made mystery de rigueur. Jim remembered her as a "lost soul," a little naive, as someone always seeking relationships—with men in particular. On a couple of occasions, after dinner, she had wanted to spend the night with him. Even though his marriage was shaky, and he would divorce his wife after the war, he did not want any complications at the time. His caution had another reason too: He sensed that she was very apt to mistake kindness for an "interest."

Sometime in April 1967, their relationship came to an abrupt end. It was a day when he had an urgent "pouch" to prepare for the Pentagon, and he was working with his Vietnamese counterpart. Philippa had called three or four times on the phone and left messages. When her fifth call came, and the messenger knocked once again on the major's door, Jim decided to pick up. "I was not very gracious. I told her she was *not* my top priority—in a word, 'get lost.' She said she just wanted to stop by that night to say hello. I said I was too busy. She started crying . . . and I hung up."[30] Leiter said he felt badly. In retrospect (and even then) he realized that he meant more to her than she had to him.

Toward the end of our interview, I gambled and mentioned that Philippa had written to her mother that she wanted to marry him. He fell silent, looked hard at me, blinked, then closed his eyes for a second. When he recomposed himself he said, "I'm very surprised. I had no idea."[31] Yet, as he searched his memory, he began to recall clues of deep affection she had left along the way. Unlike André Gascht, James Leiter did not feel guilty. In fact, it strengthened his feelings that having known Philippa, even for as little as he had, was a gift of grace. "She was such a breath of fresh air in my crazy world in Vietnam. My life is richer for having known her."[32]

❖ ❖ ❖

The war distressed Philippa on a nonmilitary plane as well. Everywhere she began to observe the degrading treatment of the black GI by his white counterpart. She was stunned. Here, where everyone was united in fighting the "communist bastards," she had thought there would be no manifestations of the American Dilemma. Philippa was also surprised to witness the general disrespect of the South Vietnamese soldier by U.S. officers. Racism became a recurring theme in her articles. She believed this contributed significantly to the failure of Vietnam: the inability of the American to communicate with the Vietnamese.

Her treatment by white GIs was no better. Whether they perceived her as a Vietnamese woman, a Negro, or Hispanic, she was always conscious of their disdain. To them, she simply was not white. And it was this that finally inspired her last novel, *Dau Tranh*, which she began on her second trip to Vietnam during the spring of 1967.

Dau Tranh held all the promise of a personal catharsis. Jeanne, an American mulatto, is an erstwhile child prodigy and musician, traveling and working in Vietnam as a foreign correspondent. For dramatic effect, the protagonist is illegitimate, but other than that, she is Philippa:

> Dau tranh [the novel begins]—the Vietnamese word for struggle flashed through Jeanne's mind. It was such a poignant and perceptive word that conveyed far

> more passion and subtlety than the stale English translation could ever possibly express.
>
> She was in a state of struggle. *Dau tranh!* That agonized cry of pain and defiance that typified all the movements for liberation in Vietnam also expressed her own drama. She had inherited the handicaps of illegitimacy and a skin that was not white. The first burden could be concealed, but the second never could. Her skin was light enough for her to be accepted as a second-class white in Rhodesia, Kenya, or South Africa, and its color made no difference in Europe. But to Americans it was the most important of all characteristics. It categorized one as a person to be insulted, to be treated as a pariah, to be deprived of respect in all deeper human relationships. The same white Americans who were supposed to be bringing democracy to Vietnam were incapable of practicing it themselves in any context that went deeper than the superficial.[33]

Suddenly, almost every letter home to Jody discussed this: "Do you know, after my clothes were stolen, who showed sympathy? NO WHITE AMERICAN. A Vietnamese girl gave me a dress. An American Negro gave me a dress. A European Catholic gave me a dress. BUT NO WHITE AMERICAN even gave me a button to put on a dress."[34]

And in another letter, arguing with her mother:

> Concerning bravery: I would like you to appreciate the fact that accepting a segregated form of life and work is not bravery *at all.* Men like William Mann and Leonard Holsey who . . . brave the slights of the race-and color-prejudiced American white who *resents* them having a position even moderately compensating their talents—THESE ARE THE TRUE HEROES. The American white man *resents* a colored person being in any role but servant or entertainer. . . . The Negro military officers over here are some of the finest men I've met . . . far superior to white officers of the same rank. . . . By the way, do you think this internecine color combat is unnoticed by the subtle Vietnamese?[35]

Her experiences in Vietnam forced Philippa to reexamine the civil rights struggle in America, a movement she had often eschewed. "This whole period has been healthy," she wrote Jody, "because it has accelerated change in the racial status quo. It is about time. How patient can one be after so many years of waiting?"[36]

And finally, in Philippa's very last letter home:

> Half of my encounters with THE WHITE AMERICAN are abortive, negative or unpleasant. They are the worst advertisement for America's supposed democracy I ever saw. The Negro American calls the White American "Charlie." The White American calls the brown-skinned Viet Cong "Charlie." "Charlie" means "the enemy." . . .
>
> Now if George, instead of letting himself be segregated all his life, had had the guts to go forth into integration and *try* to thrust his way into white companies and white neighborhoods, he would have *found out* why the Stokely Carmichaels are necessary now—as a pressure valve. . . . I am not going to cravenly accept segregation. Nor will I bring any child up into segregation.[37]

Leonard Holsey wrote George and Jody a painful letter after Philippa's death. It contained perhaps the most provocative comment about their daughter's racial "reawakening": "I am sure Philippa matured in Vietnam. . . . Although an extremely clever person, she learned what the American white man is really like only after she got here."[38]

Philippa's anger could no longer be held captive by grief.

Postlude

On Tuesday morning, May 9, Josephine and George received a cable from their daughter explaining that she would arrive on Friday. That same day they also received a letter from her.

At 2 p.m. their phone rang. George was sitting in his favorite chair, reading. Jody answered the phone. It was a friend of theirs from UPI. He had just read over the wire that Philippa had died in a helicopter crash. Jody screamed and hung the phone up.

Four hours later they received a telegram from the State Department corroborating the phone call. George had not moved from his chair. He was in a state of shock. Occasionally he would say, "It wasn't supposed to be this way. I wanted to die first. She's so young."[1]

The CBS Evening Radio News treated her death as a national story. Philippa was the tenth American correspondent and only the second woman journalist to die in Vietnam. All across America, television programs were interrupted at various times to announce the passing of an American journalist in Vietnam. Newspapers in Europe, Africa, the Far East, and South and Central America reported her death.

During the next several days, the Schuylers' telephone rang constantly. It was Jody who handled all of them; George was unable to speak.

Hundreds of telegrams and letters poured in from all over the world. One letter was addressed, simply, *Philippa Schuyler, Harlem.* "We loved her too," cabled Ella Fitzgerald. William Loeb coined the phrase "the Beautiful American" to describe her. Children in a fourth-grade class from Laconia, New Hampshire, who were part of an adoption program, wrote: "We were thinking of a name for our fourth baby when we read about your daughter. As she was helping children we decided to have a girl baby baptized Philippa. . . ."[2] Lineke, Dennis, Ernie, Gerd, John Garth, Henry Cabot Lodge, Sammy Davis Junior, priests, ambassadors, kings, presidents, all sent condolences. The list was endless.

Even people who barely knew her wrote: "I have wondered whether I, a stranger, should write to you about your daughter's death," began one letter from a Virginia Thorndike, M.D., to George. "I don't wish to intrude on your grief. I saw Philippa for about 15–20 minutes when she was seven or eight years old. I was working for *Life* at the time. Somehow, I have never forgotten her, among the hundreds, perhaps thousands of people I've talked with, since. She lives in my memory."[3]

On September 24, when Philippa had been scheduled to give a recital in Town Hall, a memorial concert was performed instead. Her book on Vietnam, *Good Men Die*, was published posthumously. (Her other book on Vietnam, *Dau Tranh*, was never finished.) And Jody began a book of her own—poems about her daughter.

In the days following her death, a Philippa Duke Schuyler Memorial Foundation was founded to raise money for nonmilitary causes in South Vietnam—seeds for farming; bows, rosin, and music for the students at the Saigon Conservatory. Josephine flew to Vietnam toward the end of October, not only to deliver the materials but also to look for Philippa's last effects. Leonard Holsey was still in Vietnam and it was he who gave her a wrapped package left by her daughter. Josephine opened it and to her surprise found Philippa's fatigues. She wept in Mr. Holsey's arms.

Back home, Jody spent many of her days writing about her daughter. George slowly began putting his estate in order, anticipating his own death. The apartment lay quiet now, Philippa's urn standing stoically on the mantelpiece.

❖ ❖ ❖

At 2 a.m. Thursday morning, May 2, 1969, George awoke with a start. His sleep had been fitful: He was worried about money, about his wife's health, about life. The second anniversary of Philippa's death was approaching in one week. The enormous pain during the last two years had often been unbearable, but the two of them had managed, somehow. Jody seemed to be improving little by little and was just beginning to pull herself out of another depression. She had just finished her book of poems—a tribute to their daughter Philippa—and made the last revisions on her autobiography, *From Texas to Harlem with Love*.

But her health had not been good.[4] For the last thirteen days, Jody had refused any food, maintaining she was following a health-food fast. It would be good for her, she insisted, and perhaps the donnatal the doctor had prescribed for her upset stomach would not be necessary. She had not been sleeping well either, although on Wednesday night after George forced her to take some soup, she had slept better.

Worried, he decided to look in on his wife. He slipped into his paisley bathrobe and matching slippers, presents from his daughter from a Christmas long gone, and walked slowly toward his wife's bedroom. Light was coming from underneath the door. He gently knocked and walked in. Josephine was standing in the middle of the floor, staring at the faded, heavy brocaded blue drapes. She turned abruptly toward her husband; he noticed how drawn and pale, how weary and defeated she looked, her eyes almost as faded as the drapes. He remembered those eyes of forty-two years ago, and how they had first shocked him by their radiance.

They talked briefly. Josephine said something about getting new drapes. George agreed but drew her toward the bed and tucked her in.

Three hours later Schuyler awoke again. He decided to check on his wife. As he approached her bedroom, he thought how strange that her light was still on. He had turned it off, hadn't he? He tried the door; it seemed jammed. He pushed harder and with some effort managed to force it open. He stood in the doorway—the way his wife had those many years ago when he had first laid eyes on her—and stared. Josephine stared back. Time froze. George blinked. Jody did not. She had wedged a faded blue drapery between the transom sash and the bedroom door and hanged herself.

❖ ❖ ❖

"Yes," wrote George to his good friend William Loeb, days after Josephine's suicide. "I truly expected that both Josephine and Philippa would outlive me, and now it is the reverse. But I doubt that I shall be here very long. I have tried to fight the good fight for what I have considered right, but now the long battle has worn me down. It is hard to hold one's head high and to carry on under crushing burdens of responsibility."[5]

Schuyler's first Christmas without his daughter or his wife was a lonely one. Hoping for some comfort from his lifelong friend, the typewriter, he banged out a few lines:

THOUGHTS AT YULETIDE—1969

How happy and content were we
In our Convent Avenue airie
Gathering from great distance
Exchanging reminiscences
Of accomplishments within the year.
Acquaintances made far and near;
Adventures in contrasting climes
Exotic food and drink betimes;
From hurtling jets to bounding jeep
To steamers swimming in the deep.
What Yuletide gifts from Up Above,
Encircled by the bonds of Love!
The airie now is almost bare.
With one alone to sorrow there;
With memories of things long past,
And counting days that cannot last.
The books gaze down from serried racks
On exotic global artifacts.

> And how this ancient memory burns
> As I see those two ash-filled urns.
> Their challenge is too much to bear,
> And soon now I shall join them there.[6]

But George would not join them so soon. He died eight years later, on August 31, 1977, in New York Hospital—alone.

NOTES

ABBREVIATIONS

ADVENTURES	Philippa Duke Schuyler, *Adventures in Black and White* (New York: Robert Speller and Sons, 1960).
DGS	Dennis Gray Stoll.
GSS	George Samuel Schuyler.
IUB	Indiana University School of Music Library, Bloomington.
JCS	Josephine Cogdell Schuyler.
NYHT	*New York Herald Tribune.*
NYT	*New York Times.*
PDS	Philippa Duke Schuyler.
SCRBC	Philippa Duke Schuyler Collection, Manuscripts, Archives and Rare Books Division, Schomburg Center for Research in Black Culture; New York Public Library, Astor, Lenox and Tilden Foundations.
SU	Department of Special Collections, Syracuse University Library.
YU	Yale Collection of American Literature, Beinecke Rare Book and Manuscript Library, Yale University.

PRELUDE

1. All quotes regarding the testimony of the four officers are from the U.S. Army Aviation Accident Report, May 17, 1967, obtained through the Freedom of Information Act. The report is sixty-nine pages; the government, however, has blacked out a significant portion.

2. In all likelihood, her injuries made it impossible for her to swim.

3. Lt. Frederick D. Gregory to JCS, July 1, 1967, SCRBC.
4. PDS, *Good Men Die* (New York: Twin Circle Publishing Company, 1969), p. 10.
5. Philippa was the second woman journalist to die in Vietnam.

CHAPTER 1

1. Jervis Anderson, *This Was Harlem: A Cultural Portrait, 1900–1950* (New York: Farrar Straus Giroux, 1982), p. 342.
2. Ibid., p. 343.
3. Lincoln Barnett, "Negro Girl, 2½, Recites Omar and Spells 5-Syllable Words," *NYHT*, Feb. 8, 1934, p. 18.
4. The Schuylers kept scrapbooks during the first thirteen years of Philippa's life. Years 1933–34 and 1937–40 are in SCRBC; the missing years, 1931–32, 1935–36, and 1941–44 are in SU.
5. Scrapbook, Jan. 10, 1932.
6. Scrapbook, Oct. 2, 1932.
7. Harry L. Shapiro, "Descendants of Mutineers of the Bounty," *Memoirs of the Bernie P. Bishop Museum*, Hawaii, vol. 2, no. 1 (1929). Noted Harvard anthropologist E. A. Hooten, in his book *Twilight of Man*, also alludes to the possibility of race mixture resulting in a genius (E. A. Hooten to JCS, Dec. 19, 1939, SCRBC).
8. Later in life, Philippa believed so firmly in Jody's nutritional theories that she rarely deviated from them, despite her travels to remote corners of the globe and her dining with important people who sometimes found her diet peculiar. At age sixteen, when she was giving a concert in Texas, Philippa gave a reporter more than he bargained for when he innocently asked her what she had eaten for breakfast that morning: "I had orange juice, a piece of slightly steamed fish, a salad of raw cauliflower, broccoli, carrots and avocados, and six teaspoons of cod liver oil. I have cod liver oil every day" (*Austin Statesman*, Apr. 12, 1948, SCRBC).
9. Scrapbook, Oct. 1932.
10. Scrapbook, Oct. 1933.
11. Scrapbook, Dec. 23, 1931.
12. Scrapbook, Oct. 1932.

CHAPTER 2

1. On legal documents (e.g., passports), JCS put her birthdate as June 23, 1900. There is room for speculation, however. Harry MacKinley Williams places her birth around 1898 ("When Black Is Right: The Life and Writings of George S. Schuyler" [Ph.D. diss., Brown University, 1988], p. 135).
2. GSS, *Slaves Today: A Story of Liberia* (1931; rpt., New York: AMS Press, 1969), pp. 183–84.
3. Anonymous [JCS], "The Fall of a Fair Confederate," *Modern Quarterly* (Winter 1930–31):532.
4. Ibid., pp. 533–34.
5. Ibid., p. 530.
6. Ibid., p. 531.
7. JCS, "An Interracial Marriage," *American Mercury* 62 (Mar. 1946):277.

CHAPTER 3

1. Joe B. Frantz, *Texas: A Bicentennial History* (New York: Norton, 1976). Subsequent descriptions of early Texas taken from Frantz.
2. Gaston Cogdell, interviewed by author, Feb. 1983.
3. Ibid.
4. Ibid.
5. This much-condensed history of the Cogdell family is based on author interviews with Gaston Cogdell (see n. 2), Susie May Cogdell (see n. 8), Kathleen Houston (see n. 7); Josephine's diaries; newspaper clippings housed at SCRBC; and microfilm from the University of Oklahoma, Norman.
6. JCS, diary, [1923–24], SCRBC.
7. Kathleen Houston, interviewed by author, Mar. 1983.
8. Susie May Cogdell, interviewed by author, Feb. 1983.
9. JCS, "An Interracial Marriage," *American Mercury* 62 (Mar. 1946):274.
10. Anonymous [JCS], "The Fall of a Fair Confederate," *Modern Quarterly* (Winter 1930–31):528.
11. JCS, diary, [1923–24], SCRBC.
12. Ibid., also JCS, "Fall," p. 529.
13. JCS, diary, [1923–24], SCRBC.
14. Susie May Cogdell, interview.
15. Gaston Cogdell, interview.

CHAPTER 5

1. It would be years before he would team up with his brother Stewart to write a now famous column.
2. Joseph Alsop, Jr., "Harlem's Youngest Philosopher Parades Talent on Third Birthday," *NYHT*, Aug. 3, 1934, p. 15.
3. Lincoln Barnett, "Negro Girl, 2½, Recites Omar and Spells 5-Syllable Words," *NYHT*, Feb. 8, 1934, p. 18.
4. Scrapbook, Mar. 2, 1934. A present-day child-development professional studying Jody's scrapbooks and reading the newspaper accounts would readily conclude that Philippa at three was very well advanced in her motor skills, particularly in hand coordination, yet within the accepted norm. But they would seriously quarrel with the Schuylers' assessment that she was not a prodigy, a word they eschewed in trying to promulgate the theory that any interracial child, through hybrid vigor and proper attention, could be as intelligent as theirs. Students of cognitive development, however, would hardly recognize a "normal" three year old as the subject of these interviews and accounts.

A normal three year old can follow two-step commands or directions, use three-word sentences, say first and last names, tell about an immediate experience (within a few hours), name an object when shown it, tell how common objects are used, ask "what" and "where" questions, point to and name body parts (head, eyes, toes, but nothing like hips), use words ending in "ing," grasp pencil and crayon (in the fist—there is, as yet, no "pincer" grasp, so most three year olds cannot pick up small objects and manipulate them), invent new words and play with words (i.e., to bow a shoe and nonsense rhyme if they sound right together), have a sense of gender identity, and know one or two colors (most common is red).

Normal children at age three cannot write, read, or spell. Very few know the alphabet; if they do it is by rote with many mistakes, and not all the way through. They have no sense of logic, time sequence, object permanence, or causality—reality is based on their changing perspective. To understand a three year old, Arnold Gesell wrote almost a half century ago, one must recognize his almost complete ignorance of the wide world beyond the nursery. Philippa obviously was years ahead of the norm in her cognitive and intellectual abilities. (From Nina Callahan, personal report to author, 1987.)

5. As an adult, Philippa would sit at the piano ten to twelve hours a day. She could concentrate, equally well, on writing. Some examples of her very early poetry:

THE SUN—EASTER POEM (March 28, 1935)

The sun is lifting his lid
The sun is leaving his crib
The sun is a walking baby
Who will bring the dawn, maybe
Thump, thump, thump out of the earth!

THE MOON OVER TENEMENTS (Sept. 1935)

The moon tonight
Is so bright
I think it must have used a cloud
To polish its face

SWEET MOTHER (Sept. 1935)

You are so sweet
Like a bunch of grapes
Wake up! Wake up!
Or I will make a wine of you
And drink you up.

6. Scrapbooks, 1936 and 1937.
7. Scrapbook, Dec. 1934.
8. Scrapbook, June 1935 (*NYHT*, June 29, 1935).
9. "Prodigious Crop," *Time*, Aug. 26, 1935, p. 27.
10. H. L. Mencken to GSS, Jan. 7, 1933, SCRBC.
11. Louis G. Gregory to GSS, Jan. 27, 1934, SCRBC.
12. JCS, "My Daughter Philippa," *Sepia* (May 1959), p. 9.
13. Scrapbook, Christmas week 1934.
14. Scrapbook, Aug. 12, 1934.
15. Scrapbook, Mar. 2, 1934.
16. Scrapbook, June [15], 1934.
17. Scrapbook, Feb. 22, 1936.
18. Scrapbook, Jan. 1935.
19. Scrapbook, Jan. 19, 1936.
20. The latter had a subsequent name change, becoming "Nigerian Dance." This was the tune George concluded was conjured out of the African lullabies he sang his daughter as a child.

21. Scrapbook, June 1936.
22. "Harlem Prodigy," *Time*, June 22, 1936, p. 40.
23. J. A. Rogers, *Sex and Race: A History of White, Negro, and Indian Miscegenation in the Two Americas*, vol. 2, *The New World* (New York: J. A. Rogers, 1942), p. 408.
24. Carolyn Mitchell, executrix of the Schuyler estate, is responsible for Philippa's first solo recital in New Jersey.
25. Scrapbook, Nov. 1936.
26. Scrapbook, Apr. 1936.
27. Scrapbook, Mar. 1936.
28. Scrapbook, Feb. 25, 1936.
29. Ibid.

CHAPTER 6

1. John Broadus Watson, *Psychological Care of Infant and Child* (New York: Norton, 1928).
2. Ibid., pp. 81–82.
3. Ibid., p. 44.
4. Ibid., p. 115.
5. Ibid., p. 3.
6. Ibid., pp. 5–6.
7. Ibid., p. 12
8. Ibid., p. 15.
9. Ibid., p. 9.
10. Ibid., p. 15.
11. Ibid., p. 40.
12. Ibid., p. 84.
13. Scrapbook, Sept. 1934.
14. Scrapbook, Nov. 2, 1934.
15. Scrapbook, Jan. 1, 1936.
16. Watson, *Psychological Care*, p. 96.
17. Scrapbook, July 2, 1934.
18. Scrapbook, Dec. 1936.
19. Ibid.
20. Ibid.
21. Scrapbook, July 2, 1934.
22. Eugene D. Genovese, *Roll, Jordon, Roll: The World the Slaves Made* (New York: Vintage Books, 1976), p. 120.
23. Scrapbook, Oct. 1936.
24. GSS to PDS, [1936], SCRBC.
25. Scrapbook, June 1935.
26. Elton Fax, interviewed by author, Feb. 1983.

CHAPTER 7

1. GSS, *Black and Conservative: The Autobiography of George S. Schuyler* (New Rochelle, NY: Arlington House, 1966).
2. Ibid., p. 4.
3. There is a question as to George's true biological father. See Harry MacKinley

Williams, "When Black Is Right: The Life and Writings of George S. Schuyler" (Ph.D. diss., Brown University, 1988), p. 2ff.

4. GSS, *Black and Conservative*, p. 8.

5. Ibid., p. 18.

6. Ibid., p. 28.

7. Many an African-American home proudly displayed lithographs of the Ninth and Tenth Cavalries charging up San Juan Hill, and black newspapers featured stories about such luminaries as Major Charles Young, First Lieutenant John E. Green, and the older Benjamin O. Davis. Green and Davis had risen from the ranks. Young had graduated from West Point in 1889, but during the four years he was at the academy not a single fellow cadet ever spoke to him. Aside from some chaplains, these were the only three black commissioned officers in the American army at that time.

8. Organized by an act of Congress in 1867, the all-black Twenty-fifth U.S. Infantry was an old, distinguished regiment. It was known for its sharpshooters and marching bands—its esprit de corps was as high as its élan. The Twenty-fifth had fought valiantly in all the important battles of the Spanish-American War and the Philippine Insurrection. It was involved in the Brownsville affair of 1907, during which some of its men, goaded by white Texans, had shot up the town, for which they were dishonorably discharged by Theodore Roosevelt.

9. GSS, *Black and Conservative*, p. 49.

10. Ibid., p. 89.

11. Ibid., pp. 89–90.

12. Ibid., p. 90.

13. GSS, "Black Warriors," *American Mercury* 21 (Nov. 1930):294–96.

14. JCS, diary, [1927?], SCRBC.

15. Fax, interview, Feb. 1983.

16. GSS, *Black and Conservative*, p. 99.

17. Schuyler would use this as fodder for future articles. Ten years later, he would describe it in his article "Memoirs of a Pearl Diver," one of Mencken's favorites. *American Mercury* 22 (April 1931):487–96.

18. GSS, *Black and Conservative*, pp. 128–30.

19. Ibid., pp. 130–32.

20. Ibid., p. 133.

21. He started as a jack-of-all-trades for ten dollars per week, and was appointed managing editor in December 1926, a job he held until the *Messenger* folded in July 1928.

CHAPTER 8

1. Thomas Fortune had once been editorial writer for the *New York Sun*, Lester Walton had been a feature writer for the *New York World*, Eugene Gordon for the *Boston Post*, and Noah Thompson for the *Los Angeles Express*.

2. GSS, *Black and Conservative: The Autobiography of George S. Schuyler* (New Rochelle, NY: Arlington House, 1966), p. 174. Among the numerous artifacts that George brought home with him from Africa was an elephant-toe bracelet. Sliced from a toe, hollowed out, polished and studded with silver, it was as rare then as now. He gave it to Amelia Earhart, who wore it as a talisman on her triumphant solo flight across the Atlantic the following year.

3. Harry MacKinley Williams refers to a letter from this same U.S. vice consul to

Schuyler, after they had both returned from Africa. The vice consul salutes George for being a "helluva man [for] the way you went through country, the women, and the palm wine" ("When Black Is Right: The Life and Writings of George S. Schuyler" [Ph.D. diss., Brown University, 1988], p. 222).

4. GSS, *Slaves Today: A Story of Liberia* (1931; rpt., New York: AMS Press, 1969), pp. 5–6.

5. GSS, *Black and Conservative*, p. 201.

6. Ibid., p. 230.

7. At the same time, George was also writing works of fiction (serialized in various black weeklies under several pen names). John A. Williams has written that Schuyler's achievement over the two-year period from 1936 through 1938 exceeds, at least in quantity, even the writings of Charles Dickens (GSS, John A. Williams, Foreword to *Black Empire*, by GSS, ed. Robert A. Hill and R. Kent Rasmussen [Boston: Northeastern University Press, 1991], p. xi).

8. GSS to JCS, n.d., SCRBC.

9. GSS to JCS, Oct. 20, 1935, SCRBC.

CHAPTER 9

1. JCS, "The Education of a Musician," *Interracial Review* 26 (Sept. 1953):151.

2. Scrapbook, June 1937.

3. Ibid.

4. After six years (in 1943), Philippa was barred from further participation in the Philharmonic contests. The judges based their decision on the somewhat circular argument that her outstanding abilities precluded other children from winning. The black newspapers took a different view: They wrote that her barring was racially provoked.

5. Deems Taylor, Foreword to *Adventures*, p. xiii.

6. Scrapbook, May 1938.

7. Delilah Jackson, conversation with author, 1987.

8. "The Shirley Temple of American Negroes," *Look*, Nov. 7, 1939, p. 4.

9. This and subsequent descriptions of the World's Fair are from Lawrence Elliott, *Little Flower: The Life and Times of Fiorello La Guardia* (New York: William Morrow, 1983).

10. "Philippa's Day at the Fair," *Time*, July 1, 1940, p. 48.

11. Milton Bracker, "Child Composer, 8, Is Honored at Fair," *NYT*, June 20, 1940, p. 28.

12. Ibid.

13. Ibid.

14. Elliott, *Little Flower*, p. 198.

15. Ibid., p. 225.

16. Joseph Mitchell, "Evening with a Gifted Child," *New Yorker*, Aug. 31, 1940, pp. 8–31. Mitchell's article, along with several of his other profiles, was later published in book form as *McSorley's Wonderful Saloon*, and recently reprinted (1993) in *Up in the Old Hotel*. The account, on the following pages, including all quoted conversations, is condensed from Mitchell's original interview.

17. Almost thirteen years later Phil would write Mitchell when inviting him to her New York Town Hall debut: "I can still remember eating ice cream with you! Your charming story went around the world. I recently met up with it in Havana. I went into a bookstore across from the Plaza Hotel and there it was! At this late date, accept my

gratitude please for your thorough coverage of what must have seemed to you a funny little girl!" (May 1, 1953, SCRBC).

18. James Mumford, conversation with author, 1987.
19. Hylan Lewis, conversation with author, May 1987.
20. Ibid.

CHAPTER 10

1. Scrapbook, Aug. 1938.
2. Scrapbook, Aug. 1939.
3. Scrapbook, May 1938.
4. Daisey Cogdell to JCS, "April 27, 1939. From the hospital," SCRBC.
5. Edna Porter to JCS, July 7, 1938, SCRBC.
6. Scrapbook, June 1937.
7. Scrapbook, July 30, 1937.
8. Scrapbook, 1938.
9. Pauline Apanowitz Styler, interviewed by author, Apr. 14, 1987.
10. Antonia Brico to JCS, June 1, 1941, SCRBC.
11. Styler, interview.
12. Although it is not unknown for children of nine or ten to threaten suicide, it is highly uncommon. No doubt Phil had heard Josephine use this technique on George (perhaps even on her) and was imitating her mother, as she often did.
13. Scrapbook, July 10, 1941.
14. A year earlier, in February 1941, Philippa was named to the Honor Roll of Race Relations, sharing her distinction, inter alia, with Albert Einstein, William Grant Still, Jack Benny, and Justice Hugo Black.
15. Beryl Williams, "Scaling New Heights," *Calling All Girls*, Oct. 1943, p. 82.
16. Scrapbook, Nov. 1942 (*Pittsburgh Courier*, Nov. 21, 1942).

CHAPTER 11

1. Six hundred American blacks had earned their wings by the end of the war.
2. Dominic J. Capeci, Jr., *The Harlem Riot of 1943* (Philadelphia: Temple University Press, 1977), p. 51.
3. At about the same time, he wrote a pamphlet entitled "Why We Are Against the War" for the Negroes Against the War Committee.
4. Harry MacKinley Williams, "When Black Is Right: The Life and Writings of George S. Schuyler" (Ph.D. diss., Brown University, 1988), p. 324.
5. Gaston Cogdell, interview, Feb. 1983. There were two exceptions: Jewell, the widow of Jody's favorite brother, Gaston; and Susie May, the ex-wife of another brother, Buster. Although Susie May—who knew about the marriage from the beginning—would not meet her niece for another two decades, Aunt Jewell came to New York during the end of 1934, when Phil was barely three, and visited Harlem. George was away on assignment. The notation in the scrapbook with subtle but palpable pride reads: "She took you on her lap and gazed a long time into your face. Then she said, 'She hasn't an ugly feature. Every feature is pretty. I see you in her smile, the side of her face, and her hands.' This is the first of your relations in Texas to see you. She had never heard a child read of spell at your age." ([Nov. 1934], SCRBC.)

When five years later, *Look* profiled Phil, Aunt Jewell saw the article and wrote Jody a worrisome letter: "I looked at P_______ a *long* time. She is wonderful and so pretty—and pretty hair—and so smart. I know you are proud of her. . . . I am very much afraid if Gertrude [Jody's sister-in-law] sees it, she'll recognize the *name*—as when you got your passport you sent *that* name [Schuyler] and Gertrude told me about it. And you know Gertrude—she'll put 2 and 2 together. In hopes she'll not see it . . ." (Nov. [10], 1939, SCRBC).

6. W. E. B. Du Bois, *Souls of Black Folk* (Millwood, NY: Kraus-Thomson Organization Limited, 1973), p. 3.

7. GSS, "Our White Folks," *American Mercury* 12 (Dec. 1927):392.

8. In January of 1931, while George was on his way to Liberia, his only other published fiction book appeared: *Black No More*, subtitled *Being an Account of the Strange and Wonderful Workings of Science in the Land of the Free, A.D. 1933–1940.* Although it appeared long after Jules Verne's *Journey to the Center of the Earth* (1864) and H. G. Wells's *Time Machine* (1895), *Black No More* is one of the earlier full-length science-fiction novels published in the twentieth century (it predates Aldous Huxley's *Brave New World* by a year). There was very little writing in the field until 1926 when the pulp magazine *Amazing Stories*, devoted exclusively to science fiction, was started. Schuyler's novel, like Huxley's, uses science fiction as an instrument of social criticism and it may well be the first full-length novel to combine the two genres.

It was well received at first by literary critics, black and white alike, although it fell out of favor when black critics began to question Schuyler's racial loyalties: His novel was accused of denying racial identity and of showing revulsion toward the Negro masses. He was branded an assimilationist.

9. In this same vein, George would later write a letter to a staffer at the *Courier* condemning his own serial, "The Black Internationale," which he wrote under the pseudonym Samuel I. Brooks. The series concerns a militant and ruthless internationale that reclaims Africa as a black nation, wiping out anyone, regardless of race, who stands in their way: "I have been greatly amused by the public enthusiasm for 'The Black Internationale,' which is hokum and hack work of the purest vein," GSS wrote. "I deliberately set out to crowd as much race chauvinism and sheer improbability into it as my fertile imagination could conjure. The result vindicates my low opinion of the human race" (GSS, *Black Empire*, ed. Robert A. Hill and R. Kent Rasmussen [Boston: Northeastern University Press, 1991], p. 200).

10. Recent scholarship on Schuyler disagrees with this statement, in particular Rasmussen and Hill in their 1991 publication *Black Empire*, which is an annotated republication of two "novellas" written by Schuyler under the name of Samuel I. Brooks—"The Black Internationale," and "Black Empire" (serialized for the *Courier*, between 1936 and 1938). Rasmussen and Hill write: "Ideologically, while [Schuyler] is principally remembered today as a militant anti-communist, during the thirties he was staunchly anti-capitalist. Schuyler evolved into a conservative after World War II, but during the thirties he was definitely a radical. If he ended up as a critic and opponent of the civil rights movement in the sixties, he was in the thirties both an activist and a militant advocate of civil rights. And if later he became an apologist for continued imperialistic rule in Africa, in the prewar years he was an articulate critic of imperialism. Likewise, while he has been thought of, and with good reason, as anti-Garvey, he was in the thirties a committed Pan-Africanist. Finally, although he prided himself on being a hard-nosed realist, he was also a radical idealist and scientific utopian. *Black Empire* reveals Schuyler as a complex radical thinker, whose ideological journey toward conservatism was more labyrinthine and problematic than has been appre-

ciated. As a writer and a thinker, he defies easy classification" (GSS, *Black Empire*, pp. 261–62).

11. Michael Peplow, *George S. Schuyler* (Boston: Twayne Publishers, 1980), p. 104.

12. On March 31, 1931, nine Negro boys were indicted in Scottsboro, Alabama, on charges of having raped two white girls on a coal car of a freight train passing through North Alabama. After long sessions and legal maneuvers and two reversals by the Supreme Court, most of the defendants were freed. The Communist Party had taken over the defense from the NAACP and made it its cause célèbre.

Angelo Herndon had been recruited in Cincinnati and sent down to Atlanta by the Reds to lead marches of hungry and unemployed workers on the City Hall and Capitol. He was promptly nabbed, jugged, tried and sentenced to a Georgia chain gang; then released on bail. The Communists paraded him at meetings all over the country.

13. GSS, "The Separate State Hokum," *Crisis* 42 (May 1935):135.

14. H. L. Mencken to PDS, Oct. 14, 1946, SCRBC.

15. Lewis, quoting Clarke, conversation, May 1987.

CHAPTER 12

1. JCS to GSS, Mar. 1944, SCRBC.

2. India McIntosh, "Prodigy to Play 1st Orchestral Number Today," *NYHT*, Aug. 2, 1944, p. 14.

3. In February 1944, George was named managing editor of the New York office of the *Pittsburgh Courier*. His expanded duties at the *Courier* forced him to resign as business manager of the *Crisis*, a job which had also mandated much of his travel over the past seven years.

4. JCS to GSS, [1944], SCRBC.

5. Ibid.

6. *Adventures*, p. 2.

7. This quote and the next, ibid., p. 3.

8. McIntosh, "Prodigy to Play."

9. Nora Holt, "Philippa, Musical Prodigy, Seems to Improve with Age," *Amsterdam News*, Aug. 12, 1944, p. 4-A.

10. *Adventures*, p. ix.

11. Scrapbook, Apr. 8, 1945.

12. Program Scrapbook, SU (*NYT*, Apr., 1945).

CHAPTER 13

1. *Adventures*, p. ix.

2. Lena Cogdell to JCS, Feb. 2, 1945, SCRBC.

CHAPTER 14

1. "Rumpelstiltskin" had debuted under the baton of Dean Dixon (Phil's erstwhile teacher) on April 6, 1946, at Hunter College, the American Youth Orchestra performing.

2. Jean Tennyson, a well-known (white) opera singer who had studied with Mary Garden in Paris, was executive vice president of the stadium concerts. She had known the Schuylers for many years, and took a special interest in Philippa. Unofficially, she

became Philippa's "sponsor"—paying for her piano lessons during some of the leaner years, often giving Phil "gifts" of two or three hundred dollars, arranging concerts with VA hospitals throughout America, and renting Town Hall in 1956 for Philippa's third appearance there (and probably for her second appearance as well, in 1954). Ms. Tennyson, along with Mrs. Charles Guggenheim, chairman of the stadium concerts, were responsible for Phil's appearance at this Lewisohn Stadium concert.

3. Program Scrapbook, SU (*NYT*, July 13, 1946).
4. Program Scrapbook, SU (Robert Hague, "Youth Rules at Stadium" [July 1946]).
5. "Goings On about Town," *New Yorker*, July 13, 1946, p. 7.
6. Thor Johnson to PDS, Aug. 30, 1946, SCRBC.
7. Arthur Fiedler to PDS, Aug. 7, 1946, SCRBC.
8. Doris Sperber, "Pops Concert," *Boston Herald*, June 26, 1946, p. 23.
9. "Original Girl," *Time*, Mar. 25, 1946, p. 62.
10. Ibid.
11. Carl Van Vechten, *Pittsburgh Courier*, May 10, 1947.
12. Virgil Thomson, "University Festival," *NYHT*, May 4, 1947, p. 6.
13. Later, Philippa would separate these two movements from her symphony and entitle them *Sleepy Hollow Sketches*. She billed them as her third major orchestral work.
14. But the scrapbook experience had had its effect, and Philippa guarded her public image carefully. One change was that she now annotated her press clippings. In the March 1946 issue of *Ebony*, for example, where the interviewer wrote (probably to make Philippa appear more real): "She loathes the daily technical piano routine necessary to keep her fingers flexible," Phil scribbled, "I do not loathe my practicing! I enjoy it!" When they mentioned a "boyfriend" she had met in Mexico, and discussed her first "date," she underlined that she was too young for dates and never had a boyfriend. Much of the rest of the article was similarly blue-penciled.
15. "Teen Age Prodigy," *PM Magazine*, July 21, 1946, pp. M4–M5.
16. Frances Ullmann to JCS, Oct. 1, 1946, SCRBC.
17. JCS to Carl Van Vechten, Oct. 6, 1946, SCRBC.
18. GSS to JCS, n.d., SCRBC.
19. Louise K. Hines, "Child Artist's Concerts Help Finance Education," *Baltimore Afro-American*, May 1947. Josephine may have been referring to Jean Tennyson but more likely to a previous teacher, Herman Wasserman, who was willing to teach Philippa for free provided he receive a percentage of her earnings after she turned twenty-one.
20. JCS to GSS, Mar. 27, 1948, SCRBC.
21. *Austin Statesman*, Apr. 12, 1948, SCRBC.
22. John Rosenfield, *Dallas Morning News*, Mar. 27, 1948.
23. JCS to GSS, Mar. 27, 1948, SCRBC.
24. Buster Cogdell to PDS, Dec. 1948, SCRBC.
25. PDS, "My Black and White World," *Sepia* (June 1962), p. 13.

CHAPTER 15

1. GSS to JCS, [Fall 1940], SCRBC.
2. GSS to JCS, Mar. 3, 1931, SCRBC.
3. GSS to JCS, Oct. 13, 1935, SU.
4. GSS, *Black and Conservative: The Autobiography of George S. Schuyler* (New Rochelle, NY: Arlington House, 1966), p. 286.

5. JCS, diary, n.d., SCRBC.
6. GSS to JCS, July 25, 1949, SCRBC.
7. Ibid.
8. JCS to GSS, June 2, 1949, SCRBC.
9. JCS to GSS, June 16, 1949, SCRBC.
10. GSS to JCS, July 3, 1949, SCRBC.
11. GSS to JCS, July 4, 1949, SCRBC.
12. GSS to JCS, July 7, 1949, SCRBC.
13. JCS to GSS, July 8, 1949, SCRBC.
14. GSS to JCS, July 16, 1949, SCRBC.
15. Ibid.
16. GSS to JCS, July 3, 1949, SCRBC.
17. GSS to JCS, July 7, 1949, SCRBC.
18. GSS to JCS, [Summer 1949], SCRBC.
19. Lucille Elferbein, from unknown Rhode Island newspaper, May 1950, IUB.
20. JCS to GSS, [Feb. 1950], SCRBC.
21. GSS to PDS, Feb. 21, 1950, SCRBC.
22. JCS to GSS, Feb. 3, 1950, SCRBC.

CHAPTER 16

1. *Adventures*, pp. 29–30.
2. PDS to JCS, Sept. 4, 1952, SCRBC.
3. JCS to Carl Van Vechten, Nov. 6, 1952, SCRBC.
4. *Adventures*, p. 52.
5. Tarot card vignette from *Adventures*, pp. 54–58.
6. Organ vignette from *Adventures*, pp. 66–68.
7. Fax, interview, Feb. 1983.
8. Ibid.
9. *Adventures*, p. 78.
10. Ibid.
11. Ibid.
12. PDS to JCS, Apr. 26, 1955, SCRBC.

CHAPTER 17

1. Program: Scarlatti, Sonata in A Minor, Sonata in C; Rameau-Godowsky, *Tambourin;* Bach, Chromatic Fantasy and Fugue; Mendelssohn, *Variations sérieuses*, op. 54; Schubert, Impromptu in G-flat, op. 90, no. 3; Chopin, Scherzo in C-sharp Minor, op. 39; Beethoven, Sonata in C, op. 53 ("Waldstein"); Ravel, *Jeux d'eau* and *El Alborada del Gracioso;* Stravinsky, *Petrouchka.*
2. Harriet Johnson, "Philippa Schuyler's Recital," *New York Post*, May 13, 1953, p. 72.
3. R. P., "Philippa Schuyler in Town Hall Bow," *NYT*, May 13, 1953, p. 33.
4. Mary Craig, "New York Concerts, Town Hall, Philippa Schuyler," *Musical Courier* 147 (June 1953):11.
5. George Sokolsky, "An Interlude with Genius," *New York Journal-American*, May 20, 1953, p. 22.
6. Francis Perkins, "Philippa Schuyler, Pianist," *NYHT*, May 13, 1953, p. 20.

7. Ralph Kammerer, "Philippa Schuyler . . . Town Hall (Debut)," *Musical America* 73 (June 1953):24.

8. PDS to JCS, July 29, 1953, SCRBC.

9. PDS to Olive Abbott, [Sept.] 1953, SCRBC.

10. Usually, but not always, this meant that the artist paid all the expenses and collected the revenue, while the agency received a modest fixed fee. The gamble was that if the reviews were sufficiently favorable, the agent might add the artist to his "stable."

11. *Adventures*, p. 102.

12. PDS to JCS, [Feb. 1953], SCRBC.

13. *Adventures*, p. 136.

14. Gerd Gamborg, interviewed by author, July 1983.

15. Ibid.

16. *Adventures*, p. 111.

17. Rudolph Dunbar, unknown London newspaper [1953].

18. *London Times*, Nov. 30, as quoted in *Pittsburgh Courier*, "Finns and Dutch Praise Pianist," Dec. 5, 1953.

19. Various music reviews, 1953, SCRBC.

20. André Gascht, *L'Afrique et le monde*, Dec. 17, 1953.

21. Various letters, PDS to JCS, and JCS to PDS, 1953, SCRBC.

22. PDS to JCS, Tampere, Finland, Nov. 14, 1953, SCRBC.

23. PDS to JCS, [Fall 1953], SCRBC.

24. Program: Bach-Busoni, Chaconne in D Minor; Bach-Liszt, Prelude and Fugue in A Minor; Scarlatti, Sonatas in A Minor, C, D Minor; Beethoven, Sonata in F Minor, op. 57 ("Appassionata"); Brahms, Intermezzi, op. 117, nos. 2 and 3; Rhapsody in E-flat, op. 119; Ravel, *Sonatine;* Griffes, Sonata in D Minor; Chopin, Scherzo in B-flat Minor, op. 31.

25. "Recital Is Given by Miss Schuyler," Harold Schonberg, *NYT*, Apr. 30, 1954, p. 27.

26. *Musical America*, [1954], n.p.

27. Francis Perkins, "Philippa Schuyler," *NYHT*, Apr. 30, 1954, p. 16.

28. Ibid.

29. Mary Craig, "Philippa Schuyler, Pianist," *Musical Courier* 149 (May 15, 1954):17.

30. Deems Taylor to PDS, Apr. 30, 1954, SCRBC.

31. Klaus George Roy, "Debut at Jordan Hall," *Christian Science Monitor*, Nov. 8, 1954, p. 7.

32. Klaus George Roy, telephone interview by author, Feb. 18, 1989. African Americans make up less than 1 percent of all classical musicians playing in this country's major orchestras.

CHAPTER 18

1. The Ethiopian royal family traced its roots back to the ninth century B.C. when the biblical queen of Sheba was reputed to have made her visit to King Solomon. Although he had promised not to touch her if she refrained from walking off with any of his property, the dusky princess left Jerusalem pregnant with Solomon's child. The boy, born in Ethiopia and raised at King Solomon's court, became Menalik I, king of Axum, the first "Lion of Judah." Haile Selassie, reigning Lion, was a direct descendant of Menalik I.

2. *Adventures*, p. 129.
3. Ibid.
4. PDS to JCS, [Nov. 1955], SCRBC.
5. *Adventures*, p. 156.
6. Anonymous to GSS, Dec. 1955, SCRBC.
7. PDS to JCS, [Dec. 1955], SCRBC.
8. PDS to JCS, [Dec. 1955], SCRBC.
9. *Le Droit de vivre*, Jan. 1, 1956.
10. *Adventures*, pp. 142–43.

CHAPTER 19

1. If one views Philippa's condition as a Mixed-Personality Disorder rather than a trough in a bipolar cycle, then it becomes less puzzling why her depressive episodes did not seriously reduce her ability to work.

People with bipolar disorders often cycle through manic and depressive stages. Typically, a depressive episode precludes concentrated periods of work. A Mixed-Personality Disorder, however, does not necessarily diminish one's productivity.

Phil may have had a Mixed-Personality Disorder, or at least features of it.

Individuals with personality disorders are characterized by personality traits that are maladaptive and inflexible, often leading to subjective distress or impairment in social or occupational functioning. The deeply ingrained maladaptive patterns usually have an onset in adolescence or early adulthood.

2. PDS, "Appassionata, or Scherzo of the Hearts," unpublished manuscript, pagination varies, SCRBC.
3. Ibid.
4. Ibid.
5. Ibid.
6. Gamborg, interview, July 1983.
7. Fax, interview, Feb. 1983.
8. Yohanan Ramati, interviewed by author, Nov. 1983.
9. Eschilia Cosi, interviewed by author, May 18, 1983.
10. Clinton Gray-Fisk, "In My Opinion—The London Concert World: Philippa Schuyler," *Musical Opinion* 80 (Feb. 1957):263.
11. Sister C. G. Carroll to author, Mar. 27, 1984.
12. Klaus George Roy, telephone interview, Feb. 18, 1989.
13. Lineke Snijders van Eyk, interviewed by author, June 19, 1983.
14. Theo Snijders van Eyk, interviewed by author, June 19, 1983.
15. Mary Craig, "On the Cover: Young Artist Abroad," *Musical Courier* 153 (Jan. 1, 1956):2.

CHAPTER 20

1. Later, Ben would receive many awards, including various honorary doctorates, a Commonwealth Certificate for his contributions to art, and the Order of the British Empire.
2. *Adventures*, p. 90.
3. Ben Enwonwu to PDS, Oct. 24, 1957, SCRBC.
4. Rudolph Dunbar to PDS, Nov. 19, 1957, SCRBC.

5. Ramati, interview, Nov. 1983.
6. Ibid.
7. Ibid.
8. Ramati vignette from *Adventures*, pp. 91–92.
9. Enwonwu vignette from *Adventures*, pp. 93–94.
10. Rudolph Dunbar to JCS, Dec. 9, 1957, SCRBC.
11. *Adventures*, pp. 94–98.
12. PDS to JCS, Dec. 5, 1957, SCRBC.

CHAPTER 21

1. Frederick Franck, *African Sketchbook* (New York: Rinehart and Winston, 1961), pp. 94, 96.
2. Robert Caputo, "Ethiopia: Revolution in an Ancient Empire," *National Geographic* 163 (May 1983):633.
3. Franck, *African Sketchbook*, p. 103. According to Franck, eucalyptus trees were introduced to Ethiopia by Menelik II in the nineteenth century.
4. Ibid., pp. 127–28.
5. *Adventures*, p. 181.
6. Ibid.
7. Ibid., p. 188.
8. PDS to JCS, [Jan. 1958], SCRBC.
9. Shana Alexander, "The Serengeti: The Glory of Life," *National Geographic* 169 (May 1986):585.
10. *Adventures*, p. 194.
11. PDS to JCS, [Jan. 1958], SCRBC.
12. Franck, *African Sketchbook*, p. 57.
13. Ibid.
14. *Adventures*, p. 198.
15. Ibid.
16. PDS to JCS, [Jan. 1958], SCRBC.
17. Amy Garvey was not overly pleased with Philippa on this trip, and wrote Josephine: "Philippa . . . should be travelling with a secretary-companion-cum-maid. . . . She should not have her way in everything. . . . We were able to have a very brief chat. . . . She needs to take greater care of herself personally. She must not loose [*sic*] her glamour" (Jan. 28, 1958, SCRBC).
18. *Adventures*, p. 206.
19. In the shallow estuary of the Niger, where less than a century earlier slave ships of many nations lay at anchor, Philippa noticed a few derricks. Oil companies began drilling in the Niger Delta that year—but very few people surmised the country's immense potential as one of the world's major producer of low-sulfer crude—and no one could predict that the oil price boom of the 1970s would bring a $5-billion surplus to Nigeria's treasury by the middle of that decade. In 1958, Nigeria's economy was still based on subsistence agriculture, palm oil, and some open-pit mining of tin and columbite. As one reporter put it, Nigeria was a Cinderella totally unprepared for the royal ball.
20. In 1950, eight years earlier, the country had only eight hospitals (426 beds) and fifteen physicians, of whom only two were Liberians. For a population of 750,000, there were only 253 schools and 220 miles of public roads—mainly around Monrovia and the

Firestone rubber plantations. (Alan C. G. Best and Harm J. de Blij, *African Survey* [New York: Wiley, 1977], p. 196.)

21. Ibid., p. 200.

CHAPTER 22

1. PDS, "The Music of Modern Africa," *Music Journal* 18 (Oct. 1960):18.
2. Ibid., p. 63.
3. Ibid., p. 18.
4. On a later trip to Africa, when she visited Madagascar, Philippa would note that here, too, women and men danced and made music together.
5. PDS to JCS, n.d., SCRBC.
6. PDS, "Music of Modern Africa," pp. 61–62.
7. *Adventures*, pp. 184–85.
8. PDS, "Music of Modern Africa," p. 62.
9. Ibid., p. 18.

CHAPTER 23

1. PDS to JCS, [Summer 1953], SCRBC.
2. PDS, "Sophie Daw," unpublished manuscript, unpaginated, SCRBC.
3. "Sophie Daw" is, first, a picaresque novel, reminding one of the episodic *Tom Jones* of Henry Fielding or Daniel Defoe's *Moll Flanders*. The novel is greatly influenced by the Naturalist school and their bleak social Darwinism. It shares George Gissing's emphasis on grinding poverty and Hardy's belief in a fate that ultimately controls man. It owes a debt to Emile Zola and to the American Naturalists. Stephen Crane's *Maggie: A Girl of the Streets* could have well been the model for "Sophie Daw."
4. In Nepal, female babies die from neglect because parents value sons over daughters; in Sudan, girls' genitals are mutilated to ensure virginity until marriage; and in India, young brides are murdered with impunity by their husbands when parents fail to provide enough dowry—all expressions of women's inferior status in its rawest form.
5. *Adventures*, p. 163.
6. PDS to JCS, [May 1960], SCRBC.
7. Lori Heise writes: "Immediate risks include hemorrhage, tetanus and blood poisoning from unsterile and often primitive cutting implements . . . damage to adjacent organs, and shock from the pain of the operation, which is carried out without anesthesia. Not uncommonly, these complications result in death." Longterm effects include chronic urinary tract infections, pelvic infections that lead to infertility, painful intercourse, and vicious scars that can cause tearing of tissues and hemorrhage during childbirth.

Ms. Heise's report (and statistics) on female circumcision and sexual surgery is harrowing. "According to the World Health Organization, more than 80 million women have undergone sexual surgery in Africa alone. The practice is also performed in the southern part of the Arabian peninsula, around the Persian Gulf, among Muslim groups in Malaysia, and on the island of Java. Recently, it has resurfaced in Europe among immigrant groups after having been abandoned less than a century ago as a 'cure' for nymphomania, hysteria, depression and epilepsy." In a recent phone conversation I had with Ms. Heise, she reports, rather shockingly, that more and more upper-middle-class families in Africa are sending their girls away to European hospitals in order to have the sexual surgery performed under optimal and sterile conditions. She also corroborated Philippa's perceptions of the Sudan: in parts of that country, female circumcision is performed on *over 90 percent* of the women ("Crimes of Gender," *World Watch* 2 (Mar.–Apr. 1989):12–21).

8. *Adventures*, pp. 118–19.

CHAPTER 24

1. A typical concert might have included Mussorgsky's *Pictures at an Exhibition*; a Chopin nocturne, fantasy, or scherzo; a Beethoven sonata; perhaps a Ravel; a short Scarlatti or Soler sonata; a Bach prelude and fugue or his chaconne; Griffes' sonata; Philippa's transcription of Gershwin's *Rhapsody in Blue* (by the end of 1960 she had completed two other Gershwin transcriptions: Concerto in F and *An American in Paris*); and one or two of her own pieces. "Rumpelstiltskin" often appeared on her printed programs (or was performed as an encore). By the end of 1959 she had composed three new pieces based on African themes, which she called alternately *Suite Africaine*, *African Suite*, or *Negro Suite*. They soon became a standard part of her programmed repertoire.

2. "On the Cover: Philippa Schuyler, American Pianist," *Musical Courier* 159 (May 1959):12.

3. Ewha had over nineteen thousand students. To Phil's amazement, she learned that its first students, in the 1880s, came from the economic underclass, because the daughters of the rich were neither allowed higher education nor permitted to walk outside of their homes unchaperoned. Ewha's graduates had gone on to make profound marks on Korean society—the first woman lawyer; the organizer of its first national opera company, and many others.

4. *Adventures*, p. 275.

5. John Mackenzie to PDS, Feb. 5, 1960, SCRBC.

6. PDS to JCS, Jan. 30, 1959, SCRBC.

7. Later, Philippa would portray Albert in a very uncomplimentary light in *Adventures*.

8. *Adventures*, p. 236.

9. PDS to JCS, Feb. 7, 1959, SCRBC.

10. *Adventures*, p. 242.

11. Ibid., p. 247.

12. Frederick Franck, *African Sketchbook* (New York: Rinehart and Winston, 1961), p. 162.

13. Ibid., p. 178.

14. *Adventures*, p. 255.

15. Gwendolyn Shackleford to JCS, Feb. 2, 1959, SCRBC.

16. William Dunbar to JCS, Feb. 26, 1959, SCRBC.

CHAPTER 25

1. PDS to JCS, Feb. 25, 1959, SCRBC.

2. PDS to JCS, Feb. 20, 1959, SCRBC.

3. Phil performed Gershwin's *Rhapsody in Blue* and his Concerto in F.

4. Queen Elisabeth was also a musician, and had studied violin with the greats: Eugene Ysaÿe, Jacques Thibault, and Georges Enesco; she had also played duets with

Yehudi Menuhin, and with Albert Einstein when, in 1934, he was in exile on the Belgian littoral. She established a concert series at Brussels's Palais des Beaux-Arts (where Phil had performed in 1953) and the Concours Reine Elisabeth, an international competition which catapulted many a promising musician to fame. Among its winners in the piano competition have been Leon Fleisher (1952), Vladimir Ashkenazy (1956) and Malcolm Frager (1960).

5. *Adventures*, p. 124.

6. Ibid.

7. Ibid., p. 125.

8. André Gascht, interviewed by author, Dec. 3, 1988.

9. Even her erstwhile conducting teacher, Dean Dixon, had given up on America. Like Philippa, he had met with early successes—stints with the NBC Summer Symphony, summer concerts with the New York Philharmonic, Philadelphia Orchestra, and Boston Symphony. However, "taking a cold, hard look at his life, Dixon realized that these successes, which would have been a springboard for a young white conductor, actually marked the end of the line for a black American. 'It suddenly dawned on me that these first opportunities were a gesture in [what was becoming a] confrontation between black and white America. I suppose I could have enlisted help . . . but I wanted my music, not my color, to open doors.' So Dixon left the United States for Europe, saying later, 'I had kicked myself out of America, and even if I hadn't, they weren't interested in helping me. Because being an American Negro *in my field* [means acknowledging that we Negroes possess] abilities that America says we don't.'" (Noah André Trudeau, "When the Doors Didn't Open," *Hi Fi/ Musical America* [May 1985]:58.)

10. PDS, diary, Mar. 30, 1959, SCRBC.

11. PDS to JCS, Feb. 12, 1959, SCRBC.

12. Carl Van Vechten to JCS, Jan. 31, 1959, SCRBC.

13. Eric Salzman, "Music: Maturing Artist," *NYT*, June 8, 1959, p. 31.

14. Mary Craig, "Carnegie Hall . . . ," *Musical Courier* 160 (July 1959):10.

15. Ralph Kammerer, "Summer Concerts in New York," *Musical America* 79 (July 1959):26.

16. Salzman, "Music."

17. Kammerer, "Summer."

18. Carl Van Vechten to PDS, June 8, 1959, SCRBC.

19. Deems Taylor to PDS, [June 1959], SCRBC.

CHAPTER 26

1. *Adventures*, p. 36.

2. Ernie Pereira, "Nothing Could Stop Philippa," *Hong Kong Tiger Standard*, Jan. 11, 1959, pp. 1, 7.

3. Ibid., p. 1.

4. Ibid., p. 7.

5. Ibid.

6. Ibid.

7. PDS to GSS, Feb. 1, 1959, SCRBC.

8. GSS, "Khruschev's African Foothold," *American Mercury* 88 (Mar. 1959):59.

9. PDS to JCS, Feb. 20, 1959, SCRBC.

10. Ibid.

11. PDS, *Jungle Saints* (Rome: Herder, 1963), p. 41.

12. PDS to JCS, Aug. 16, 1962, SCRBC.

13. In 1962, Philippa wrote to herself: "It seems a shallow mockery to me to be carrying a USA passport. It seems to burn in my hands . . . like the badge of shame of a third-class citizen."

14. PDS, diary, Apr. 15, 1960, SCRBC. In Jessie Fauset's 1933 novel, *Comedy: American Style*, Teresa, the mulatto protagonist is described in terms so reminiscent of this note, as to be uncanny: "Emotionally . . . she was a girl without country. . . . She had become, and she would always remain, individual and aloof, never a part of a component whole."

15. Gamborg, interview, July 1983.

16. Cosi, interview, May 18, 1983.

17. PDS to JCS, May 12, 1960, SCRBC.

CHAPTER 27

1. *Honolulu Star*, Aug. 28, 1958.

2. Although by now somewhat distanced from the family scene, George worried about his daughter. For a while, he had been sensing her emotional deterioration. But it was not something he talked about easily to either Phil or Josephine. But while Philippa was in Hawaii that August, he wrote her hosts the Ehlers (Schuyler had known Sabina Ehler since Phil's days at New York University's Clinic for the Gifted). "George sent me a letter," Mrs. Ehler confided to me years later, "asking if I could help her as she was so unhappy" (Mar. 14, 1986).

3. PDS to JCS, Mar. 21, 1960, SCRBC.

4. *Adventures*, p. 121.

5. Gascht, interview, Dec. 3, 1988.

6. *Adventures*, pp. 121–22.

7. Ibid., p. 122.

8. Ibid.

9. Gascht, interview.

10. Ibid.

11. Ibid.

12. Ibid.

13. Ibid.

14. Ibid.

15. PDS to JCS, Apr. 15, 1960, SCRBC.

16. JCS to PDS, [Apr.] 1960, SCRBC.

17. Gascht, interview.

18. Ibid. André finally married in 1964, at the age of forty-three. His mother died the same year as Philippa, in 1967.

19. Tonino Ciccolella to PDS, Sept. 7, 1960, SCRBC.

20. Lineke Snijders van Eyk to PDS, Nov. 15, 1959, SCRBC.

21. Ernie Pereira to PDS, June 8, 1961, SCRBC.

22. PDS to JCS, [June or July] 27, 1963, SCRBC.

CHAPTER 28

1. PDS, notes to *Who Killed the Congo?* SCRBC.

2. PDS, *Who Killed the Congo*? (New York: Devin-Adair, 1962), p. 127.

3. Brian Urquhart, *Hammarskjold* (New York: Harper and Row, 1972), p. 391.
4. PDS, *Who Killed the Congo?* pp. 7–9.
5. PDS, notes to *Who Killed the Congo?* SCRBC.
6. PDS, *Who Killed the Congo?* p. 217.
7. PDS to JCS [July 1960], SCRBC.
8. Ibid.
9. PDS to JCS, [Aug.? 1960], SCRBC.
10. PDS, *Who Killed the Congo?* p. 161.
11. Robert MacNeil, *The Right Time at the Right Place* (Boston: Little, Brown and Company, 1982), pp. 53–55.
12. PDS, *Who Killed the Congo?* p. 19.
13. The commission had found that on January 13, 1961, a mutiny occurred in the Thysville military camp over pay and lodging. The mutineers had opened Lumumba's cell door, but, fearing a trap, he had refused to come out. Even before that, the Léopoldville regime had determined to send Lumumba from Thysville, where he was being guarded on Hammarskjold's orders by a Ghanaian U.N. contingent, to a "safer place." Tshombe had been approached with the view of sending Lumumba to Katanga. Not cherishing that idea in the least, he had been noncommittal.

Kalonji had urged that Lumumba be sent to Bakwanga, where the Balubas could be counted on to avenge themselves on his person for the events of August. But the Bakwanga airport was under U.N. control, and the Léopoldville government decided to send Lumumba to Elisabethville without Tshombe's consent, or knowledge of the resident U.N. chief. With him were sent two of his close collaborators.

The prisoners were transported on January 17 in an Air Congo DC-4 with an all-Baluban escort. During the five-hour flight the guards beat up the prisoners so brutally that the Belgian crew became nauseated, and after attempting unsuccessfully to intervene, finally locked themselves in the flight deck.

The enquiry concluded that, most likely, Lumumba was already in critical condition when he arrived. On landing, the plane was ordered to taxi to a remote part of the field. There, Godefroi Munongo, Katanga's iron-willed interior minister, boarded with a handful of Katangese soldiers. Munongo approached Lumumba, and after a few remarks, took a bayonet from one of the soldiers and plunged it into the prisoner's chest. While Lumumba lay on the ground dying, a certain Captain Ruys, a Belgian mercenary serving with the Katanga army, ended his suffering by putting a bullet through his head. The body, it is known, was taken to the deep-freeze in the laboratories of the Union Minière.

It became known many years later that the CIA—under Allen Dulles, who had called Lumumba "a Castro, or worse"—hatched a plot to assassinate him by poisoning his food or his toothpaste. The poison was reputedly delivered from CIA labs to agents in the Congo, but by that time Lumumba had been murdered.
14. Urquhart, *Hammarskjold*, pp. 508–9.
15. PDS, *Who Killed the Congo?* p. 272.
16. Urquhart, *Hammarskjold*, pp. 588–89.
17. Finally forced to abdicate, Tshombe went into exile in Europe. He returned in 1964, and in July of that year President Kasavubu named him prime minister of a government of national reconciliation. He served for fifteen months before being dismissed. Accused (1966) of treason against the government, Tshombe went into exile in Spain and was sentenced to death in absentia. In June 1967, a plane in which he was

flying was hijacked to Algeria, where he was first jailed and then held incommunicado until his death in 1969.

18. Philippa disagreed with this statistic. She pointed out that over four hundred native Congolese Catholic priests had "bootlegged" their college education via seminaries.

19. PDS, notes for *Who Killed the Congo?* SCRBC.

20. It is an intriguing question: How did one man become the undisputed "owner" of so much of central Africa, a huge territory which he had neither conquered or inherited? It all started with a Dickensian character by the name of John Rowlands. Born into abject poverty in Wales, the boy entered the notorious St. Asaph's Workhouse at the age of five, where he was subjected to vicious treatment by a half-demented schoolmaster. John was fourteen when he finally ran away and engaged himself as a cabin boy on a sailing ship bound for the New World.

Jumping ship in New Orleans, he met, on a street corner, Henry Morton Stanley, an ordained minister turned merchant. Childless, Stanley informally adopted John, gave him his name, and introduced him into his home, but shortly thereafter died, leaving no provisions for his protégé.

Young "Stanley" enlisted in the Confederate Army, was captured in the Battle of Shiloh, and suffered two months of harsh imprisonment in Camp Douglas (Chicago). After the war, he finally found his calling: Henry Morton Stanley, Jr., became a roving journalist, rapidly gaining a reputation as a vivid descriptive writer.

His journalist career was colorful. He covered Lord Napier's punitive expedition in Abyssinia; witnessed for the *New York Herald* the extravagant opening of the Suez Canal, and then wandered through Jerusalem and Constantinople on his way to the battlefields of the Crimea.

In early 1871, he leaped to the *New York Herald*'s challenge to "Find Livingston," the great Scottish missionary, explorer, and abolitionist, presumably lost somewhere in equatorial Africa.

Starting out from Zanzibar, Stanley reached Ujiji, Tanganyika, eight months later after an incredibly difficult journey. There, against overwhelming odds, he found Dr. Livingston. Their reported greeting may be apocryphal, but it is known to every schoolchild.

Africa got into Stanley's blood. Between 1874 and 1877, he made a great journey across central Africa, tracing the course of the mighty Congo River. Intrigued by Stanley's findings that the region had great economic potential, King Leopold II of Belgium engaged him in 1878 to establish the king's authority in the Congo Basin. To do this, Stanley founded a number of stations along the Congo and signed treaties with several African rulers. King Leopold named the territory the Congo Free State, and proclaimed a series of lofty purposes for his mission in Africa, including a "philanthropic system for liberating the natives."

Nothing could have been further from the truth, and it was not long before the forced labor, official thievery, and other atrocities of Leopold's administration in the Congo became known internationally. Ultimately the whole civilized world would be incensed. But it would take a long time, and possibly 6 to 12 million Congolese would die in the vast forced labor camps. Initially there was a great deal of indifference. Slavery, which treated the African not very differently from a beast of burden, had been universally abolished only recently. Civil rights were the perogative of a very few in select countries, and these alleged crimes were taking place somewhere in the remote recesses of "Darkest Africa."

Several men devoted their lives to exposing the horror. Among them the Englishman Edmund Morel, who in 1904 published a book called *Red Rubber*, which depicted the inhuman conditions attending the gathering of this precious commodity from tapping the rubber vine growing in the fetid, partially submerged forests. Joseph Conrad visited the Congo in 1890 as a commander of a tin steamer plying the Upper Congo River. He stayed several months, almost perishing from dysentery and fever. The exposé he wrote about it called *Heart of Darkness* is one of the most extraordinary short novels ever written. And Roger Casement, British consul in the Congo Free State, caused, in 1903, the conditions in the Upper Congo to become a public scandal. (Always on the side of the oppressed, Casement, ironically, was to be hanged by the British in 1910 for his participation in the Irish Easter Rebellion.)

But eventually the world press took up the call. Congo reform, the elimination of King Leopold's abuses became the cause célèbre of the emerging century. After an exhaustive debate in the Belgian parliament spearheaded by the Socialists, Leopold was deprived of his personal ownership of the Congo, and the vast country became, in 1908, a colony of Belgium.

21. Kwame Nkrumah to PDS, Apr. 19, 1963, SCRBC.

22. Gwendolyn Carter, *NYHT*, book review section, June 17, 1962.

23. As a result of siphoning of funds, the copper-mining industry is in particularly bad shape. The once fabulous but now decaying smelters of Katanga are operating at less that 10 percent of their peak capacity.

24. Colin M. Turnbull, *The Lonely African* (New York: Simon and Schuster, 1962), p. 53.

CHAPTER 29

1. GSS, "Views and Reviews," *Pittsburgh Courier*, Sept. 22, 1934, p. 10.

2. Harry MacKinley Williams, "When Black Is Right: The Life and Writings of George S. Schuyler" (Ph.D. diss., Brown University, 1988), p. 91.

3. Alfonso Zaratti, *Circle of Love* (Italy: Casamari Abbey Publishing Co., 1969), p. 536.

4. Various undated letters, Ernie Pereira to PDS, and PDS to Ernie Pereira, SCRBC.

5. PDS to Doug Maze, *Conversations at Large*, radio interview, Charlotte, NC, 1966.

6. PDS, *Jungle Saints* (Rome: Herder, 1963), pp. 183–84.

7. *Adventures*, pp. 153–55.

CHAPTER 30

1. Jean E. Cazort and Constance Tibbs Hobson, *Born to Play: The Life and Career of Hazel Harrison* (Westport, CT: Greenwood, 1983), p. 118.

2. Houston, interview, Mar. 1983.

3. Hubert Roussel, "Philippa Schuyler in a Piano Recital of Extraordinary Level," *Houston Post*, Apr. [3 or 4], 1962.

4. Houston, interview.

5. Kathleen Houston to JCS, Feb. 20, 1962, SCRBC.

6. Houston, interview.

7. Ibid.

8. Ibid.
9. Ibid.
10. PDS to JCS, Jan. 30, 1959, SCRBC.
11. Ibid.
12. Although he was aware of the Monterro plan, George knew very little about it. There is some indication, however, that father and daughter had a serious argument during this time, and it appears to have been over an extra set of programs that Phil had printed up for her Philharmonic recital listing "Felipa Monterro" as the pianist. In light of the fact that she was trying to adopt an Iberian heritage for her alter ego, and her program contained pieces only by Spanish, Portuguese, and Latin American composers, it is possible Philippa thought Miss Monterro could "pull a fast one." George, no doubt, thought the idea mad, as well as insulting.
13. PDS to JCS, [Feb. 1963], SCRBC.
14. PDS to JCS, Feb. 11, 1963, SCRBC.
15. PDS to JCS, July 10, 1963, SCRBC.
16. Lineke Snijders van Eyk to PDS, Oct. 12, 1962, SCRBC.
17. Madame Beltjens to PDS, Sept. 27, 1962, SCRBC.
18. Roger Nonkel to PDS, Jan. 21, 1963, SCRBC.
19. PDS to JCS, Apr. 4, 1963, SCRBC.
20. The Birch-Monterro liaison was largely to earn money. It is somewhat puzzling, however, that Phil chose this avenue. Not only was her father one of the few black members of the society at the time, and she, therefore, easily recognized; but for years Phil had deplored George's extremism, claiming it as one of the causes for her failure in America.
21. PDS, "Terror in Angola," *American Opinion* (Apr. 1963): 29–35. The article was accompanied by a small photograph of the author. It may have been a printer's error, but the woman bore no resemblance whatsoever to Philippa. In fact, Miss Monterro might be considered a rather ugly white woman with her sagging chin and hookish nose—hardly the beauty that Philippa was. Even if it had been a printer's error, Philippa's intentions all along had been to send a highly disguised version of herself: "At last I have gotten a photo made that I can send to Amer. Opinion as Felipa Monterro," she wrote Jody. "I tried before, but strangely, the other photos looked like me! I enclose a small one of the photo. The large one has the unruly strands of hair removed. Don't you think it looks like Felipa Monterro?? Maybe using that name is making me look like her!" (n.d., SCRBC).
22. PDS to JCS, May 13, 1963, SCRBC.
23. Like her father, Philippa considered becoming a paid propagandist for the Portuguese, not only to earn a little extra money but also to give her new identity a cause, in fact, a chauvinistic one, which would validate her newly sought after heritage even more. "The Portuguese don't give a hoot what *name* I do something for them under," she wrote Jody. "They don't care if I use Adolf Hitler, if it has the result of making their cause better known" (May 13, 1963, SCRBC).
24. *TAT* (Zurich), Apr. 30, 1963.
25. *Tages-Anzeiger* (Zurich), Apr. 29, 1963.
26. *Neue Züricher Nachrichten*, May 2, 1963.
27. *Neue Züricher Zeitung*, May 3, 1963.
28. PDS to JCS, Apr. 26, 1963, SCRBC. Defending herself even more, she wrote home: "I actually made some *Money* off the concert. I made 55 dollars. . . . Something brought the people to the concert. It must have been my *name* and the program. What

else could it have been? . . . If I had played a Mozart, Haydn, Beethoven program, I'll bet nobody would have come, because so many people play these things. . . . I've already tried playing conventional programs in Europe and I had mixed success. Sometimes I got wonderful reviews, and sometimes I got criticism—but my name did not become a *household word* in association with Beethoven, or Bach, etc. . . . Do you know when Rubinstein first lifted himself up from anonymity to being considered an important artist, he did it with Spanish music?"

29. There is an interesting letter from Josephine to Carl Van Vechten during this period (Dec. 1963, YU). In the habit of collecting as many of his goddaughter's concert programs as possible, he was surprised when he did not receive any from Jody, who over the years had dutifully forward them. They had run out, she wrote him rather sheepishly; you know how disorganized the Italians are. In fact, neither she nor Philippa wanted Van Vechten to know what was going on.

30. PDS to JCS, [July 1963], SCRBC.

31. PDS to JCS, May 17, 1963, SCRBC.

32. PDS to JCS, Aug. 3, 1963, SCRBC.

33. Ibid.

34. One of the publishers (Oriel Schadel) who knew Philippa fairly well was shocked when he learned from me, over twenty years later, that these two were one and the same person. A careful reading of the book also reveals that Philippa identified herself to the reader as white. On page 182, when she asks to meet Father Tempels, Muteba replies, "Well, that might be a little difficult . . . you see . . . he rarely meets or associates with whites."

35. PDS to JCS, Aug. 3, 1963, SCRBC.

36. Cosi, interview, May 18, 1983.

37. Ibid.

38. Ibid.

39. Ibid.

40. The album included Mussorgsky, *Pictures at an Exhibition;* Turina, *Jeudi Saint à Minuit;* Copland, *Scherzo humoristique;* Gershwin, *An American in Paris;* Middleton, Andante Satirico; Liszt, Hungarian Rhapsody no. 6; Infante, *Sevillian Variations;* Casanovas, *Voyage around Cape Horn.*

CHAPTER 31

1. Two hotel bills have been preserved. They read: "Hotel am Zoo, Oct 10 to 14, Herr Raymond—Zimmer 53; Frau Schuyler—Zimmer 40." The "40" is crossed out and "53" written over. Preserved is also the record of a tarot card session, written on Hotel am Zoo stationary: "Does Maurice Raymond love me?" "Does he intend to marry me?" "Should I marry Maurice Raymond?" On the verso is more writing: "Philippa Raymond," "Philippe Raymond," "Philippine Raymond," all juxtaposed to "Philippa Schuyler" and "Felipa Monterro."

2. PDS to JCS, Oct. 12, 1963, SCRBC.

3. Ibid.

4. JCS to PDS, [1963], SCRBC.

5. Maurice Raymond to PDS (translation), Feb. 3, 1964, SCRBC.

6. PDS to JCS, [Feb. 1964], SCRBC.

7. PDS to JCS, Feb. 29, 1964, SCRBC.

8. Ibid.

9. Ibid.
10. Ibid.
11. Ibid.
12. PDS to JCS, Feb. 24, 1964, SCRBC.
13. "Felipe Monterro: Pianiste internationale de passage à Lyon," *L'Echo*, Feb. 18, 1964.
14. PDS to JCS, Mar. 8, 1964, SCRBC.
15. Ibid.
16. PDS to JCS, Apr. 1, 1964, SCRBC.
17. Maurice Raymond to JCS, May 21, 1964, SCRBC.
18. Rupert A. Lloyd to JCS, Sept. 11, 1964, SCRBC.
19. Ibid.

CHAPTER 32

1. The Town Hall recital was a benefit for the John La Farge Committee of the Catholic Interracial Council. The program included: Chopin's Scherzo in B-flat Minor, op. 31; Beethoven's *Appassionata;* three works by Liszt: Spanish Rhapsody, Hungarian Rhapsody no. 6, *Rigoletto Variations;* Yohanan Ramati's *Memories of Poland;* and three of her own works: *Sanga; Chisambaru the Nogomo* (premier); and *White Nile Suite* (premier).
2. Raymond Ericson, "Music: Town Hall Recital," *NYHT*, Sept. 14, 1964.
3. PDS to JCS, Oct. 29, 1964, SCRBC.
4. DGS, *I Give Myself Away* (London: Hodder and Stoughton, 1936).
5. Denis Craig [DGS], *Man in Ebony: A Novel* (London: Victor Gollancz, 1950). Reviews are from dustjacket of 1950 edition.
6. PDS to JCS, Nov. 4, 1964, SCRBC.
7. PDS to JCS, May 1, 1964, SCRBC.
8. PDS to JCS, Nov. 4, 1964, SCRBC.
9. DGS to PDS, Nov. 14, 1964, SCRBC.
10. DGS to PDS, Nov. 21, 1964, SCRBC.
11. DGS to PDS, Nov. 14, 1964, SCRBC.
12. DGS to PDS, Feb. 24, 1965, SCRBC.
13. DGS to PDS, Feb. 4, 1965, SCRBC.
14. Deems Taylor to ISM, Mar. 9, 1965, SCRBC.
15. DGS to PDS, Feb. 8, 1965, SCRBC.
16. DGS to PDS, Feb. 12, 1965, SCRBC.
17. PDS to DGS, [Feb. 1965], SCRBC.
18. DGS to PDS, Feb. 17, 1965, SCRBC.
19. DGS to PDS, Feb. 26, 1965, SCRBC.
20. Ibid.
21. DGS, *Comedy in Chains: A Novel of South India* (*1939–1941*), (London: Victor Gollancz, 1944), p. 40.
22. Ibid., p. 33.
23. March 21, Westminster Theatre, London; March 25, Iran Society, Iranian Embassy; April 8, American Embassy Theatre; April 21, Stroud; April 24, Royal Philharmonic Orchestra; April 25, Odeon Theatre; May 3, Gandhi Foundation.
24. Marcel Gros to PDS, Jan. 12, 1965, SCRBC.
25. DGS to PDS, Aug. 7, 1965, SCRBC.

CHAPTER 33

1. Georges Apedo-Amah to PDS, Jan. 11, 1965, SCRBC.
2. PDS to JCS, Apr. 15, 1965, SCRBC.
3. Ibid.
4. Ibid.
5. John MacKenzie to author, Aug. 2, 1983. Mackenzie ends his letter with a poignant statement. "I do not wish to be unkind nor uncharitable to Josephine who nurtured such a fine flower which bloomed too unseen, perhaps wasted its fragrance on the desert air, and certainly bloomed far too briefly."
6. DGS to PDS, June 7, 1965, SCRBC.
7. DGS to JCS, May 4, 1965, SCRBC.
8. Ibid.
9. DGS to JCS, May 25, 1965, SCRBC.
10. PDS to JCS, May 27, 1965, SCRBC.
11. DGS to PDS, May 12, 1965, SCRBC.
12. Patricia reportedly described Philippa as "a golddigger, a bitch, the most insensitive and selfish woman [I have] ever met" (DGS to PDS, Aug. 18, 1965, SCRBC).
13. PDS to JCS, May 27, 1965, SCRBC.
14. John MacKenzie to PDS, June 15, 1965, SCRBC.
15. DGS to PDS, July 30, 1965, SCRBC.
16. DGS to PDS, Aug. 17, 1965, SCRBC.
17. DGS to PDS, Aug. 22, 1965, SCRBC.
18. DGS to JCS, Aug. 24, 1965, SCRBC.
19. "NDS," interviewed by author, Jan. 14, 16, 1991.
20. Lineke Snijders van Eyk, interview, June 19, 1983.
21. DGS to JCS, Sept. 8, 1965, SCRBC.
22. PDS to JCS, Sept. 5, 1965, SCRBC.
23. PDS to JCS, Sept. 6, 1965, SCRBC.
24. DGS to PDS, Oct. 24, 1965, SCRBC.

CHAPTER 34

1. In a small notebook, written in an equally diminutive hand—presumably a "secret diary"—Philippa recorded her feelings immediately preceding and following her illegal abortion. This chapter is culled from that diary (SCRBC).
2. Georges Apedo-Amah to PDS, Nov. 6, 1965, SCRBC.
3. Gascht, interview, Dec. 3, 1988.
4. Oriel Schadel, interviewed by author, May 15, 1983.

CHAPTER 35

1. In 1967, when the *Times* of London printed an obituary of PDS, which Dennis felt sure would have upset her, he sent the following corrigendum to the editors: "You refer to . . . the distinguished American pianist Philippa Schuyler . . . as 'a Negro.' Miss Schuyler's mother is a white Texan and her father is descended from the Dutch General Schuyler, a hero of the American War of Independence. He is partly Madagascan, and Miss Schuyler was therefore partly Polynesian. The error doubtless arose from the fact that her father edits an American Negro newspaper, and she herself has fre-

quently toured Africa as a concert pianist and written brilliantly about the country. . . . It is with great respect that I pay my last tribute, quoting Deems Taylor's words—'Superb. A second Myra Hess'" (May 18, 1967).

2. Georges Apedo-Amah to PDS, Mar. 10, 1966, SCRBC.

3. "NDS," interview, Jan. 14, 16, 1991.

4. Joseph and Mary Myers, interviewed by author, Aug. 18, 1978.

5. J. A. C. Dunn, "The Prodigy: 'I'm Not Sure What I Am.'" *Charlotte Observer*, June 25, 1966, p. 20D.

6. PDS to the Myers, June 6, 1966, IUB.

7. Carl Rowan, "New Group Seeks to 'Save' Africa," *Evening Star*, Feb. 23, 1966, p. A-23.

8. GSS, *Black and Conservative: The Autobiography of George S. Schuyler* (New Rochelle, NY: Arlington House, 1966), p. 341.

9. Ibid.

10. GSS, *Manchester (NH) Union Leader*, Nov. 10, 1964, p. 1.

11. Ernie Pereira to PDS, [Jan. 1966], SCRBC.

12. PDS to JCS, Feb. 26, 1966, SCRBC.

13. PDS to JCS, July 26, 1966, SCRBC.

14. Ernie Pereira to PDS, Sept. 23, 1966, SCRBC.

CHAPTER 36

1. Virginia Elwood-Akers, *Women War Correspondents in the Vietnam War, 1961–1975* (Metuchen, NJ: Scarecrow Press, 1988), p. 76.

2. PDS to JCS, Sept. 8, 1966, SCRBC.

3. PDS to JCS, Sept. 12, 1966, SCRBC.

4. PDS, notes for *Good Men Die*, SCRBC.

5. PDS to JCS, Sept. 15, 1966, SCRBC.

6. PDS, *Good Men Die* (New York: Twin Circle Publishing Company, 1969), pp. 215–16.

7. PDS, "Dau Tranh," unpublished manuscript, unpaginated, SCRBC.

8. When Philippa arrived in Vietnam during the fall of 1966, the number of female correspondents had grown from a mere six or seven the year before to over two dozen—still a small percentage of the more than four hundred accredited journalists there at the time. Their general makeup had also changed. Not only were the women younger than their counterparts had been during the first half of 1966, they were also not as fervently anticommunist as their older colleagues had been, Philippa being an obvious exception. The death of Philippa would leave Patches Musgrove as the only woman war correspondent in Vietnam at the time who spoke out in favor of American involvement in the war.

9. Elwood-Akers, *Women War Correspondents*, p. 74.

10. A quarter of a century later, when I finally tracked down "Mr. Leiter" (not his real name) after several years of search, he recalled a somewhat different story of their first encounter—that they had met at a press conference. He remembered Philippa as "a very outgoing person. It did not take long to establish a rapport. . . . She was a very attractive young lady. One thing I did not like about her was that ratty wig she wore quite often [part of her Vietnamese guise]. It distracted from her appearance." "James Leiter," interviewed by author, Sept. 1990.

11. PDS, "Dau Tranh," SCRBC.

12. PDS, *Good Men Die*, p. 70.

13. "Leiter," interview.
14. Ibid.
15. Ibid.
16. Ibid.
17. Ibid.
18. Ibid.
19. When I asked Leiter what three words he would use to describe Philippa, he answered, "Troubled, very interesting, and driven—especially the last."
20. JCS to PDS, Jan. 29, 1967, SCRBC.
21. JCS to PDS, Feb. 5, 1967, SCRBC.
22. PDS, notes to *Good Men Die*, SCRBC.
23. Elwood-Akers, *Women War Correspondents*, p. 102.
24. PDS, "What Is Really Going On," *Manchester (NH) Union Leader*, Apr. 21, 1967, p. 1 ff.
25. "Leiter" remembers this incident differently: that although he heard her perform it was not at her concert but later, at the embassy, where she performed several pieces.
26. PDS to JCS, Feb. 21, 1966, SCRBC.
27. "Leiter," interview.
28. PDS to JCS, Feb, 21, 1966, SCRBC.
29. Ibid.
30. "Leiter," interview.
31. Ibid.
32. Ibid.
33. PDS, "Dau Tranh," SCRBC.
34. PDS to JCS, n.d., SCRBC.
35. PDS to JCS, Apr. 9, 1967, SCRBC.
36. PDS to JCS, [May 1967], SCRBC.
37. PDS to JCS, May 6, 1967, SCRBC.
38. Leonard Holsey to JCS and GSS, June 11, 1967, SCRBC.

POSTLUDE

1. Fax, interview, Feb. 1983.
2. The Fourth Grade, St. John School, to JSC and GSS, May 22, 1967, SCRBC.
3. Virginia Thorndike to GSS, May 10, 1967, SCRBC.
4. It is rumored that Jody had just been told she had cancer.
5. GSS to William Loeb, May 5, 1969, YU.
6. GSS, YU.

INDEX